Nitrogen Fixation

Nitrogen Fixation

Volume I:
Free-living Systems and
Chemical Models

Edited by

William E. Newton
Charles F. Kettering Research Laboratory
Yellow Springs, Ohio

and

William H. Orme-Johnson
University of Wisconsin-Madison

University Park Press
Baltimore

UNIVERSITY PARK PRESS
International Publishers in Science, Medicine, and Education
233 East Redwood Street
Baltimore, Maryland 21202

Copyright © 1980 University Park Press

Typeset by American Graphic Arts Corporation.

Manufactured in the United States of America by
The Maple Press Company.

Library of Congress Cataloging in Publication Data

Kettering International Symposium on Nitrogen Fixation,
    3d, Madison, Wis., 1978.
    Nitrogen fixation.

    Includes index.
    1. Micro-organisms, Nitrogen-fixing—Congresses.
2. Nitrogen—Fixation—Congresses. I. Newton, William
Edward, 1938–   II. Orme-Johnson, William H.
III. Title.
QR89.7.K47 1978   589   79-28285
ISBN 0-8391-1560-1 (v.1)
ISBN 0-8391-1561-X (v.2)

Contents

Contributors List

Frederick M. Ausubel
Department of Biology
Harvard University
Cambridge, Massachusetts 02138

Winston J. Brill
Department of Bacteriology and Center
    for Studies of Nitrogen Fixation
University of Wisconsin
Madison, Wisconsin 53706

Barbara K. Burgess
Charles F. Kettering Research
    Laboratory
Yellow Springs, Ohio 45387

A. Burns
Charles F. Kettering Research
    Laboratory
Yellow Springs, Ohio 45387

R. H. Burris
Department of Biochemistry
University of Wisconsin
Madison, Wisconsin 53706

Frank C. Cannon
A.R.C. Unit of Nitrogen Fixation
University of Sussex
Brighton, BN1 9RQ
England

Joseph Chatt
A.R.C. Unit of Nitrogen Fixation
University of Sussex
Brighton, BN1 9RQ
England

James L. Corbin
Charles F. Kettering Research
    Laboratory
Yellow Springs, Ohio 45387

Ray Dixon
A.R.C. Unit of Nitrogen Fixation
University of Sussex
Brighton, BN1 9RQ
England

Robert R. Eady
A.R.C. Unit of Nitrogen Fixation
University of Sussex
Brighton, BN1 9RQ
England

Mechthild Filser
A.R.C. Unit of Nitrogen Fixation
University of Sussex
Brighton, BN1 9RQ
England

Vincent P. Gutschick
Los Alamos Scientific Laboratory
Los Alamos, New Mexico 87545

Huub Haaker
Department of Biochemistry
Agricultural University
De Dreijen 11, 6703 BC Wageningen
The Netherlands

Bryan Hainline
Department of Biochemistry
Duke University Medical Center
Durham, North Carolina 27710

R. W. F. Hardy
Central Research & Development
    Department
Experimental Station
E. I. Dupont de Nemours and
    Company, Inc.
Wilmington, Delaware 19898

Keith O. Hodgson
Department of Chemistry
Stanford University
Stanford, California 94305

Hsu Chi-Ching
Nitrogen Fixation Research Group
Department of Chemistry
Kirin University
Changchun, Kirin, China

C. W. Huang
Department of Chemistry
University of Southern California
Los Angeles, California 90007

Kaaren A. Janssen
Department of Biology
Harvard University
Cambridge, Massachusetts 02138

Jean L. Johnson
Department of Biochemistry
Duke University Medical Center
Durham, North Carolina 27710

Harold P. Jones
Department of Biochemistry
Duke University Medical Center
Durham, North Carolina 27710

J. B. Jones
Department of Chemistry
University of Southern California
Los Angeles, California 90007

Christina Kennedy
A.R.C. Unit of Nitrogen Fixation
University of Sussex
Brighton, BN1 9RQ
England

Jon A. Kirby
Laboratory of Chemical Biodynamics
Building #3
Lawrence Berkeley Laboratory
University of California
Berkeley, California 94720

Melvin P. Klein
Laboratory of Chemical Biodynamics
Building #3
Lawrence Berkeley Laboratory
University of California
Berkeley, California 94720

Colja Laane
Department of Biochemistry
Agricultural University
De Dreijen 11, 6703 BC Wageningen
The Netherlands

G. I. Likhtenstein
Institute of Chemical Physics
Academy of Sciences USSR
142432 Moscow oblast
Chernogolovka, USSR

S. Lough
Charles F. Kettering Research
    Laboratory
Yellow Springs, Ohio 45387

David J. Lowe
A.R.C. Unit of Nitrogen Fixation
University of Sussex
Brighton, BN1 9RQ
England

Lu Jiaxi
Nitrogen Fixation Research Group
Fujian Institute of Research on the
    Structure of Matter
Chinese Academy of Sciences
Fuzhou, Fujian, China

Paul W. Ludden
Department of Biochemistry
University of California
Riverside, California 92521

Douglas MacNeil
Department of Bacteriology and Center
    for Studies of Nitrogen Fixation
University of Wisconsin
Madison, Wisconsin 53706

Tanya MacNeil
Department of Bacteriology and Center
    for Studies of Nitrogen Fixation
University of Wisconsin
Madison, Wisconsin 53706

Leonard Matz
Department of Biochemistry
Agricultural University
De Dreijen 11, 6703 BC Wageningen
The Netherlands

C. E. McKenna
Department of Chemistry
University of Southern California
Los Angeles, California 90007

M.-C. McKenna
Department of Chemistry
University of Southern California
Los Angeles, California 90007

Mike Merrick
A.R.C. Unit of Nitrogen Fixation
University of Sussex
Brighton, BN1 9RQ
England

T. Nakajima
Department of Chemistry
University of Southern California
Los Angeles, California 90007

William E. Newton
Charles F. Kettering Research
   Laboratory
Yellow Springs, Ohio 45387

H. T. Nguyen
Department of Chemistry
University of Southern California
Los Angeles, California 90007

D. E. Nichols
Division of Chemical Development
Tennessee Valley Authority
Muscle Shoals, Alabama 35660

Michael J. O'Donnell
A.R.C. Unit of Nitrogen Fixation
University of Sussex
Brighton, BN1 9RQ
England

W. J. Payne
Department of Microbiology
University of Georgia
Athens, Georgia 30602

John R. Postgate
A.R.C. Unit of Nitrogen Fixation
University of Sussex
Brighton, BN1 9RQ
England

K. V. Rajagopalan
Department of Biochemistry
Duke University Medical Center
Durham, North Carolina 27710

Raymond L. Richards
A.R.C. Unit of Nitrogen Fixation
University of Sussex
Brighton, BN1 9RQ
England

Gerard E. Riedel
Department of Biology
Harvard University
Cambridge, Massachusetts 02138

Gary P. Roberts
Department of Bacteriology and Center
   for Studies of Nitrogen Fixation
University of Wisconsin
Madison, Wisconsin 53706

Alan Robertson
Laboratory of Chemical Biodynamics
Building #3
Lawrence Berkeley Laboratory
University of California
Berkeley, California 94720

J. J. Rowe
Department of Microbiology
University of Georgia
Athens, Georgia 30602

Gerard Scherings
Department of Biochemistry
Agricultural University
De Dreijen 11, 6703 BC Wageningen
The Netherlands

Vinod K. Shah
Department of Bacteriology and Center
   for Studies of Nitrogen Fixation
University of Wisconsin
Madison, Wisconsin 53706

B. F. Sherr
Department of Microbiology
University of Georgia
Athens, Georgia 30602

Barry E. Smith
A.R.C. Unit of Nitrogen Fixation
University of Sussex
Brighton, BN1 9RQ
England

Joseph P. Smith
Department of Biochemistry
University of Wisconsin
Madison, Wisconsin 53706

Edward I. Stiefel
Charles F. Kettering Research
   Laboratory
Yellow Springs, Ohio 45387

Mark A. Supiano
Department of Bacteriology and Center
  for Studies of Nitrogen Fixation
University of Wisconsin
Madison, Wisconsin 53706

Albert C. Thompson
Building #29, Room 107
Lawrence Berkeley Laboratory
University of California
Berkeley, California 94720

Roger N. F. Thorneley
A.R.C. Unit of Nitrogen Fixation
University of Sussex
Brighton, BN1 9RQ
England

K. R. Tsai
Nitrogen Fixation Research Group
Laboratory for Catalysis Research
Xiamen University
Xiamen, China

Louise Van Zeeland-Wolbers
Department of Biochemistry
Agricultural University
De Dreijen 11, 6703 BC Wageningen
The Netherlands

Cees Veeger
Department of Biochemistry
Agricultural University
De Dreijen 11, 6703 BC Wageningen
The Netherlands

D. R. Waggoner
Division of Chemical Development
Tennessee Valley Authority
Muscle Shoals, Alabama 35660

C. C. Walker
A.R.C. Unit of Nitrogen Fixation
University of Sussex
Brighton, BN1 9RQ
England

Gerald D. Watt
Charles F. Kettering Research
  Laboratory
Yellow Springs, Ohio 45387

Scot Wherland
Charles F. Kettering Research
  Laboratory
Yellow Springs, Ohio 45387

P. C. Williamson
Division of Chemical Development
Tennessee Valley Authority
Muscle Shoals, Alabama 35660

M. G. Yates
A.R.C. Unit of Nitrogen Fixation
University of Sussex
Brighton, BN1 9RQ
England

Preface

These volumes constitute the Proceedings of an international symposium that was held on June 12–16, 1978, in Madison, Wisconsin (USA). Three principal benefactors made this Symposium possible. The Steenbock Symposia Committee of the Department of Biochemistry, the University of Wisconsin-Madison, provided a grant as a part of a regular program of annual symposia established in honor of the eminent biochemist, the late Professor Harry F. Steenbock, under the benefaction of Mrs. Evelyn Steenbock, whom we sincerely thank. The meeting was thus the Seventh Harry F. Steenbock Symposium. The Charles F. Kettering Foundation, Dayton, Ohio, and the Kettering Research Laboratory, Yellow Springs, Ohio, were established by the renowed scientist-inventor, Charles F. Kettering, and have organized several previous gatherings devoted to plant science. This meeting was the Third Kettering International Symposium on Nitrogen Fixation, the previous two being held in Pullman, Washington, in 1974 and Salamanca, Spain, in 1976. We thank Mr. Robert G. Chollar, President and Chairman of the Board of the C. F. Kettering Foundation, for his continued support of the concept of an interdisciplinary meeting in the area of nitrogen fixation. Finally, the Chemical Development Division of the Tennessee Valley Authority, under its programmatic interests in industrial nitrogen fixation and fertilizer economics, also provided support. Our thanks are extended to Mr. Charles H. Davis and his colleagues at TVA for their timely interest and generosity.

We are particularly pleased to note that three names widely associated with the application of science and technology to the betterment of mankind's condition have joined in this enterprise, which we hope has stimulated (and will continue to do so) interdisciplinary discussion on this vitally important contemporary problem.

The organization of this meeting was in the hands of a series of committees under our overall direction. We thank our colleagues on the International Program Committee, M. J. Dilworth (Murdoch University, Australia), R. W. F. Hardy (DuPont Company, USA), and J. R. Postgate (A.R.C. Unit of Nitrogen Fixation, UK), for their invaluable advice and counsel. R. H. Burris and W. J. Brill (University of Wisconsin-Madison) and H. J. Evans (Oregon State University) also contributed to this part of the endeavor, as did the many participants who wrote with programming suggestions. All are thanked.

The hard work of local arrangements and day-to-day organization fell in varying proportions to our colleagues on the Local Committee, Karen Davis, Vicki Hudson Newton, Bill Hamilton, Cynthia Touton, Nanette Orme-Johnson, Catherine Burris and Bob Burris, without whose efficiency and fidelity we would have been lost.

The interest in and importance of this area of research is demonstrated by the increasing demand for participation in these meetings. In 1974, about 200 scientists were in attendance, whereas in 1976 nearly 300 people participated. Attendance in 1978 was in excess of 400. This attendance trend is paralleled by the enormous progress made in the science in the four years since the Kettering-Pullman Symposium. These volumes speak very clearly to this point. We hope that this record of an outstanding symposium will benefit all persons interested in this area of research endeavor.

William E. Newton
Yellow Springs, Ohio

William H. Orme-Johnson
Madison, Wisconsin

Nitrogen Fixation

Section I
Global Nitrogen and Carbon Economy

Nitrogen Fixation, Volume I
Edited by W. E. Newton and W. H. Orme-Johnson
Copyright 1980   University Park Press   Baltimore

The Global Carbon and Nitrogen Economy

R. W. F. Hardy

Why initiate this International Symposium on Nitrogen Fixation with a session on the global carbon and nitrogen economy? A variety of reasons exist, ranging from the academic justification of scientific interest to the practical justification of meeting the additional needs for fixed nitrogen to support the expanding food needs within acceptable energy and environmental constraints.

How much N_2 is fixed annually? What are the strengths and weaknesses of the data basis? The necessity of balancing the efflux (denitrification) by equating it with the influx (N_2 fixation) is unsatisfactory. Will the new acetylene inhibition of N_2O reduction technique (Balderston, Sherr, and Payne, 1976; Yoshimari, Hynes, and Knowles, 1977) facilitate direct measurement of denitrification? What will be the effect of inhibitors of nitrification and/or denitrification on the nitrogen cycle? These and many other questions regarding the global nitrogen economy are of scientific importance. Much additional investigation is needed to provide the answers.

The need for additional food and feed will produce a major impact on the global nitrogen and carbon economy. It has been suggested (Hardy, 1977; Hardy, Havelka, and Quebedeaux, 1978) that cereal grain production will have to double from 1.3 to 2.6 billion tons per year during the fourth quarter of this century, and grain legume production may have to quadruple from 0.13 to 0.5 billion tons per year during the same period, in order to meet the food expectations of the expanding population. These increases in production will require the provision of large additional quantities of fixed nitrogen, estimated at 75–150 million tons per year.

Several options for provision of this additional fixed nitrogen may be feasible. In the case of cereal grains, the generalized options include: 1) expansion of fertilizer nitrogen production from 40 to 160 million tons by the Haber-Bosch process, accompanied by the development of alternative feedstocks, such as coal (Waggoner, this volume); 2) improvement in the efficiency of use of fertilizer nitrogen by crops from the current average of 50% to 75% through, for example, agents such as nitrification inhibitors (Huber et al., 1977); 3) invention of a zero direct energy–consuming process for the manufacture of fertilizer nitrogen, such as an oxidative process; and 4) discovery of an economical biological N_2 fixation process that will provide adequate fixed nitrogen for high yields. In the case of grain legumes, the options include: 1) enhancement of the *Rhizobium*-legume N_2-fixing process; and 2) development of an economical system that enables high legume yields in response to fertilizer nitrogen (Garcia and Hanway, 1976).

Energy and environmental constraints as well as economics will be important factors in selecting the preferred options. Currently, about 1% of the world fossil energy consumption is used to produce fertilizer nitrogen and about 1–2 billion tons of plant carbohydrate may be consumed for biological N_2 fixation—in the latter case, an amount equal to the world cereal grain and grain legume production combined. Can we continue to afford, on an energy basis, to fix nitrogen by the currently available energy-wasteful synthetic and biological processes?

Within the past three years, a concern has been raised regarding the possible alteration of stratospheric ozone by the additional N_2O produced by denitrification of the expanded fixed nitrogen needed for crop production (Crutzen and Ehalt, 1977). About a year ago, it was indicated that a doubling in N_2O concentration would produce a 15–20% decrease in ozone, but consideration of additional rate data now leads to an estimate of at most 10% in ozone reduction for a doubling of atmospheric N_2O (Crutzen, 1977). In addition, it has been shown that nitrification of ammonia in fields may yield as much N_2O as denitrification of nitrate (Bremner and Blackmer, 1978). Clearly, expansion of our relatively elementary information on the global carbon and nitrogen economy is needed for both scientific and practical objectives. The "N_2-fixer" must be cognizant of this broader area in order for his/her work to be of the greatest impact.

REFERENCES

Balderston, W. L., B. Sherr, and W. J. Payne. 1976. Blockage by acetylene of nitrous oxide reduction by *Pseudomonas perfectomarinus*. Appl. Environ. Microbiol. 31:504–508.

Bremner, J. M., and A. M. Blackmer. 1978. Nitrous oxide: Emission from soils during nitrification of fertilizer nitrogen. Science 199:295–296.

Crutzen, P. J. 1977. Ozone variations caused by changing atmospheric concentrations of nitrous oxide: A discussion of some remaining uncertainties. In: Denitrification Seminar. The Fertilizer Institute, 1015 18th Street, N.W., Washington, D.C. 20036.

Crutzen, P. J., and D. H. Ehalt. 1977. Effects of nitrogen fertilizer on the stratospheric ozone layer. Ambio 6:112–117.

Garcia, R., and J. J. Hanway. 1976. Foliar fertilization of soybeans during seed-filling period. Agron. J. 68:653–657.

Hardy, R. W. F. 1977. Increasing crop productivity: Agronomic and economic considerations on the role of biological nitrogen fixation. In: Report of the Public Meetings on Genetic Engineering for Nitrogen Fixation, pp. 77–106. U.S. Government Printing Office, Washington, D.C.

Hardy, R. W. F., U. D. Havelka, and B. Quebedeaux. 1978. Increasing crop productivity: The problem, strategies, approach, and selected rate-limitations related to photosynthesis. In: D. O. Hall, J. Coombs, and T. W. Goodwin (eds.), Proceedings of the Fourth International Congress on Photosynthesis, pp. 695–719, The Biochemical Society, London.

Huber, D. M., H. L. Warren, D. W. Nelson, and C. Y. Tsai. 1977. Nitrification inhibitors—new tools for food production. BioScience 27:523–529.

Yoshimari, T., R. Hynes, and R. Knowles. 1977. Acetylene inhibition of nitrous oxide reduction and measurement of denitrification and nitrogen fixation in soil. Soil Biol. Biochem. 9:177–183.

Nitrogen Fixation, Volume I
Edited by W. E. Newton and W. H. Orme-Johnson
Copyright 1980   University Park Press   Baltimore

The Global Nitrogen Budget—Science or Seance?

R. H. Burris

It is the privilege of age to talk on a subject like this one that defies adequate quantitation. The organizers reason—let the young fellows collect the solid data, but offer the geriatric set the opportunity to talk harmlessly about global impacts, nitrogen budgets, and the philosophy of N_2 fixation.

The global nitrogen cycle can be drawn in as simple or as complicated a form as you wish. Figure 1 shows it in rather simplistic fashion, indicating major nitrogen sinks in the atmosphere, the land, and the sea and showing interactions. Burns and Hardy (1975) prefer to consider five sinks: the primary rocks, the sedimentary rocks, the deep sea sediments, the atmosphere, and the soil-water pool. They clearly recognize that the biological action is centered in the soil-water (i.e., terrestrial and sea) pools. The assessments of Burns and Hardy (1975) and of Delwiche (1970, 1977) are thorough, carefully thought out, and particularly helpful to anyone trying to assess the nitrogen cycle. The reservoir of N_2 in the atmosphere is generally recorded (Delwiche, 1977) as 3.9×10^9 teragrams (Tg), or 3.9×10^{15} metric tons. However, this is only about 2% of global nitrogen; almost 98% is held in the earth's primary rocks. Plants and animals contain only about 1×10^{10} metric tons (Garrels, Mackenzie, and Hunt, 1975; Delwiche, 1977) or one four-hundred-thousandth of the nitrogen in the atmosphere. It is apparent that, with a balance of 400,000 parts versus 1 part, the complete cycling of the atmospheric pool through the plant-animal pool is not going to be very rapid. The organic and inorganic matter of the soil carries a substantial pool of the terrestrial nitrogen. It is estimated to be 3.3×10^{11} metric tons, or roughly one ten-thousandth of the nitrogen in the atmosphere (Delwiche, 1977).

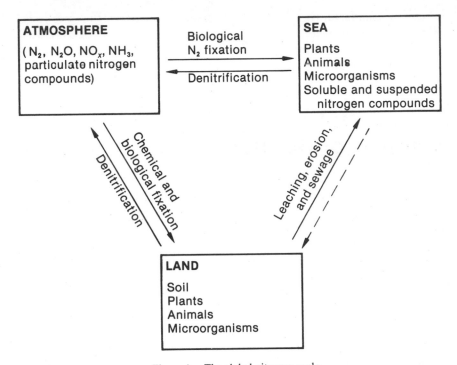

Figure 1.   The global nitrogen cycle.

How does nitrogen make the transition between the atmospheric and the terrestrial pools? The transition comes through chemical fixation, natural fixation by lightning discharges and other atmospheric reactions, and biological N_2 fixation (see Figure 1). Biological fixation can be attributed to symbiotic and nonsymbiotic N_2 fixation systems. The fixed nitrogen in the symbiotic system goes directly into the plants from the captive microorganisms, whereas the free-living bacteria retain most of their fixed nitrogen until they die and decompose. At this point the fixed nitrogen is returned to the soil, where it becomes available to plants.

Plants support animals, and the animals do not cut much of a figure in the overall nitrogen cycle. Studies of animal metabolism receive most of the research largess in this country and elsewhere because government agencies are run by mammalian chauvinists. However, the nitrogen cycle will be grinding away and largely ignoring the animals on this planet long after the bureaucrats have gone to their reward.

There are losses from the terrestrial pool back to the atmospheric pool (Delwiche and Bryan, 1976). A considerable variety of microorganisms are capable of using nitrate and nitrite as oxidants; in the process they reduce these compounds to N_2, N_2O, and other nitrogen oxides that are lost as gases to the atmosphere. This loss is deplorable, but it does serve a useful purpose for organisms that are dependent on nitrate and nitrite to support their oxidative metabolism, and it does keep the nitrogen cycle turning so that the terrestrial pool is not unduly enriched at the expense of the atmosphere. Depletion of the atmosphere is not a major concern, but too much fixed nitrogen in the terrestrial pool could cause problems.

Finally, there is the nitrogen pool in the sea. Data on the size of this pool are particularly imprecise, but the total pool of soluble and suspended nitrogen may be near 1×10^{10} metric tons of nitrogen, or about three times the terrestrial pool (Delwiche, 1977). The ocean sediments are another matter. They have been estimated to be 200 times as great as the dissolved and suspended pool, or about 2×10^{14} tons of nitrogen. Note that the sea receives fixed nitrogen from sewage and solid wastes and from the leaching and erosion of nitrogenous compounds from the soil (see Figure 1). Minimal amounts of nitrogen are returned from the sea to the land as wastes from marine products. The atmospheric pool, however, is in direct and substantial contact with the pool in the sea via N_2-fixing and denitrifying marine organisms. Quantitative data are soft on the flux in both directions. Burns and Hardy (1975) have suggested that there is fixation of about 1 kg of nitrogen/ha annually in the sea to yield an annual total of 36×10^6 metric tons. Although this is not very active fixation, I nevertheless think the estimate is high. If you have pulled a plankton net through the ocean and tested the ability of your meager catch to reduce C_2H_2, you realize that most of the ocean is a biological desert rather than the bountiful bonanza belt that it usually is pictured. It has its areas of upwelling, high nutrient levels and high productivity, but these areas are limited in extent and atypical of the ocean in general.

Why are meaningful data on the nitrogen cycle so hard to accumulate? First, the areas to be analyzed are vast. It involves a major effort to gather data even on a limited section of the land or ocean surface of the earth. Second, the various sections of the earth are heterogeneous in soil type, climate, and nutrient levels. Third, productivity varies seasonally, and it is difficult to integrate productivity rates over the entire year.

One only has to fly over the countryside in spring or fall to see the heterogeneity of the soil laid bare by the plow. We pride ourselves on our rich soils in the Midwest, but the cultivated soils run from heavily leached sands to highly productive loams, from yellow to black. The heterogeneity within single fields forms a dramatic patchwork. And what of the uncultivated areas? Some have reasonably good soils but too much slope, whereas others are predominantly rocky. Not much time is spent measuring the productivity of the untilled areas, but they must appear in the balance sheets with estimates of their inputs and outputs.

At least the land generally stays put. This is not true of a body of water. Again, if you examine our lakes from the air in summer, you can observe striking areas of algal growth contrasted with blue areas of algal deficits on the same lake. A few hours later, a shift in the winds may have altered the entire pattern. As if this were not discouraging enough, in a body of water you must deal with three dimensions, and changes in the vertical direction are much more dramatic than those on land. To compound the difficulties still further, the vertical migration may be quite species specific, so that migration of the N_2-fixing blue-green algae may occur when the general population shows little vertical movement, or vice versa (Burris and Peterson, 1978).

So we have problems in quantitation. As Burns and Hardy (1975) stated, ". . . values represent best estimates frequently derived from conflicting literature reports." Delwiche (1977) observed, "Figures are uncertain in any case and subject to revision as new data are obtained." Can we count on improvements in the data? I'm sure that the data base will improve dramatically. More people are now inclined to take a broad view and to ask questions about the biogeochemistry of important nutrient elements. For many years only a few people like G. Evelyn Hutchinson asked the big questions. Now many people are interested in how our ecosystems are faring. The regulatory agencies want data, and have their own people or investigators under grant or contract collecting them. The task becomes no less complex with time, but the manpower assigned to the task is growing rapidly and methodology is improving. The International Biological Program (IBP) had its detractors, but its biome studies did support examination of a variety of ecosystems to determine balances of carbon, nitrogen, and other elements. These data have been collected in a series of volumes and some parts of the data are still under analysis.

A particularly encouraging development is the organization of SCOPE under the International Council of Scientific Unions (ICSU), a nongovernmental organization that has adhering societies in many countries. These include 66 National Academies of Science, 18 International Unions, and 12 other scientific associations. Fairly recently, ICSU has organized SCOPE, the Scientific Committee on Problems of the Environment. Among its seven areas of interest is that designated as Biogeochemical Cycles. Under this title, SCOPE has formed a Nitrogen Unit at the Royal Swedish Academy of Sciences in Stockholm. A publication on the nitrogen, phosphorus, and sulfur cycles has been sponsored and has appeared as "SCOPE 7" (Svensson and Söderlund, 1976). The nitrogen cycle group in Stockholm is sponsoring data collection on a global scale to shore up the inadequate data base for the nitrogen cycle. The group is planning a new integrated nitrogen project, including literature surveys on simulation models of nitrogen cycling, nitrogen in agriculture, and organic matter in agriculture. The Australian contingent may attempt to establish a nitrogen budget for the entire Australian continent. A West African workshop in 1978 emphasized nitrogen cycling in West African ecosystems, and training courses on applied aspects of the nitrogen cycle were organized. The interest of SCOPE, with its international representation, in the specific area of N_2 fixation is very encouraging. Broad representation may aid in resolving some of the difficulties in assessing and reconciling the data from many countries. We all wish SCOPE the best of success in clarifying the quantitative aspect of the nitrogen cycle, and I trust everyone asked will help the group to gather data.

What is the methodology for establishing a nitrogen budget? Judging from the discrepancies in the data from different writers, one concludes that investigators use a crystal ball, consult a sage in a cave, go into a trance, or, to modernize the process, ask a computer for a random number. There is unquestionably a good deal of nonsense published. As an example, let me quote from unnamed University of California authorities who, in 1975, described N_2 fixation in "Tahoe's Troubled Waters." They stated, "Astonishingly, both pure rain and snow contain higher concentrations of nitrate than Lake Tahoe, so certainly some nitrate enters the lake as precipitation. A far more significant source, however, is the surrounding vegetation, which converts atmospheric nitrogen to nutrient nitrogen at the rate of many tons per acre per year." Many is hardly a quantitative term, but, if we modestly assume five is many, then the group reports 10,000 pounds of nitrogen fixed annually per acre. Astonishing indeed, that the rocky slopes of the Sierras should hold the world's record for the annual rate of N_2 fixation.

The more commonly employed method for estimating pools and flux in the nitrogen cycle is to consult the literature, choose the data that seem to make sense, and then construct a table to express the balance. This furnishes a satisfactory starting point, but the matter should not be left there. How can investigators improve the data base? Rather obviously, the same way they improve any data base—they go to the field to collect the best quantitative data they can. Fortunately, methodology has been improved in recent years. C_2H_2 reduction provides a simple, cheap, rapid, and highly sensitive method that has been widely applied in the field. There are problems associated with the method, but despite its limitations it yields far better data than were available earlier. The simplicity of the method permits it to be used on many samples, and this is important when sampling a heterogeneous medium like the soil or a body of water. My pet complaint is that the method often is used without any determination of a factor to convert C_2H_2 to N_2 reduction (Peterson and Burris, 1976). It is not just that nitrogenase has an inherent capability to divert electrons from N_2 reduction to produce H_2. It is much more complicated than this. I suggest that the simplistic idea that when C_2H_2 is present all electron flow is directed to its reduction and when N_2 is present all electron flow is directed to its reduction should be discarded. Simultaneous measurements of the production of ammonia, C_2H_2, and H_2 show that the balance in electron flow to generate these individual products varies drastically as the total electron flux through the system is changed. General adoption of a $3:1$ ratio for the $C_2H_2:N_2$ reduction ratio clearly is inappropriate. A ratio of $4:1$ is usually nearer to the true value, but a proper value for every set of experimental conditions should be determined.

It is encouraging that more people are adopting $^{15}N_2$ as a tracer both for standardization of the C_2H_2 reduction technique and for direct measurement of N_2 fixation. Commercially available isotope-ratio mass spectrometers now routinely provide measurements of high precision and accuracy. They make it possible to use compounds only modestly enriched with ^{15}N and still obtain valid analytical results. ^{15}N-depleted material can be employed effectively, and isotope fractionation can be measured.

The group at the USDA laboratories at Beltsville has employed isotope dilution to measure N_2 fixation by plants growing in large cylinders buried in the field. ^{15}N-enriched inorganic nitrogen sources, usually nitrate, are mixed with soil and sugar and are left for a period sufficient for the nitrate to be used in the synthesis of bacterial cells. The ^{15}N is in organic compounds when the cylinders are planted. The plants grow on the ^{15}N-enriched compounds in the cylinder, but their associated bacteria also have the option of fixing N_2 from the atmosphere. Any normal N_2 fixed from the air and assimilated by the plants effects an isotope dilution of the ^{15}N in the plants. Analysis of the plants at the termination of the experiment reveals increases in total dry matter and total nitrogen, and isotope dilution analysis reveals the fraction of nitrogen in the plant that came from the soil and from atmospheric N_2. The method is somewhat cumbersome, but it permits measurement of N_2 fixation in plants grown under conditions approaching those in the field, and it reveals clearly how much of the plants' nitrogen comes from the soil and how much from the air. The method should be used more widely since it can counter the statements in the literature quoted like Mosaic law to allege that soybeans take 25% of their nitrogen from the air and 75% from the soil. This statement is repeated as if the soil and its nutrient level had no influence on the ratio of use of N_2 and fixed nitrogen.

Some index of the nitrogen cycle relative to the carbon cycle can be derived. The carbon cycle has attracted considerable attention, but rather large discrepancies still appear among estimates by various individuals. Woodwell et al. (1978) have been making a conscientious effort to bring reason into the quantitation of the carbon cycle, and they evaluate the earth's primary production at 78×10^9 metric tons of carbon annually. If we assume a nitrogen content of 0.2% in this material, that would be equivalent to 156 million tons of nitrogen assimilated in the annual cycle, a not unreasonable figure.

Denitrification values are more difficult to derive than values for N_2 fixation (Delwiche and Bryan, 1976). The gaseous products of denitrification are produced slowly from dispersed sources and are difficult to capture and analyze. Hence, most data are derived from confined systems, and these may be quite different from natural systems in their rates of denitrification. The usual solution to these analytical difficulties is to conclude that the nitrogen cycle seems to be in balance, and that one is therefore justified in equating annual N_2 fixation and denitrification. This conclusion has no analytical basis and may be quite unjustified.

What projections can be made regarding the nitrogen cycle? Certainly chemical fixation of N_2 has been increasing and will continue to increase. World capacity now is around 60 million metric tons per year and actual output is near 50 million metric tons. Over 40 million metric tons of this is used as fertilizer. Prices rose dramatically a few years ago, but now have dropped back somewhat below $200 per ton of anhydrous ammonia. Capacity is adequate and there is currently little pressure for a price increase. This situation could change with any marked change in the price of gas or oil. An increase in capacity for chemical N_2 fixation can be anticipated in the Middle East, where funds for plant construction are available and large quantities of natural gas are still flared.

There is talk about the pressure for more food and for more biological N_2 fixation, but few areas have shown dramatic increases in acreage planted to legumes. There have been small increases in legumes in Britain and large increases in soybean plantings in the U.S. and in Brazil. Elsewhere, crop patterns are rather stable. An appealing story for increasing the use of plants such as the winged bean can be developed, but food and agricultural customs change slowly, and legumes other than the soybean have sparked no agricultural revolution of late. Legumes have a marvelously adapted N_2-fixation system, and they should be exploited more fully. Legumes also can be worked into rotations that will reduce soil erosion; erosion is becoming increasingly serious as the soil is mined to counter the balance of payment deficit engendered by oil importation (Brink, Densmore, and Hill, 1977).

I look for an increasing interest in and an increasing use of plants nodulated by actinomycetes. The alder already is being more widely accepted for pulp and lumber in the Pacific Northwest. It fixes N_2 vigorously, and its wood is acceptable for a variety of applications. The recent isolation of a pure culture of the endophytic actinomycete by Callaham, del Tredici, and Torrey (1978) has sparked interest in studies of this group of N_2 fixers. With this developing interest will come a better assessment of the contribution of this group to the fixed nitrogen pool.

In recent years the estimates of global biological N_2 fixation have been increasing, with a recognition that contributions are being made by groups or organisms that earlier were not recognized as N_2 fixers or were ignored. Delwiche (1970) estimated that about 100 million metric tons of nitrogen were fixed annually by biological agents. Burns and Hardy (1975) increased this estimate to about 175 million metric tons. At the 1976 symposium on N_2-fixing blue-green algae in Uppsala, E. A. Paul (1978) chaired a session that explored fluxes of nitrogen on a global basis. The participants used the Burns and Hardy (1975) estimates as a starting point and then combined their wisdom and intuition to modify the estimates. The group considered the Burns and Hardy estimates of 140 kg of nitrogen/ha too high for average N_2 fixation by cultivated legumes, and they estimated that half of this value would be more realistic. Likewise, the consensus was that 15 kg of nitrogen fixed annually/ha of grasslands was two to three times the real value. Several people at the Uppsala conference had measured N_2 fixation in forest systems and disagreed with the Burns and Hardy estimate of 10 kg of nitrogen fixed annually/ha; their data more commonly showed fixation of 1–3 kg of nitrogen/ha, although there was one Douglas fir value as high as 5 kg and one oak-hickory association with 12 kg of nitrogen fixed/ha. A reduction from 10 to 3 kg would reduce total fixation from 40 to 12×10^6 metric tons of nitrogen fixed in forests and woodlands annually.

These specific modifications reduced the estimate of global biological N_2 fixation from the 175×10^6 metric tons of Burns and Hardy (1975) to about 122×10^6 metric tons. Assuming the 122×10^6 metric tons of nitrogen and the 40×10^6 metric tons of chemically fixed fertilizer nitrogen are assimilated, a total of 162×10^6 metric tons is obtained, in close agreement to the 156×10^6 metric tons estimated from the carbon cycle. These revised estimates give a value for total N_2 fixation that is probably no more believable than that of Delwiche (1970) or Burns and Hardy (1975). The value merely indicates that, with the data presently available, there is plenty of room for disagreement. The experimental data are not sufficiently solid to permit one to take a firm stand, so we will arbitrarily accept the consensus of the Uppsala group and modify our position as better data accumulate.

I have high hopes that the SCOPE subcommittee will explore the nitrogen cycle in greater depth than has been done before and will promote the research and information gathering that will be necessary to establish defensible data on nitrogen pool sizes and flux. Then the results can be further sharpened with time, and a suitable base for prediction of agricultural productivity can be established.

How can biological N_2 fixation be increased? First, by planting more hectares to legumes and by using the best bacterial strains, plant cultivars, and agronomic practices to assure maximum fixation. However, farmers will not plant more legumes unless they are assured a profitable market for them. Second, the free-living blue-green algae and the *Azolla-Anabaena azollae* association can be exploited more extensively in rice husbandry (Talley, Talley, and Rains, 1977). These organisms are active N_2 fixers, and they have been shown to work effectively in extensive rice-growing areas. Their use can be extended and the techniques for their application improved. Third, we can look for N_2-fixing organisms that grow freely in the soil or in association with roots, and adopt practices that will enhance their N_2 fixation. For example, we can adopt cropping practices that will leave adequate residues in the soil to furnish energy to support growth and N_2 fixation by the organisms. There are many microorganisms in soils capable of fixing N_2, but they usually lack adequate supplies of available energy. If we could conveniently and economically supply such energy, it would enhance N_2 fixation. In addition to these primary ways for increasing biological N_2 fixation, alders and other actinomycete-nodulated plants could be used more extensively, as well as a variety of little-publicized N_2 fixers. Finally, farmers should be trained to keep a sharp eye for minimizing losses of nitrogen by denitrification, leaching, and erosion.

REFERENCES

Brink, R. A., J. W. Densmore, and G. A. Hill. 1977. Soil deterioration and the growing world demand for food. Science 197:625–630.

Burns, R. C., and R. W. F. Hardy. 1975. Nitrogen Fixation in Bacteria and Higher Plants. Springer-Verlag, Berlin.

Burris, R. H., and R. B. Peterson. 1978. N_2-fixing blue-green algae: Their H_2 metabolism and their activity in freshwater lakes. In: U. Granhall (ed.), Environmental Role of Nitrogen-fixing Blue-green Algae and Asymbiotic Bacteria. Ecol. Bull. (Stockholm) 26:26–40.

Callaham, D., P. del Tredici, and J. G. Torrey. 1978. Isolation and cultivation *in vitro* of the actinomycete causing root nodulation in *Comptonia*. Science 199:899–902.

Delwiche, C. C. 1970. The nitrogen cycle. Sci. Am. 223:136–146.

Delwiche, C. C. 1977. Energy relations in the global nitrogen cycle. Ambio 6:106–111.

Delwiche, C. C., and B. A. Bryan. 1976. Denitrification. Annu. Rev. Microbiol. 30:241–262.

Garrels, R. M., F. T. Mackenzie, and C. Hunt. 1975. Chemical Cycles and the Global Environment: Assessing Human Influences. Kaufman, Inc., Los Altos, California.

Paul, E. A. 1978. In: U. Granhall (ed.), Environmental Role of Nitrogen-fixing Blue-green Algae and Asymbiotic Bacteria, p. 282. Ecol. Bull. (Stockholm), Vol. 26.

Peterson, R. B., and R. H. Burris. 1976. Conversion of acetylene reduction rates to nitrogen fixation rates in natural populations of blue-green algae. Anal. Biochem. 73:404–410.

Svensson, B. H., and R. Söderlund (eds.) 1976. Nitrogen, Phosphorus and Sulphur—Global Cycles. SCOPE Report 7. Ecol. Bull. (Stockholm), Vol. 22.

Talley, S. N., B. J. Talley, and D. W. Rains. 1977. In: A. Hollaender (ed.), Genetic Engineering for Nitrogen Fixation, p. 259. Plenum Publishing Corp., New York.

Woodwell, G. M., R. H. Whittaker, W. A. Reiners, G. E. Likens, C. C. Delwiche, and D. B. Botkin. 1978. The biota and the world carbon budget. Science 199:141–146.

Nitrogen Fixation, Volume I
Edited by W. E. Newton and W. H. Orme-Johnson
Copyright 1980   University Park Press   Baltimore

Energy Flows in the Nitrogen Cycle, Especially in Fixation

V. P. Gutschick

The thermodynamically minimal energy that plants or microbes must spend to fix nitrogen is considerable, and necessary activation energy losses in enzymatic paths increase the cost. Efficiency of respiration, which is high in legume-*Rhizobium* symbioses and low in free-living fixers, is even more of a crucial determinant of energy costs. The roles of legumes and of free-livers in ecosystems are strongly conditioned by their resulting energy intensiveness. On the whole, in the nitrogen-starved biosphere, fixers are allowed about 2.5% of total photosynthate directly or indirectly. The strategy of the biosphere as a whole is to recycle the less costly nitrogen of dead plants as nitrate. *Energy* is lost in oxidation of ammonia (the first product of recycling) to nitrate, and *nitrogen* itself is lost in microbial reactions at the "low end" of the nitrogen cycle; both losses are explicable as "commissions" to microbes for direct and indirect services to individual plants and to the biosphere.

Dynamic responses of ecosystems to local conditions present some expected patterns (e.g., tightness of cycling in climax ecosystems) and some quantitative puzzles, especially in transfers of ammonia by air and in the oceanic energy budget for fixation. Individual legumes respond to nitrogen stress by fixing N_2, but neither energy costs nor their photosynthetic competitiveness seem to have roles in this "strategy."

Human response to the needs of crops for nitrogen includes diversion of about 1.6% of world energy to fertilizer synthesis. Worries about ultimate fuel shortages have sparked research, especially for increasing biofixation, but several caveats apply because crop yields are lowered by the energy

This work was performed under the auspices of the U.S. Department of Energy.

drain on plants. Decreasing fertilizer losses, which currently average 50%, is one worthy goal of research; two practical routes to do so differ notably in a total energy picture (fossil plus photosynthetic).

The following discussion quantifies the points just made, and points out some specific research needs. (Eight captioned figures used in the oral presentation of this paper are omitted from this text for lack of space. Copies are available on request from the author.)

INTERNAL ENERGY ECONOMY OF MICROBES

Minimal Energy Requirements

The nominal reaction in N_2 fixation is

$$N_2 \text{ (gas)} + 3\,H_2O \text{ (liquid)} \rightarrow 2NH_3 \text{ (aqueous)} + 3/2\,O_2 \text{ (gas)}$$

Ammonia is the form that all fixed nitrogen achieves, and is the highest point in energy. The reactants N_2 and H_2O are considered to be costless, that is, maintained in high concentrations by natural, abiological processes, although N_2 is, in fact, partly maintained by biological denitrification (see below). The *least* energy input for fixation is the free-energy change, ΔG, for the reaction above. ΔG is positive for a process that must be driven, and negative for a spontaneous, usually energy-liberating process. Its value depends on concentrations of reactants and products, but a standard $\Delta G°$ is used as a close approximation. For fixation, $\Delta G°$ is about $+340$ kJ per mole of NH_3. Fixation can be driven by coupling to the oxidation of glucose with $\Delta G° = -3140$ kJ/mole. With perfect energy transfer, 0.11 mole of glucose must be burned to fix 1 mole of ammonia, or 1.44 g of glucose per gram of nitrogen fixed. Much more glucose is used in living organisms.

Energy Use by Legumes

Minchin and Pate (1973) carefully measured all carbohydrate flows in pea seedlings 21 to 30 days old, inoculated with the appropriate symbiotic *Rhizobium*. The roots used 18.8 g of glucose per gram of N_2 fixed. Minchin and Pate "charged" only the 14.8 g respired as the cost of fixation, for an energy efficiency of 10%. I agree that perhaps one-half of root respiration is for fixation, and one-half of root and all of nodular growth should be charged as necessary "overhead" for fixation (but see "Response by Legumes" on scavenging this overhead). This accounting yields a cost of 12.0 g of glucose/g of N_2 fixed and an efficiency of 12%. Fixation is therefore costly. If all of a legume's nitrogen (2.8% by weight) came from fixation, then 150 g of primary photosynthate might yield only 82 g of final dry weight, after 28 g are respired for fixation and 41 g for other biosyntheses. Comparing this to 100 g of dry weight if fixed nitrogen were supplied "free," we see that the cost in yield is about 18%. This cost is ameliorated in several ways: 1) because nitrogen shortage limits growth on most land, any gram of N_2 fixed allows an extra 36 g of plant to grow, offsetting the crude energy cost by three to one; 2) legumes fix only 25% to perhaps 85% of their needs over the growing season (Hardy and Havelka, 1975), and the rest is obtained from the soil by recycling past plant growth; 3) even recycled nitrogen as nitrate costs almost as much to reduce to ammonia as does N_2, if this is done in the roots by nitrate reductase with glucose respiration (Minchin and Pate, 1973). Although ammonia that is free of energy cost is an initial product in nitrogen recycling, it is soon oxidized bacterially to nitrate. This is fortunate for the bacteria that gain the energy and also for all higher plants, because ammonia is immobilized in soil by adsorption (on clays, especially), whereas nitrate is free to migrate to roots.

Energy Use by Enzymes Alone

Investigations using mutants of free-living bacteria (reviewed by Ljones, 1974) have shown that nitrogenase uses 10–12 ATP molecules and three reductants (carrying two hydrogens each, denoted as [2H]) per molecule of N_2 reduced. For each N_2 reduced, an average of 1.3 molecules of H_2 is liberated in a "useless" but probably obligatory side reaction (Schrauzer, 1976; discussion by Evans et al., 1977). Each H_2 costs only one-third as much in reductant and ATP as an N_2, so the total drain is 40% of N_2-dedicated consumptions. In many bacterial species two or three ATP molecules are recovered from the 1.3 H_2 molecules later. Total use per N_2 molecule thus averages 13.5 ATP molecules and 4.2 reductants [2H].

Anaerobic respiration supplies 2 ATP molecules and 2 reductants per glucose molecule. Allowing for microbial facility in rebalancing the ATP:reductant ratio, as with nitrate terminal oxidation, then nearly 3.3 molecules of glucose would suffice per N_2, or 1.63 per NH_3—an efficiency of 6.7%. Aerobic respiration could supply 38 ATP per glucose, or one-third as many reductants, such that 0.69 molecules of glucose would suffice per N_2—an efficiency of 32%. Aerobism is tolerable only if nitrogenase is protected from free O_2, a system perfected only in *Rhizobium*-legume symbioses (and perhaps in *Actinomycete-Alnus* symbioses—carbohydrate consumption in the latter, as discussed during this symposium, may be so low as to merit the term "perfected"). Halving the above efficiency to account crudely for cellular overhead gives 16% efficiency, near the estimates for whole legumes.

Energy Use by Free-living Microbes

Certain species of bacteria living free in the soil can fix N_2. They fall into all metabolic classes. Some are photosynthetic autotrophs, all of which fix N_2

only anaerobically and with efficiencies that are yet unmeasured. Most important are heterotrophs, for which Mulder and Brotonegoro (1974) reviewed the empirical efficiencies. *Azotobacter*, a typical free-living aerobe, uses 100 g of glucose per gram of N_2, for an efficiency of only 1.4%. The efficiency of aerobic respiration is offset by the wasteful burning of glucose (apparently) near the nitrogenase to protect the enzyme from O_2. Consequently, *Azotobacter*'s metabolic rate is the highest measured in any living matter. Anaerobes also do poorly, because of their low respiratory efficiency. General surveys of soil heterotrophs (Virtanen and Miettinen, 1963) show 0.3%–4% efficiencies. On leaves of plants (Ruinen, 1974) the average may be near 3%.

Blue-green algae of many species also fix N_2, usually when growing photosynthetically (Stewart, 1974). Perhaps efficiencies are good because fixation is done in protected heterocysts, but measurements are lacking.

Why is Fixation Inefficient?

Fixation's primary efficiency of 32%, before overhead, compares favorably with the 34% maximal efficiency of photosynthesis (Thorndike, 1976), perhaps the only comparably difficult biological reaction. However, photosynthesis is more of a feat. The number of steps involved is larger, and losses of activation energies, E_a, needed to prevent back-reactions are very well minimized—especially in the splitting of water to H^+ and O_2 by photosystem II, which is 75% as energy costly as direct photolysis to H and OH radicals, even with work to separate products added to the former. Fixation has fewer steps but greater E_a losses. Fixation does have more difficult work, the splitting of the $N\equiv N$ bond, one of the strongest bonds, at room temperature. Furthermore, evolutionary pressure for efficient fixation is likely to be less than that for efficient photosynthesis; perhaps partly in consequence, only procaryotes have been "chosen" for the task.

ENERGY AND NITROGEN FLOWS
IN THE GLOBAL NITROGEN CYCLE

Energy Cost of the Cycle

Annual biological nitrogen fixation on land has been estimated (most recently but not definitively) by Burns and Hardy (1975) to be 140 teragrams (Tg; million metric tons). I estimate that perhaps 80% is fixed by symbiotic fixers, which is the same proportion that Burns and Hardy estimate for cultivated land: cultivated and wild legumes *do* abound, in both species (Heywood, 1971) and numbers (discussion by Lopes, 1978). The remaining 20% is done by free-livers at a global average intensity of 2 $kg \cdot ha^{-1} \cdot yr^{-1}$, an estimate that is near the upper limits of persistent rates in the few measurements in the wild on algae (Fogg et al., 1973; Stewart et al., 1978) and on bacteria (Döbereiner, 1974). At estimated costs of 12 and 50 grams of glucose, respectively, per gram of N_2 fixed, the total cost of fixation is $1344 + 1400 = 2744$ Tg of glucose. This equals 2.5% of the estimated primary photosynthesis on land of 110,000 Tg (Rodin, Bazilevich, and Rozov, 1975). [A more recent estimate by Whittaker and Likens of 132,000 Tg as dry matter is reported by Woodwell (1978).] The legumes use this photosynthate directly, whereas free-livers use the previous years' product in dead matter.

Because fixation is quite costly, especially to the particular higher plant or microbe where it occurs, and because fixing microbes have limited ecological niches (partly in consequence), the biosphere meets most of its nitrogen demand by recycling with the help of various soil bacteria and fungi. The biosphere needs about 1050 Tg of nitrogen annually, to support a gross production (after respiratory losses) of near 70,000 Tg, which contains an average of 1.5% nitrogen by weight (my estimate, for 67% production as herbaceous growth or tree leaves and 33% as wood, averaging 2% and 0.5% nitrogen, respectively). The 140 Tg of biological fixation and 40 Tg of abiological fixation (by lighting, forest fires, and ozonization) are supplemented by an estimated 870 Tg of decay-recycled nitrogen, which is 83% of that needed. Decay itself uses primarily the energy in litterfall, respiring away much carbon content as CO_2. The nitrogen-enriched residue left after aerobic fungi and less demanding decomposers act (Burges and Raw, 1967; Dickinson and Pugh, 1974) is largely oxidized to NH_3 and CO_2; a small amount is assimilated directly for microbial protein (Campbell and Lees, 1967). NH_3, in the barely aerobic environment, is oxidized through nitrite to nitrate for further energy by other microbes.

Soil nitrate (and to a small extent, ammonia) is available for uptake by higher plants. As noted in "Energy Use by Legumes," nitrate must be re-reduced to NH_3 inside the plant, at an energy cost per mole of NH_3 nearly equal to that of fixation ($\Delta G° = +331$ kJ/mole). Thus, 870 Tg of recycled nitrate would cost the biosphere about 10,400 Tg (10%) of primary photosynthate. The actual cost is halved to 5% because some nitrate, probably one-half on a worldwide average, is reduced nearly free of energy cost in the leaves, where excess photosynthetic capacity makes reductant available. This reduction rarely competes with photosynthesis because the latter is often shut down for lack of water. The only small overhead is maintenance energy for excess photosynthetic units.

Balancing new fixation, and indeed making it necessary, is the loss of soil nitrate when denitrifying bacteria use it as oxidant anaerobically, either in deeper soil layers or in waters carrying nitrate away as leachate or runoff. Primarily N_2 and N_2O gases are liberated to the atmosphere. [Substrate balances among carbon source, nitrate, and oxygen dictate that these gases and not NH_3 are favored to form (Campbell and Lees, 1967; Delwiche, 1977).]

Need for Denitrification

First, recycling via decay is needed to keep the world's nitrogen mobile and not tied up in dead matter that would, within 4 million years at present primary production, deplete the world of mobile nitrogen. Denitrification in particular keeps nitrogen available at all locations, that is, in the air. Ammonia and nitrate carried away by surface water and precipitation to the ocean at a total of 15 + 45 = 60 Tg per year would deplete even the air in 70 million years (my estimate, using the data of Burns and Hardy, 1975; most nitrogen content in precipitation is generated over the land and not the ocean). Denitrification-fixation and other elemental redox cycles of iron, manganese, and sulfur also keep *oxygen* in the air, at the cost of biological energy flow, instead of tied up in the elemental oxides that are the thermodynamically favored forms given the earth's gross composition (Board, 1977). On a more local scale, I contend that denitrification is a "commission" to soil microflora as a whole, which are faced with excess nitrogen:carbon ratios at the end of the litter cycle; some nitrogen must be dumped for microbial viability, and for prevention of nitrite/nitrate toxicity to higher plants.

DYNAMIC, LOCAL BALANCES

Individual Ecosystems

Nitrogen cycling tends to be least loss prone and most nearly constant in time in climax ecosystems (Bonnier and Brakel, 1969; Franco, 1978), although definitive studies are lacking. In tropical climax forests primary photosynthesis is very high, whereas nitrogen cycling and its attendant energy costs are kept modest. In very rainy climes the leaching of soluble nutrients, including nitrogen, may raise the need for fixation to average or above (Jaiyebo and Moore, 1963); this may be evidenced indirectly by the preponderance of nodulated leguminous trees (Lopes, 1978). To settle questions of openness of nitrogen cycles and of energy needs for fixation/nitrate reduction, we need Hubbard-Brook–style experiments on nutrient balances, plus surveys of leaf nitrate-reductase levels and water-stress frequency, for major ecosystems.

   Fixed nitrogen is also transported aerially (as volatilized ammonia representing much energy lost) between rich but open ecosystems and poorer ones, Magnitudes of the turnovers are unknown, but the total may be 50 Tg or more annually (Burns and Hardy, 1975; Dawson, 1977; also, a simple calculation, based on 1 ppb NH_3 in air as an average and a 3-day residence time, yields 660 Tg!). This turnover process is receiving little attention to date.

Oceanic life seems to be somewhat less intensive than the land in budgeting energy for fixation. It supports photosynthesis equal to 30%–50% of that of land (Whittaker, 1971; Rodin, Bazilevich, and Rozov, 1975; Whittaker and Likens, cited by Woodwell, 1978) with fixation equal to less than 25% of that of the land (Delwiche, 1970; Burns and Hardy, 1975). This occurs even though food chain turnovers with risk of nitrogen loss are far more frequent in oceans, as evidenced by the small standing biomass:production ratio. Oceans *are* partly subsidized by fixed nitrogen in river runoff and in precipitation (see "Need for Denitrification"), but they also lack any known symbioses as efficient as those in terrestrial legumes, so that their fixation demand is very costly in energy. Energy-efficient fixing species, if any, may yet be found in the poorly characterized dark or dysphotic zones.

Responses by Legumes

Legumes are substantially, if not totally, self-sufficient in nitrogen (see "Energy Use by Legumes"). However, they readily decrease fixation and take up soil nitrate in almost exact replacement quantities. The reason for this response is unclear. It is not likely to be a saving of energy, because soybeans, for one, reduce their nitrate almost entirely in the roots at high energy cost. [The claim for dominance of root reduction was challenged by a number of symposium attendees—eds.] Energy supplies for fixation *or* nitrate reduction are apparently quite limited in legumes like soybeans, for unknown reasons. Neither superabundant nitrate nor highly favorable conditions for photosynthesis (except the $CO_2:O_2$ ratio; as seen below) will prod soybeans to flourish. For unknown reasons, photosynthesis appears to be limited genetically by high photorespiratory losses, poor leaf coverage, and so on.

Bonnier and Brakel (1969) have found evidence that legumes will nodulate and fix significant N_2 only under nitrogen stress, when an ecosystem's nitrogen cycle is open and losses are high, in early succession. Under such conditions, the large drain on photosynthate (and poor photosynthetic performance in general) is tolerable, because photosynthetically efficient potential competitors lack nitrogen for growth. Eventually, legumes fertilize the soil and presumably trigger their own demise.

Legumes apparently destroy their own nodules when the need for nitrogen is greatest (Sinclair and de Wit, 1976) by scavenging them for all available nitrogen in their structure. This limits seasonal fixation, but note that: 1) I propose that this is an optimal racing strategy (Keller, 1974). whereby one runs out of fuel just *before* the finish of a race—in short, nodules left at the end of a season are just wasted; and 2) nodular senescence *can* be delayed by increasing photosynthesis. Both increased CO_2 concentrations in air and lowered O_2 concentrations help, largely by reducing photorespiratory losses (Hardy, 1977a). Lowered oxygen pressure also exerts an independent effect, that of suppressing reproduction in favor of continued vegetative growth, for unknown reasons and by unknown mechanisms.

We might ask why photosynthetically efficient C_4 plants have no highly productive symbioses. Surely, one reason is that C_4 plants need only half as much nitrogen as C_3 plants for the same dry weight (Neyra, 1978; this also implies that they will be little help, if any, as agricultural protein sources). C_4 plants are even penurious in their energy use for reducing recycled nitrate and do most in their leaves. At low soil nitrate levels, however, they are as helpless as nonleguminous C_3 plants. Even their twofold larger litterfall to support heterotrophic fixers is of little avail, because such fixers are poor competitors for carbon and energy in the litter and will fix N_2 only in the limited anaerobic zones of soil.

AMENDED SYSTEMS IN AGRICULTURE

Much of agricultural management is directed toward the nitrogen supply. In modern, intensive agriculture, high yields are sustained only with synthetic ammonia-based or ammonia-derived fertilizers. The required industrial synthesis costs much energy, mostly in the form of natural gas, because it is subject to the same thermodynamic strictures mentioned under "Minimal Energy Requirements" for the biological synthesis of ammonia. Current industrial synthesis is 54% energy efficient, remarkably high but with little hope of improvement. The 1.6% of worldwide fossil energy that we devote to nitrogenous fertilizers can increase only to perhaps 8% in 50 years. Fossil fuel shortages climaxing at that time will pose a severe threat to intensive agriculture, although these shortages are capable of being overcome by careful planning (even allowing some political failures) and intensive, selective research (Gutschick, 1977). Increased biological fixation can only lift a part of the burden from fossil fuels, because legumes have limited dietary roles and limited yields, and genetic engineering of nonlegumes to fix their own nitrogen is a distant prospect, although worthy of selective research. *Almost* all avenues to increase nitrogen fixation deserve exploitation. However, any that reduce crop yields, such as simply increasing biological fixation, are less affordable because of increased risk of famine (Hardy, 1977b). Similarly, the use of less fertilizer but of more currently wild land could cause a serious depletion of the wild gene pool needed at times for crop breeding (Gutschick, 1978). In addition, there are social, economic, and even scientific constraints that limit other avenues.

In fertilizer, fossil energy embodied in ammonia or nitrate merges directly with the flow of photosynthetic energy. Available nitrogen "catalyzes" the retention of photosynthetic energy as net growth. In addition, large concentrations of free nitrate are soon diverted by the small but voracious population of denitrifying bacteria to generate metabolic energy that is generally the prime limitation of all soil microbial growth (Burges and Raw, 1967); nitrogen per se is only secondarily limiting in most soils. In other soils where nitrogen is limiting, added nitrogen can stimulate the decomposition of carbonaceous, energy-embodying litter. The added nitrogen "catalyzes" microbial metabolism, especially the uptake of nitrogen for microbial protein synthesis (immobilization of soil nitrogen, as it is called).

Nitrate from fertilizers is usually made available in large, short-time pulses to both higher plants and soil microbes. The former are decent competitors only up to modest concentrations of nitrate, after which uptake begins to saturate. At the same concentrations, microbial uptake may be stimulated above kinetic first order, as microbes are aroused from energy starvation. An average of 50% of the nitrogen applied as fertilizer is thus lost (versus 10%–20% of soil nitrogen lost on wild land) to increased denitrification and leaching (Huber et al., 1977). Chemical suppressors of nitrifying bacteria, applied with ammoniacal fertilizer, spread out the pulse of nitrate generation over longer time spans and ultimately considerably increase the uptake efficiencies of crops. So also does delivery of fertilizer in many small doses at time intervals (Ingestad, 1977), a practice likely to find only limited agricultural use because the fossil energy (and dollar) cost to deliver fertilizer by tractor even once to a field is a significant fraction of the energy used to synthesize fertilizer. These and other practices for upgrading efficiency of fertilizer use promise even greater benefits to food production than improved efficiency of industrial ammonia synthesis.

ACKNOWLEDGMENT

I thank Lou Ellen Kay (Los Alamos) for constructive criticism, and for artwork in slides for the talk based on this paper.

REFERENCES

Board, P. A. 1977. Redox regulation of atmospheric oxygen and its consequences. Nature 270:591–592.
Bonnier, C., and J. Brakel. 1969. Lutte Biologique Contre la Faim. J. Duculot, Gembloux.
Burges, A., and F. Raw. 1967. Soil Biology. Academic Press, London.
Burns, R. C., and R. W. F. Hardy. 1975. Nitrogen Fixation in Bacteria and Higher Plants. Springer, New York.

Campbell, N. E. R., and H. Lees. 1967. The nitrogen cycle. In: A. D. McLaren and G. H. Peterson (eds.), Soil Biochemistry, Vol. 1, pp. 194–215. Marcel Dekker, New York.

Dawson, G. A. 1977. Atmospheric ammonia from undisturbed land. J. Geophys. Res. 82:3125–3133.

Delwiche, C. C. 1970. The nitrogen cycle. Sci. Am. 223:136–146.

Delwiche, C. C. 1977. Nitrous oxide and denitrification. Paper presented at the Denitrification Seminar of the Fertilizer Institute, 26–27 October, San Francisco.

Dickinson, C. H., and G. J. F. Pugh. 1974. Biology of Plant Litter Decomposition. Academic Press, London.

Döbereiner, J. 1974. Nitrogen-fixing bacteria in the rhizosphere. In: A. Quispel (ed.), The Biology of Nitrogen Fixation, pp. 86–120. North-Holland, Amsterdam.

Evans, H. J., T. Ruiz-Argüeso, N. Jennings, and J. Hanus. 1977. Energy coupling efficiency of symbiotic nitrogen fixation. In: A. Hollaender (ed.), Genetic Engineering for Nitrogen Fixation, pp. 333–354. Plenum Publishing Corp., New York.

Fogg, G. E., W. D. P. Stewart, P. Fay, and A. E. Walsby. 1973. The Blue-Green Algae, pp. 323–342. Academic Press, London.

Franco, A. A. 1978. Contribution of the legume-*Rhizobium* symbiosis to the ecosystem and food production. In: J. Döbereiner, R. H. Burris, and A. Hollaender (eds.), Limitations and Potentials for Biological Nitrogen Fixation in the Tropics, pp. 65–74. Plenum Publishing Corp., New York.

Gutschick, V. P. 1977. Long-term strategies for supplying nitrogen to crops. Los Alamos Scientific Laboratory report LA-6700-MS. May. Los Alamos Scientific Laboratory, Los Alamos, New Mexico.

Gutschick, V. P. 1978. Energy and nitrogen fixation. BioScience 28:521–525.

Hardy, R. W. F. 1977a. Rate-limiting steps in biological photoproductivity. In: A. Hollaender (ed.), Genetic Engineering for Nitrogen Fixation, pp. 369–399. Plenum Publishing Corp., New York.

Hardy, R. W. F. 1977b. Increasing crop productivity: Agronomic and economic considerations on the role of biological nitrogen fixation. In: A. Hollaender (ed.), Report of the Public Meeting on Genetic Engineering for Nitrogen Fixation, 5–6 October, Washington, D.C. U. S. Government Printing Office, Washington, D.C.

Hardy, R. W. F., and U. D. Havelka. 1975. Photosynthate as a major factor limiting nitrogen fixation by field ground legumes, with emphasis on soybeans. In: P. Nutman (ed.), Symbiotic Nitrogen Fixation, pp. 421–439. Cambridge University Press, London.

Heywood, V. H. 1971. The leguminosae—a systematic purview. In: J. B. Harborne, D. Bouiter, and B. L. Turner (eds.), Chemotaxonomy of the Leguminosae, pp. 1–29. Academic Press, London.

Huber, D. M., H. L. Warren, D. W. Nelson, and C. Y. Tsai. 1977. Nitrification inhibitors—new tools for food production. BioScience 27:523–529.

Ingestad, T. 1977. Nitrogen and plant growth; maximum efficiency of nitrogen fertilizers. Ambio 6:146–151.

Jaiyebo, E. O., and A. W. Moore. 1963. Soil nitrogen accumulation under different covers in a tropical rain-forest environment. Nature 197:317–318.

Keller, J. B. 1974. Optimal velocity in a race. Amer. Math. Mo. 81:474–480.

Ljones, P. 1974. The enzyme system. In: A. Quispel (ed.), The Biology of Nitrogen Fixation, pp. 617–638. North-Holland, Amsterdam.

Lopes, E. S. 1978. Ecology of legume-*Rhizobium* symbiosis. In: J. Döbereiner, R. H. Burris, and A. Hollaender (eds.), Limitations and Potentials for Biological

Nitrogen Fixation in the Tropics, pp. 173–190. Plenum Publishing Corp., New York.

Minchin, F. R., and J. S. Pate. 1973. The carbon balance of a legume and the functional economy of its root nodules. J. Exp. Bot. 24:259–271.

Mulder, E. G., and S. Brotonegoro. 1974. Free-living heterotrophic nitrogen-fixing bacteria. In: A. Quispel (ed.), The Biology of Nitrogen Fixation, pp. 37–85. North-Holland, Amsterdam.

Neyra, C. A. 1978. Interactions of plant photosynthesis with dinitrogen fixation and nitrate assimilation. In: J. Döbereiner, R. H. Burris, and A. Hollaender (eds.), Limitations and Potentials for Biological Nitrogen Fixation in the Tropics, pp. 111–120. Plenum Publishing Corp., New York.

Rodin, L. E., N. I. Bazilevich, and N. N. Rozov. 1975. Productivity of the world's main ecosystems. In: Productivity of World Ecosystems, pp. 13–26. National Academy of Sciences (U.S.A.), Washington, D.C.

Ruinen, J. 1974. Nitrogen fixation in the phyllosphere. In: A. Quispel (ed.), The Biology of Nitrogen Fixation, pp. 121–167. North-Holland, Amsterdam.

Schrauzer, G. N. 1976. Biological nitrogen fixation: model studies and mechanism. In: W. E. Newton and C. J. Nyman (eds.), Proceedings of The First International Symposium on Nitrogen Fixation, Vol. 1, pp. 79–116. Washington State University Press, Pullman.

Sinclair, T. R., and C. T. de Wit. 1976. Analysis of the carbon and nitrogen limitations to soybean yield. Agron. J. 68:319–324.

Stewart, W. D. P. 1974. Blue-green algae. In: A. Quispel (ed.), The Biology of Nitrogen Fixation, pp. 202–237. North-Holland, Amsterdam.

Stewart, W. D. P., M. J. Sampaio, A. O. Isichei, and R. Sylvester-Bradley. 1978. Nitrogen fixation by soil algae of temperate and tropical soils. In: J. Döbereiner, R. H. Burris, and A. Hollaender (eds.), Limitations and Potential for Biological Nitrogen Fixation in the Tropics, pp. 41–63. Plenum Publishing Corp., New York.

Thorndike, E. H. 1976. Biological energy and ecosystems. In: Energy and Environment: A Primer for Scientists and Engineers, pp. 24–43. Addison-Wesley Publishing Company, Inc., Reading, Massachusetts.

Virtanen, A. I., and J. K. Miettinen. 1963. Biological nitrogen fixation. In: F. C. Steward (ed.), Plant Physiology: A Treatise, Vol. III, pp. 539–669. Academic Press, Inc., New York.

Whittaker, R. H. 1971. Communities and Ecosystems. Macmillan Publishing Company, Inc., New York.

Woodwell, G. M. 1978. The carbon dioxide question. Sci. Am. 238:34–43.

Nitrogen Fixation, Volume I
Edited by W. E. Newton and W. H. Orme-Johnson
Copyright 1980   University Park Press   Baltimore

Dentrification: A Plea for Attention

W. J. Payne, J. J. Rowe, and B. F. Sherr

Workers in dentrification are devoted to gaining a better understanding of the mechanisms of gathering, distributing, and using nitrogen. We want to exploit the phenomena to fullest benefit in agriculture. However, a wider devotion than occurs presently is needed. Despite continual fixation, nitrogen is *the* limiting element for plant growth in a great many agricultural and natural environments, yet we seem relatively unconcerned, and allot only the barest fraction of our resources to studying the leakage of nitrogen back into the atmosphere. We should be as concerned about loss as about gain.

BACTERIOLOGY

The global scope of chemical release of nitrogen is negligible. We are interested here specifically in biological denitrification. Several but not all of the species in the bacterial genera listed in Table 1 (Buchanan and Gibbons, 1974) are uniquely accountable. These are ordinary types of bacteria found wherever one may sample. No other microorganisms—indeed, no other forms of life—have the capacity for denitrification. These denitrifiers are all heterotrophs except the *Thiobacillus* species. Inorganic sulfur compounds serve as electron donors for the thiobacilli, whereas all others oxidize organic compounds. Their acceptors and end products vary too. Nitrate is the usual electron acceptor. Some *Alcaligenes* species start the denitrifying process at nitrite rather than nitrate, however, and "*Corynebacterium*" *nephridii* and certain *Pseudomonas* species stop at nitrous oxide rather then dinitrogen. A study of numerical dominance in soils (Gamble,

Supported by grant #OCE77-15854 from the National Science Foundation.

Table 1. Genera of bacteria that include denitrifying species [$2NO_3^- \rightarrow 2NO_2^- \rightarrow 2NO \rightarrow N_2O \rightarrow N_2$]

Genus[a]	Shape	Genus[a]	Shape
(Achromobacter)	Gram-negative rods	(Kingella)	Gram-negative coccoid cells
Agrobacterium	Gram-negative rods	(Moraxella)	Gram-negative coccoid cells
Alcaligenes[b]	Gram-negative rods	Neisseria	Gram-negative cocci
Bacillus	Gram-positive rods; endospores	Paracoccus	Gram-negative coccoid cells
Chromobacterium	Gram-negative rods	Propionibacterium	Gram-positive rods
(Corynebacterium)[c]	Gram-positive rods	Pseudomonas[c]	Gram-negative rods
Flavobacterium[b]	Gram-negative rods	Rhizobium	Gram-negative rods
Hyphomicrobium	Gram-negative swarmer cells; germ-tube–forming budding cells	Spirillum[c]	Gram-negative spirals
		Thiobacillus	Gram-negative rods

[a] Parentheses indicate uncertainty about the taxonomy of the isolates studied.

[b] One species does not reduce nitrate but initiates denitrification at nitrite.

[c] Certain strains accumulate nitrous oxide as terminal product of denitrification.

Betlach, and Tiedje, 1977) revealed that denitrifiers comprise a significant part of the microflora and that *Pseudomonas fluorescens* biotypes are the most numerous. Other pseudomonads and *Alcaligenes* species follow. Strangely enough, none of the three most frequently studied denitrifiers, *Pseudomonas denitrificans*, *Pseudomonas perfectomarinus*, and *Paracoccus denitrificans*, appeared among the numerically dominant. Denitrifiers are also found in most all marine and fresh waters.

PHYSIOLOGY

Figure 1 presents diagrammatically the electron transport that comprises denitrification. Basically, the process represents consumption, in series, of four oxidants (as indicated at the far right). Anaerobic respiration permits the denitrifiers, many of which are unable to ferment organic substrates, to substitute nitrogen oxides for oxygen as terminal electron acceptors. *Bacillus* and *Chromobacterium* species are exceptions that can ferment even while respiring as denitrifiers. *Propionibacterium acidi-propionici* is further remarkable in that it ferments and denitrifies but does not respire aerobically. There is no obvious pattern to the distribution of denitrification among the bacteria. Many ordinary organic materials and some unusual substrates, including aromatic compounds (Williams and Evans, 1975), are

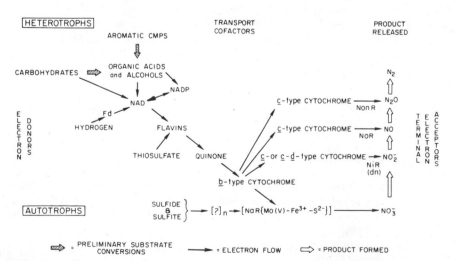

Figure 1. Composite accounting for the events known to comprise denitrification in various types of bacteria.

utilized as carbon sources for denitrifying growth. *Thiobacillus* species fix carbon dioxide as sole carbon source. Preliminary studies revealed changes in catabolic rates and pathways, when aerobic respiration gives way to denitrification in *Pseudomonas stutzeri* (Spangler and Gilmour, 1966). The tricarboxylic acid cycle is reportedly modulated, but still functions, in denitrifying *Paracoccus* (Forget and Pichinoty, 1965). Nothing is known of the nature of, or cellular sites for, such regulation.

Because there are identifiable gaseous intermediates as well as a gaseous end product, denitrifying experiments are carried out with cell suspensions or extracts, degassed with helium, in anaerobic reaction vessels. Samples are withdrawn at intervals from the head space and analyzed on columns of Porapak Q (Hollis, 1966; Barbaree and Payne, 1967).

NITRATE REDUCTION

During the initial stages of heterotrophic denitrification electron transport seems unexceptional (Payne, 1973; Pichinoty, 1973). Nitrate reduction in denitrifiers is similar to that in other microorganisms and in plants, in that other nitrate reductases also yield nitrite, are molybdoproteins, and (except for the bacterial assimilatory enzyme) also reduce chlorate and several other anions. However, it differs in that all the respiratory nitrate reductases (i.e., those produced by enterobacteria as well as the denitrifiers) are synthesized and function only under anoxic or near-anoxic conditions and are unaffected by the presence of ammonia, whereas the assimilatory reductases (produced by green plants, algae, fungi, and bacteria) are largely indifferent to oxygen and are repressed by ammonia. The nitrate reductase of denitrifiers (Forget, 1971) is apparently the simplest, with a molecular mass of about 150,000 daltons as opposed to 300,000 daltons or more for the others. The least well studied of all nitrate reductases is the assimilatory enzyme in bacteria, which is unstable in cell-free extracts. A few species, such as *Enterobacter aerogenes* and *Pseudomonas aeruginosa*, can produce and utilize both assimilatory and respiratory nitrate reductases simultaneously. Here, one functional entity in different biological matrices may be involved.

NITRITE REDUCTION

Since nitrate, followed by nitrite, reduction in all other pathways leads directly or indirectly to ammonia production, the uniqueness of denitrification is in the components of its nitrite reductase system, and consequently in the product of the enzyme's action. We observed (Payne, Riley, and Cox, 1971) that a subfraction with restricted activity, derived from extracts of the marine bacterium *Ps. perfectomarinus*, produced nitric oxide when supplied with NADH, FAD, and nitrite. Extracts from other denitrifiers perform the same function, using a unique *c-d*-type cytochrome (Yamanaka and Okunuki, 1963; Newton, 1969). There was no reductase for any other nitrogen oxide in the subfraction from *Ps. perfectomarinus* we used, and nitric oxide thus accumulated stoichiometrically. Electron paramagnetic resonance (EPR) evidence also implicates nitric oxide as the product of nitrite reduction. When a system comprising NADH, FAD, and nitrite was incubated with an extract of *Ps. perfectomarinus* from which a cytochrome-rich component had been removed, only a minimal EPR signal and no nitric oxide (by gas chromatography) were evident. When the cytochrome-rich material was restored, a typical heme nitric oxide signal was generated and persisted for the period during which nitrite reduction continued and nitric oxide was released (Cox, Payne, and Dervartanian, 1971).

SIGNIFICANCE OF NITRIC OXIDE PRODUCTION

Our hypothesis that nitric oxide is the product of nitrite reduction seems tenable even though the gas is not ordinarily released from whole cells, unlike the other intermediates, nitrite and nitrous oxide. Despite our results, nitric oxide is not universally accepted as an obligatory intermediate. Whole cells of a few species do reportedly reduce nitric oxide and grow at the expense of that reduction (Pichinoty ct al., 1978), which supports our hypothesis, but additional experiments are needed in this area. What is so important about nitric oxide? First, nitric oxide is toxic for living cells, and learning how denitrifying cells deal with nitric oxide is thus intriguing. Second, nitric oxide, unlike the other intermediates, is not normally released and does not accumulate in the medium around growing denitrifying cells. It must be held internally and very tightly. Third, and most important, nitric oxide production is the departure point that makes unchecked denitrification an agronomic defect of major proportions. No known agent will selectively inhibit denitrifying nitrite reductase and not interfere with the assimilatory kind. The only prospects tested in our laboratory, the alkyl nitrates and nitrites, are not selective at all and inhibit bacterial metabolism altogether. The search for an acceptable inhibitor is an open field for investigation with a useful agronomic pay-off potential.

NITRIC OXIDE REDUCTION—SPECIFIC BINDING AGENT

For nitric oxide to be bound intermediate in the ordinary process of denitrification, its reduction by enzymes from denitrifying bacteria must be demonstrated. Several lines of evidence can be offered. For example, exposure of a crude, undialyzed extract of *Ps. aeruginosa* to either nitrite or nitric oxide (Rowe et al., 1977) resulted in the raising of three light absorption peaks in a difference spectrum—the major one at 573 nm, a lesser one at 540 nm, and another at 425 nm. Following the major peak in the treated versus untreated difference spectra of a dialyzed extract revealed that only nitric oxide generated the peak. None of the other oxides of nitrogen (nitrate, nitrite, or nitrous oxide) nor carbon monoxide raised the peak. After a 10-min incubation, the system supplemented with nitrite also exhibited the peak, but only because much nitrite had been reduced to nitric oxide. A longer period of incubation was required for nitrate supplementation to raise the peak, indicating reduction by the residual endogenous electron donors through nitrite to nitric oxide.

When a nitric oxide–treated dialyzed extract of Ps. aeruginosa was supplied with malate as electron donor, and NAD and FAD as carriers, there was no change at 573 nm for several minutes in the system charged with nitrite (Rowe et al., 1977). In contrast, the intensity of the nitric oxide–generated absorption at 573 nm decreased rapidly. Most significantly, there was rapid and sustained production and release of nitrous oxide during this time. None was released during the incubation period from the system receiving nitrite. The function thus appears restricted to nitric oxide. If read immediately, the intensities of absorption at 573 and 540 nm were approximately linearly proportional to the quantity of nitric oxide added to the extract, as indicated in Figure 2. The peak height at 573 nm diminished with time after injection of nitric oxide, however, as endogenous electron donors were dehydrogenated. When additional nitric oxide was injected, the cycle was repeated (Figure 3). Several repetitions followed until presumably the endogenous stock was depleted. The peak height then remained constant. These observations are consistent with our earlier demonstration that a subfraction and a c-type cytochrome derived from extracts of Ps. perfectomarinus (and found to reduce nitric oxide but no other nitrogen oxide) displayed a heme–nitric oxide EPR signal while reducing nitric oxide (Cox, Payne, and Dervartanian, 1971). There was no signal before the cytochrome component was added; then, a typical heme–nitric oxide EPR signal was seen. The signal broadened as time passed and nitrous oxide release continued.

If the 573-nm material is an integral part of the denitrifying process, all denitrifiers should produce it, but only when grown under denitrifying conditions. Furthermore, bacteria other than denitrifiers should not produce the substance. The nitric oxide binding agent was indeed found in a variety of denitrifiers (Table 2), and only when they grew under denitrifying conditions. Nondenitrifiers, including organisms that respire nitrate to nitrite but no further, did not produce the nitric oxide–generated 573-nm material, nor did the nitrate-assimilating fungus Neurospora crassa. Rejection of the hypothesis that nitric oxide is an obligate intermediate in denitrification seems thus to be ill-advised on both new and old grounds.

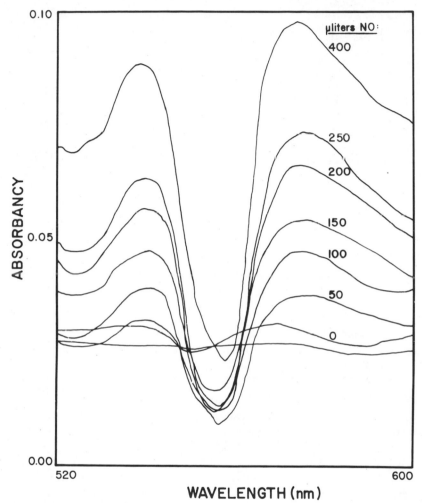

Figure 2.   Difference spectra (treated versus untreated) resulting from injection of increasing quantities of nitric oxide into anaerobic cuvettes containing dialyzed extract of tryptone-nitrate broth–grown *Pa. denitrificans* (protein, 10 mg/ml). Cells broken by passage through the French Pressure Cell; extract clarified by centrifugation at 13,300 × *g* for 20 min before overnight dialysis in the cold against 0.01M phosphate buffer at pH 7.0.

Treatment of active dialyzed extracts from *Ps. perfectomarinus* or *Pa. denitrificans* with ammonium sulfate at 50% saturation precipitated the reductive part of the complex, whereas most of the binding component remained in suspension. The suspended part that bound nitric oxide then displayed an intense absorption peak at 565 nm, rather than 573 nm, and a less intense peak as before at 540 nm (Figure 4). Addition of malate, NAD,

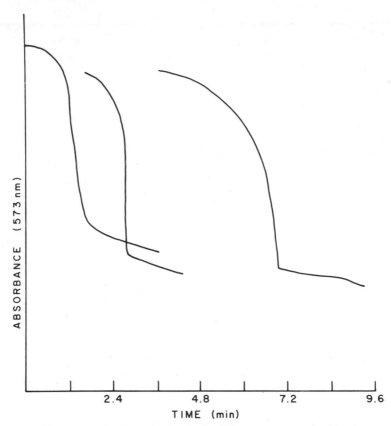

Figure 3.   Effect on peak height at 573 nm in the difference spectrum resulting from repeated injections of 100-μl lots of nitric oxide into an anaerobic cuvette containing crude extract of tryptone-nitrate broth–grown *Pa. denitrificans* (protein, 12 mg/ml).

and FAD did not result in reduction of nitric oxide by the separated binding fraction, nor was there release of nitrous oxide. When resuspended the ammonium sulfate–precipitated portion exhibited only a minimal 573-nm peak, but did reduce nitric oxide to nitrous oxide when supplied with the electron donor and the carriers. When the two fractions were mixed back together, the nitric oxide–generated peak was once again at 573 nm and the restored extract rapidly reduced nitric oxide to nitrous oxide.

When reduced with dithionite in the absence of nitric oxide, the ammonium sulfate–precipitated and resuspended fraction displayed a reduced c_{550} peak in absorption spectral scans. As added nitric oxide was reduced to nitrous oxide, this typical, reduced *c*-type spectrum disappeared. The *c-d*–type cytochrome of nitrite reductase was found not to be involved in nitric oxide reduction. Nitrite (but not nitric oxide or nitrate) oxidized the reduced *c*-type component of purified *c-d* material from *Ps. aeruginosa*. (We are indebted to Dr. D. C. Wharton for this assay.)

Our own assays and a search of pertinent literature revealed that the proper and specific oxidant for any of a variety of *c*-type cytochromes will provide a difference spectrum of the type we observed here with nitric oxide. The significance of our experimental results lies in the specificity of nitric oxide as oxidant. Attempts to purify and characterize the cytochrome involved are underway. Our working hypothesis is that the specific *c*-type cytochrome is a component of nitric oxide reductase ordinarily in reduced form. It is oxidized by the nitric oxide generated by nitrite reductase. Additional electrons are then needed to reduce the bound nitric oxide to nitrous oxide, which is released. The *c*-type cytochrome is then reduced and the cycle is repeated.

NITROUS OXIDE REDUCTION

The final step in denitrification, nitrous oxide reduction, results in the liberation of dinitrogen. Many denitrifiers release nitrous oxide transiently but later regain and reduce it. A few denitrifiers do not carry out this final step, but most do, and some can grow at the expense of nitrous oxide reduction without nitrate or nitrite present. Several denitrifiers are so sensitive to

Table 2.   Influence of growth conditions on production of the nitric oxide binding material

| | NO generated, 573-nm peak in extracts of cells grown | |
Microorganism	Aerobically (O_2)	Anaerobically (NO_3^-)
Denitrifiers		
Alcaligenes odorans	—	+
Bacillus azotoformans	—	+
Paracoccus denitrificans	—	3+
Pseudomonas aeruginosa	—	2+
Pseudomonas denitrificans	—	2+
Pseudomonas fluorescens	—	+
Pseudomonas perfectomarinus	—	2+
Nitrate respirers		
Enterobacter aerogenes	—	—
Escherichia coli	—	—
Nitrate assimilators		
Neurospora crassa[a]	—	NG

[a] Grown aerobically with nitrate as sole source of nitrogen. No growth (NG) anaerobically.

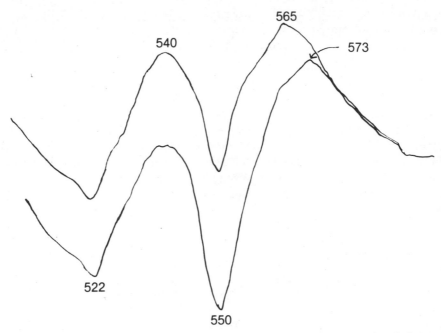

Figure 4.   Comparison of difference spectra showing the nitric oxide–generated peaks in scans of crude extract of tryptone-nitrate broth–grown *Pa. denitrificans* (lower curve) and the supernatant fluid (upper curve) from extract treated with enough ammonium sulfate to provide 50% saturation.

nitrite that cultures in media initially containing 0.5% nitrate accumulate enough nitrite to cause them to stop at, and not reduce, nitrous oxide. In media initially containing 0.1% nitrate, these same bacterial isolates release only dinitrogen. Nitrous oxide reductase activity is easily demonstrated in whole cells, but is unstable and variable in cell-free extracts. Negative experiments are not uncommon, but successes have been reported (Garcia, 1977). The enzyme system is rapidly synthesized in cultures grown on nitrate, nitrite, or nitrous oxide and is routinely demonstrable in whole cells.

ACETYLENE BLOCKAGE EFFECT

Using whole cells, we found that addition of acetylene gas to the suspension along with nitrous oxide caused blockage of the reduction of nitrous oxide (Balderston, Sherr, and Payne, 1976; Yoshinari and Knowles, 1976). Even if nitrate or nitrite was reduced, the process stopped at nitrous oxide in the presence of acetylene. As little as 4.46×10^{-5} M of acetylene effected this inhibition. The inhibitory effect was progressively less pronounced in systems receiving less acetylene. Significantly, ethylene was not produced, nor was it

or several nongaseous, triple-bonded organic compounds (benzonitrile, ace-
tonitrile, and 2-propyn-1-ol) inhibitory for nitrous oxide reduction. The
inhibition was precise and remarkably gentle.

Acetylene is apparently not harmful to the denitrifying bacteria since
suspensions exposed for more than 6 hr still vigorously reduced nitrous
oxide after the acetylene was removed. Moreover, denitrifiers grew in
nitrate-containing medium in the continual presence of acetylene and
produced all of the denitrifying enzymes, including nitrous oxide reductase.
These cultures accumulated nitrous oxide as long as acetylene was present,
but immediately reduced nitrous oxide when acetylene was removed.
Neither the synthesis nor the functioning of *E. aerogenes*-type and fungal-
type nitrate and nitrite reductases was affected in any obvious way by the
presence of acetylene during growth of some representative species
(Balderston, Sherr, and Payne, 1976).

Reduction of nitrous oxide by whole cells of *Ps. perfectomarinus* and
Pa. denitrificans was also blocked by acetylene when it was added after
reduction was in full flight (Figure 5). In addition, acetylene inhibited
nitrous oxide reduction by the bacterial populations in samples of salt
marsh mud as well as those in pure culture. Acetylene blockage thus

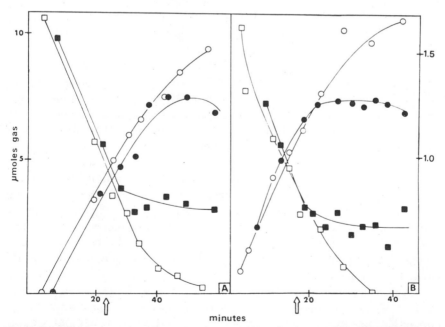

Figure 5.   Influence on rates of ongoing nitrous oxide reduction resulting from injection (at
the arrow) of enough acetylene to provide 4.46×10^{-5} M in anaerobic cuvettes containing fresh,
whole-cell suspension of nitrate-grown bacteria. (A) *Ps. perfectomarinus;* (B) *Pa. denitrificans.*

appears to offer promise for use in the measurement of ongoing denitrification rates in various types of soils. The advantage of measuring accumulation of nitrous oxide rather than dinitrogen to quantify denitrification is obvious. Living in an atmosphere that is 79% dinitrogen raises the danger of assay mistakes resulting from contamination to an almost unacceptable level. However, only very small amounts of nitrous oxide are present in our atmosphere, and its accumulation can thus serve as a useful indicator of denitrification.

The effectiveness of the practical use of acetylene blockage for measuring denitrification in soil has been compared by Tiedje (1978) with the use of ^{13}N-labeled nitrate. The short radioactive half-life of ^{13}N restricts its use to laboratories with immediate access to a cyclotron. Acetylene can be taken almost anywhere for use in assay systems for denitrification. The ^{13}N method revealed a bit more activity, but the acetylene blockage procedure was nearly as precise and thus has great promise—particularly if better means of achieving even diffusion of the gas inhibitor in the soil samples can be devised.

SUMMARY AND CONCLUSIONS

Some possibly useful leads toward control might develop from our findings. First, denitrification is a term that describes a concerted series of electron transport phenomena. It is difficult to imagine halting this flow without interfering with others that must continue. Successful antibiotics and antimicrobial agents selectively exploit some vital but susceptible operation among the biosynthetic mechanisms of the target microorganisms, but are not highly selective. A c-d–type cytochrome is the key component of denitrifying nitrite reductase. Another c-type cytochrome appears to be involved in nitric oxide reduction, and Japanese workers indicate that nitrous oxide reduction also involves c-type cytochromes. If it were possible, therefore, to devise a soil additive that would not be easily inactivated in soil or easily absorbed into plant roots, but would be selectively inhibitory for *de novo* synthesis of c-type cytochromes in bacteria, then denitrification rates could be slowed down. The c-type cytochromes do not appear to be involved in the assimilatory reduction of nitrate or nitrite, so the inhibitor would not harm this desirable process. Ideally, such an additive would also be inexpensive to prepare, easy to distribute, and subject to degradation only after several weeks in the soil.

We realize that these suggestions are presently, and may continue to be, impractical, but they do represent a positive and reasoned approach to an age-old problem. A solution would be worth the effort. Estimates circulated informally but based on analyses of many data derived from studies with ^{15}N fertilizer suggest that loss to denitrification ranges from 15% to greater than 50%, according to the condition of the soil. Our current estimates (from data not reported here) for localized rates of denitrification in soil (data are insufficient in our judgment even to guess at global rates) show that they are only slightly less than the rates of nitrogen fixation. Nitrogen fixation and incorporation into metabolism assure that the biological world stays ahead of a nitrogen deficiency, but denitrification just as certainly assures that the margin is always slim.

ACKNOWLEDGMENTS

We are grateful to Dr. Walter G. Zumft for his suggestions and criticisms.

REFERENCES

Balderston, W. L., B. F. Sherr, and W. J. Payne. 1976. Blockage by acetylene of nitrous oxide reduction in *Pseudomonas perfectomarinus*. Appl. Environ. Microbiol. 31:504–508.

Barbaree, J. M., and W. J. Payne, 1967. Products of denitrification by a marine bacterium as revealed by gas chromatography. Mar. Biol. 1:136–139.

Buchanan, R. E., and N. E. Gibbons (eds.) 1974. Bergey's Manual of Determinative Bacteriology, 8th ed. The Williams & Wilkins Co., Baltimore.

Cox, C. D., Jr., W. J. Payne, and D. V. Dervartanian. 1971. Electron paramagnetic resonance studies on the nature of hemoproteins in nitrite and nitric oxide reduction. Biochim. Biophys. Acta 253:290–294.

Forget, P. 1971. Les nitrate-réductases bactériennes. Solubilization, purification et propriétés de l'enzyme A de *Micrococcus denitrificans*. Eur. J. Biochem. 18:442–450.

Forget, P., and F. Pichinoty. 1965. Le cycle tricarboxylique chez une bactérie dénitrifante obligatoire. Ann. Inst. Pasteur (Paris) 108:364–377.

Gamble, T. N., M. R. Betlach, and J. M. Tiedje. 1977. Numerically dominant denitrifying bacteria from world soils. Appl. Environ. Microbiol. 33:926–939.

Garcia, J.-L. 1977. Étude de la denitrification chez une bactérie thermophile sporulée. Ann. Microbiol. (Inst. Pasteur) 128A:447–458.

Hollis, O. L. 1966. Separation of gaseous mixtures using porous polyaromatic beads. Anal. Chem. 38:309–316.

Newton, N. 1969. The two-haem nitrite reductase of *Micrococcus denitrificans*. Biochim. Biophys. Acta 185:316–331.

Payne, W. J. 1973. Reduction of nitrogenous oxides by microorganisms. Bacteriol. Rev. 37:409–452.

Payne, W. J., P. S. Riley, and C. D. Cox, Jr. 1971. Separate nitrite, nitric oxide and nitrous oxide reducing fractions from *Pseudomonas perfectomarinus*. J. Bacteriol. 106:356–361.

Pichinoty, F. 1973. La reduction bactérienne des composes oxygènes mineraux de l'azote. Bull. Inst. Pasteur (Paris) 71:317–395.

Pichinoty, F., J.-L. Garcia, M. Mandel, C. Job, and M. Durand. 1978. Sur l'existence de bactéries capables de croitre aux dépens de l'oxyde nitrique comme accepteur d'électrons respiratoire. C. R. Acad. Sci. (Paris) 286D:1403–1405.

Rowe, J. J., B. F. Sherr, W. J. Payne, and R. G. Eagon. 1977. A unique nitric oxide-binding complex formed by denitrifying *Pseudomonas aeuginosa*. Biochem. Biophys. Res. Commun. 77:253–258.

Spangler, W. J., and C. M. Gilmour. 1966. Biochemistry of nitrate respiration in *Pseudomonas stutzeri*. I. Aerobic and nitrate respiration routes of carbohydrate catabolism. J. Bacteriol. 91: 245–250.

Tiedje, J. M. 1978. Denitrification in soil In: D. Schlessinger (ed.), Micro-biology—1978, pp. 362–366. ASM Publications, Washington, D.C.

Williams, R. J., and W. C. Evans. 1975. The metabolism of benzoate by *Moraxella* species through anaerobic nitrate respiration. Evidence for a reductive pathway. Biochem. J. 148:1–10.

Yamanaka, T., and K. Okunuki. 1963. Crystalline *Pseudomonas* cytochrome oxi-dase. II. Spectral properties of the enzyme. Biochim. Biophys. Acta 67:394–406.

Yoshinari, T., and R. Knowles. 1976. Acetylene inhibition of nitrous oxide reduction by denitrifying bacteria. Biochem. Biophys. Res. Commun. 69:705–710.

Nitrogen Fixation, Volume I
Edited by W. E. Newton and W. H. Orme-Johnson
Copyright 1980   University Park Press   Baltimore

Assessment of Alternatives to Present-Day Ammonia Technology with Emphasis on Coal Gasification

D. E. Nichols,
P. C. Williamson, and D. R. Waggoner

FEEDSTOCKS FOR AMMONIA

Last year, we in the United States used 60% more natural gas and 100% more oil than we discovered (American Gas Association, 1978). This situation is having a serious impact on the chemical industry, which is already faced with dwindling supplies of natural gas and is heavily dependent on imported oil. Both industry and government experts are attempting to answer the question of just how long supplies of these two important resources will last. Although there are wide differences of opinion, all agree that the U. S. is short on existing reserves and that rapid development of additional reserves must take place. In order to maintain the 1975 level of demand for natural gas, reserves must be added to at the rate of 17 trillion cubic feet/year. In reality additions to the reserves have dropped from a high of 21.1 trillion cubic feet/year in 1967 to only 8 trillion cubic feet/year in 1975 (Mayfield, 1977).

Gas reserves at the end of 1976 were estimated to be about 216 trillion cubic feet and total annual consumption to be about 20 trillion cubic feet (Meyer, 1978). Considering increased production from known reserves plus gas likely to be discovered in the future, most experts predict that our gas will be depleted in the next few decades (Schanz, 1978). As a result, substantial increases in gas prices are expected. Also, gas may become unavailable to those on interstate (regulated) supplies.

How is all of this likely to affect the nitrogen fixation industry? About 95% of current ammonia capacity is based on natural gas, which is used both as fuel and the hydrocarbon feedstock. Practically all nitrogen fertilizer is produced from ammonia. In order to assure adequate supplies of food and fiber, our agricultural industry must have nitrogen fertilizer. Therefore, development of alternative feedstocks for ammonia production as well as alternative nitrogen sources deserves our highest priority efforts.

Much valuable research work is in progress on alternative methods of nitrogen fixation. Parallel efforts are also being carried out to develop alternative energy sources, such as solar energy, nuclear fusion, energy derived from thermal differences in oceans, and use of organic wastes such as cow manure. Although all of these efforts will help in the future to provide nitrogen to our agricultural industry, they will not solve the immediate need to develop an alternative feedstock for the existing ammonia-producing industry. Gas curtailments in recent years have already resulted in significant production losses.

For the short run, major programs are already underway for conversion to oil or coal for fuel (Friedman, 1978). Naphtha and heavy fuel oils can be used as feedstock for producing ammonia, but because of high cost, scarcity, and large capital investments required for conversion they are not attractive alternatives. It appears that coal is the only viable alternative for the foreseeable future. Coal reserves in the U. S. recoverable by current technology have been estimated to be sufficient for 700 years at current use rates and for approximately 100 years assuming a 6%/year increase in consumption. Sizeable portions of these reserves are located in or near the large agricultural producing regions of the nation.

COAL AS A FEEDSTOCK

The basic technology for producing ammonia from coal is now available, but is in need of further development, particularly in the areas of improved reliability and definition of economics. Existing technology was developed in Germany prior to World War II and has been used in about 20 plants, including two large plants in India that are in the commissioning phase and one large plant in South Africa that is performing satisfactorily following a 2-year period beset by difficulties. There is a need for research, development, and demonstration of coal-based technology in the U. S. to provide the information and confidence needed for the industry to embark on a multibillion dollar conversion and expansion program.

In 1975, the Tennessee Valley Authority (TVA) and the Fertilizer Institute identified the need to develop highly visible, efficient U. S. technology for production of ammonia from coal as the top priority need for the nation's fertilizer research and development. This conclusion was also reached independently by the Agricultural Research Policy Advisory

Committee and the National Academy of Science. The TVA has responded to this need by establishing an Ammonia from Coal Project at its National Fertilizer Development Center at Muscle Shoals, Alabama.

The TVA's usual approach to such a project would be to construct a pilot plant to test available technology and/or develop new technology. Pilot plant results would then be used as a basis for moving to the demonstration plant phase. Because of the urgent need for development and demonstration of the production of ammonia from coal, this approach was considered too costly and time consuming. An investigative study (Tennessee Valley Authority, 1976) showed that the quickest and most low-cost approach to the problem would be to retrofit coal gasification facilities to the TVA's small but modern ammonia plant at Muscle Shoals. The TVA plant, completed in 1972, uses steam reforming of natural gas to produce 225 tons/day of ammonia. It can be operated at a minimum capacity of 60%, or 135 tons, per day. The least-cost installation would be one that would gasify enough coal to produce 135 tons/day of ammonia and would make maximum use of the equipment in the existing plant. The new facility would produce a gas to match the composition, temperature, and pressure at the inlet of the existing low-temperature shift converter (Figure 1). This would allow flexibility to operate with 60% of the synthesis gas from·coal and 40% from natural gas, or at 60% capacity using coal only.

The gasification process to be used was chosen on the basis that it should accept a wide variety of coals, produce a gas at a composition, temperature, and pressure compatible with ammonia processes and with a minimum of undesirable contaminants, and be of reliable, modern, sufficiently advanced U. S. technology and competitive economics (Tennessee Valley Authority, 1976). After a thorough review of all available processes, the Texaco Development Corporation process was selected for the project because it best met the above criteria and has been commercially proven with oil as a feedstock. The process will be used to partially oxidize and gasify about 170 tons of coal/day.

There are about 100 natural gas–steam reforming ammonia plants in the U. S., of which about 30 have a 1000 ton/day capacity (Waitzman, 1977). The TVA demonstration facility should provide a basis for retrofitting these existing plants. In addition, much of the technology developed will be applicable to new grass roots facilities designed to produce ammonia from coal. Besides the TVA Ammonia from Coal Project, the U. S. Department of Energy (DOE) is conducting two other projects related to ammonia production (Energy Research and Development Administration, 1977). These complement the TVA project in that they involve grass roots facilities, whereas the TVA plant is primarily a retrofit project.

Four basic steps are required to produce ammonia from any feedstock: gas preparation, shift conversion, gas purification, and ammonia synthesis. The shift conversion, gas purification, and ammonia synthesis steps are

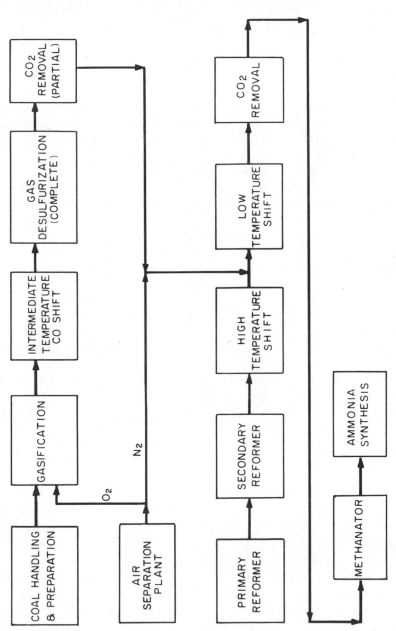

Figure 1.   Coal gasification retrofit to existing NH_3 plant.

common to most commercial processes. Significant variations occur when pure hydrogen is available (e.g., from electrolysis of water) or when by-product gases such as refinery gas or coke-oven gas are used. The first step, gas preparation, has the widest variation and uses most of the energy required by the process. This is the step that would be replaced by coal gasification. The coal-based operation has several problems not encountered with other feedstocks. Among these are higher capital costs for gasifiers, gas purification units, and air separation plants, and stringent environmental requirements. Also, reliability of the process has not yet been proven and the economics are uncertain. The TVA's project is designed to develop a reliable process and provide a basis for determining the economics of producing ammonia from coal.

THE TVA'S AMMONIA-FROM-COAL FACILITIES

The TVA conducted conceptual design studies to arrive at a basic flow scheme for a coal gasification facility that could be retrofitted to the existing ammonia plant. TVA process specifications stipulated the flow rates, product gas composition and condition, emission limitation, and certain physical features, such as instrumentation and control building, and the Texaco gasifier and its related equipment. A contract to design, procure, and construct the coal gasification and gas purification facilities was awarded to Brown and Root Development, Inc. on May 31, 1978, with erection and startup scheduled to be completed 21 months later in February, 1980.

The flow scheme is shown in Figure 1. Coal is received by rail, crushed to ½″ or less, and sent to open storage or to the coal slurry preparation area. The coal is sized to process requirements in a wet pulverizer and is pumped as a slurry (about 55% solids) to an agitated hold tank, then to one of two agitated feed tanks. Coal slurry is pumped to the gasifier at the process rate of about 8 tons/hr, dry basis. Oxygen from an air separation plant is also fed to the high-temperature, pressurized Texaco reactor at the rate required to satisfy the partial oxidation process. The raw gas from the reactor is quench-cooled and the particulate matter removed. Water and particulates from the quenching operation are recycled to the reactor. The quenched gas flows to the CO shift converter, which contains sulfur-activated shift catalyst. The shifted gas (containing about 2% CO) then flows to a COS hydrolysis unit and on to the acid-gas removal system (AGR).

Brown and Root has selected a single-stage Rectisol unit for bulk acid-gas, mainly CO_2, removal. The gases removed by the Rectisol process are sent to a Stretford unit, which decreases the sulfur content to meet the emission requirements of not more than 150 ppm H_2S and less than 11.3 pounds/hr total sulfur to the atmosphere. About 5 tons/day of elemental sulfur is recovered from this unit. The CO_2 stream to the urea unit is routed through a bed of zinc oxide to decrease the total sulfur content of that stream to less than 0.5 ppm. The process gas stream from the AGR contains a maximum of 1 ppm sulfur compounds. The sulfur is further decreased to 0.1 ppm by beds of expendable ZnO catalyst.

After final sulfur removal, nitrogen is added to obtain the 3:1 hydrogen:nitrogen ratio for ammonia synthesis, and water is added for the low-temperature shift reaction. The gas is then piped to the existing ammonia plant for low-temperature shift, final CO_2 removal, methanation of remaining carbon oxides, and finally ammonia synthesis. An incinerator is provided for use during the estimated 2-hr startup period or until the process gas stream meets specifications and is routed to the ammonia plant.

A chemical precipitation process is furnished to remove dissolved solids from the wastewater purge stream. The chemical sludge is recycled to the reactor. Ammonia is steam stripped and recovered from the water that is leaving the chemical precipitation step. Remaining organics, primarily cyanates and formates, are biologically removed in an activated sludge treatment unit. Slag is removed from the gasifier and dewatered. The solids are then trucked to a disposal area (landfill) and the water is recycled to the coal slurry preparation area.

The foregoing describes the process as contracted. Investigations are in progress with hopes of replacing the AGR system with a simpler, lower-cost process, and one that may eliminate the COS hydrolysis unit in the process gas stream. Changes here will affect other areas, particularly the sulfur recovery system. A 3-year test and demonstration program is planned for further developing the process. Changes and/or modifications will be made as necessary to accomplish the objective of demonstrating reliable technology for producing ammonia from coal.

ECONOMICS OF AMMONIA FROM ALTERNATIVE FEEDSTOCKS

U. S. ammonia producers generally fall into one of several categories: 1) those entering the business for the first time or building a completely new plant; 2) those having one or more small ammonia plants that may have a short remaining operating life; and 3) those having one or more plants, large or small, that have a substantial remaining operating life. Therefore, determination of the economics of converting from natural gas to alternative feedstocks would be heavily dependent on each producers' particular set of circumstances. For this reason, several assumptions were required to simplify the calculations made in this study.

Conceptual designs and cost estimates were prepared for 1000 short ton/day newly constructed ammonia plants for the following processes:

1. Natural gas—steam reforming
2. Naphtha—steam reforming
3. Fuel oil—partial oxidation
4. Coal—partial oxidation
5. Electrolytic hydrogen

Estimates of the capital required for each of the processes are shown in Table 1. Each of the plants has a capacity of 1000 tons/day and is rated at 330,000 tons annually. The working capital is based on 60 days' production cost. The capital investment and the production cost estimates are for mid-1978.

The equivalent energy requirements per ton of ammonia for fuel and process feedstocks range from 34.5 million Btu for natural gas to 45 million Btu for partial oxidation of coal, as shown in Table 2 (Quartulli and Turner, 1972; Slack and James, 1974; Smith and Hatfield, 1978). Thus, ammonia synthesis is a highly energy-intensive process, not only in terms of the heat energy (fuel) required, but also in terms of the chemical process feed required. In Table 3, requirements per ton of ammonia for the fuel, feedstock, and cooling water for natural gas, naphtha, and fuel oil processes were taken from Buividas, Finneran, and Quartulli (1974). The data for the electrolytic production of hydrogen were developed from information presented by Norsk Hydro (Grundt, 1977), except for the electricity requirement, which is from Smith and Hatfield (1978). The other process require-

Table 1.   Capital investment requirements for 1000-TPD ammonia plants

| | Investment[a], $ million (1000 t/day plant) | | | | |
| | | | Partial oxidation | | |
Investment item	Natural gas reforming	Naphtha reforming	Oil	Coal	Electrolytic (convention)
Battery limits NH$_3$	50.0	57.1	94.5	99.9	74.3
Site preparation	1.8	1.8	1.8	2.7	1.8
Auxiliary facilities	7.5	8.5	10.2	15.0	7.5
Support facilities	7.5	8.5	10.2	15.0	7.5
Subtotal	66.8	75.9	116.7	132.6	91.1
Product storage	3.6	3.6	3.6	3.6	3.6
Total depreciable investment	70.4	79.5	120.3	136.2	94.7
Land	0.1	0.1	0.1	0.1	0.1
Working capital	5.1	7.2	5.4	4.5	18.8
Total capital investment	75.6	86.8	125.8	140.8	113.6

Abbreviations: t, ton; TPD, tons per day.

[a] Chemical Engineering Plant Index = 218.4.

Table 2.   Energy requirements for one ton of ammonia

Fuel/feedstock	Process	Energy consumption (Btu $\times$ 10^6/t NH_3)
Natural gas	Reforming	34.5
Naphtha	Reforming	37
Fuel oil	Partial oxidation	38
Coal	Partial oxidation	38
Electricity/water	Electrolysis	36

Abbreviations: Btu, British thermal unit; t, ton.

Table 3.   Unit rates for base cases

	Units	Price/Unit ($)	$/MM Btu
Energy			
Natural gas (1,000 Btu/scf)	1,000 MCF	2.50	2.50
Naphtha (19,000 Btu/lb)	t	126.00	3.31
Fuel oil (19,000 Btu/lb)	Bbl	13.00	2.21
Coal (10,800 LHV Btu/lb)	t	27.00	1.25
Electricity	kWh	0.27	7.91
Labor			
Operating labor and supervision	manhours	8.00	
Utilities			
Cooling water makeup	M gal	0.25	
Boiler feedwater	M gal	1.00	
Materials—intermediates, catalysts, and chemicals			
		$/ton HN_3	
Natural gal—steam reforming		0.75	
Naphtha—steam reforming		1.00	
Fuel oil—partial oxidation		0.50	
Coal—partial oxidation		0.50	
Electrolysis		0.50	
Maintenance			
50% material and 50% labor (chemical processes)	5% of depreciable investment		
50% material and 50% labor (electrolytic)	2.5% of depreciable investment		
Indirect Costs			
Average capital charges	19.7 of total plant investment		
Administrative and plant overhead	75% operating and maintenance labor		
Marketing	3.00/ton NH_3		
Working Capital	60 days' direct production cost		

Abbreviations: Bbl, barrel; Btu, British thermal unit; gal, gallon; kWh, killowatt hour; lb, pound; LHV, lower heating value; M, thousand; MM, million; mcf, thousand cubic feet; scf, standard cubic feet.

Table 4.   Capital charges for base case

	Percentage
Total depreciable capital investment	
Depreciation (15-year life)	6.7
Insurance	0.5
Property taxes	1.5
	8.7
Unrecovered capital investment[a]	
Cost of Capital (capital structure assumed to be 40% of debt and 60% equity)	
Debt at 10% interest	4.0
Equity[b] at 15% return on investment	9.0
Income taxes (federal and state)[c]	9.0
Total rate applied to depreciation base	22.0
Total capital investment (total annual capital charge rate)	19.7[d]

[a] Original investment yet to be recovered or "written off."

[b] Contains retained earnings and dividends.

[c] Since income taxes are approximately 50% of gross return, the amount of taxes is the same as the return on equity.

[d] Applied on an average basis, the total annual percentage of original investment for new (15-year) plants would be 8.7 + ½ (22.0%) = 19.7%.

ments, such as boiler feedwater, labor, maintenance, etc., are from Blouin and Nichols (1977). Costs assigned to feedstock and fuel, labor, utilities, materials, and maintenance are also shown in Table 3. Capital charges were taken as 19.7 percent of the total capital investment (Table 4). The plant life was assumed to be 15 years, with straight-line depreciation. Calculations were also made for the revenue requirements for each of the processes (Tables 5 through 9). The revenue requirements represent the average price that would have to be charged under the assumptions of the base case to recover the full costs, including a normal return on investment. Operation at less than the assumed efficiency and/or rate could significantly increase the revenue requirements per ton of ammonia. Revenue requirements ranged from a low of $147/t for ammonia from the natural gas reforming process to a high of $417/t from the electrolytic process. Ammonia at $180/t from a coal-based plant was second to natural gas, followed closely by ammonia at $184/t from fuel oil and at $191/t from naphtha. Although this study shows the revenue requirements for the base case for coal to be lower than either fuel oil or naphtha, circumstances different from the base case could cause any one of these three feedstocks to be selected for a specific plant. Even ammonia from electrolytic hydrogen should be considered under

Table 5.   Natural gas–steam-reforming/ammonia production[a] (total average revenue requirements)

	No. units/t product	Unit cost ($)	Cost/t product ($)	Total annual cost ($)
Direct costs				
Materials				
Feedstock, natural gas (1,000 Btu/SCF)	20.40 MCF	2.50/MCF	51.00	16,830,000
Intermediates, catalysts, and chemicals			0.75	248,000
Utilities				
Fuel, natural gas	10.60 MCF	2.50/MCF	26.50	8,745,000
Electricity	30.00 kWh	0.027/kWh	0.81	267,000
Cooling water	2.00 M gal	0.25/M gal	0.50	165,000
Boiler feedwater	0.60 M gal	1.00/M gal	0.60	198,000
Labor				
Operating labor and super-vision	0.18 man-hr	8.00/man-hr	1.44	475,000
Maintenance labor and material (5% of depreciable plant investment)			10.70	3,531,000
General expense			1.00	330,000
Subtotal direct costs			93.30	30,789,000
Indirect costs				
Capital charges, 19.7% total capital investment			45.13	14,893,000
Administrative and plant over-head			5.09	1,680,000
Marketing			3.00	990,000
Subtotal indirect costs			53.22	17,563,000
Total revenue requirement			146,52	48,352,000

Abbreviations: Btu, British thermal unit; gal, gallon; kWh, kilowatt hour; lb, pound; M, thousand; man-hr, man-hour; mcf, thousand cubic feet; scf, standard cubic foot; t, ton.

[a] Basis: plant capacity 1,000 t/day, 330,000 t/year; life of plant, 15 years; depreciable plant investment, $70,400,000; total capital investment, $75,600,000.

special circumstances. Electrolytic hydrogen plants have been used for ammonia production for many years where inexpensive electricity is available.

The major components of the revenue requirements, representing 80% of the cost of ammonia from natural gas–steam reforming, are energy costs (feedstock and fuel) and capital charges. The effect of variation in the cost of energy, for both fuel and feedstock, on revenue requirements per ton of ammonia is shown in Figure 2. Minor inputs of heat (steam) or electrical energy are held constant. It can be seen that, if natural gas rises to around $3.60/million ft^3, coal (10,800 Btu/lb) would be competitive at the base case price of $27/t. Naphtha (19,000 Btu/lb) at the base case price of $126/t would be competitive with natural gas at $3.90/million ft³. Similarly, heavy oil at $13/barrel would be competitive with natural gas at $3.70/million ft³. The capital charges actually used by a producer would be based on the investment requirements and the capital structure of the firm. Both of these components can vary considerably. Figure 3 shows the effect of 30% increases and decreases in the capital charges. The changes in revenue requirements are significant. At the higher capital charge rate, the revenue requirement of ammonia from coal approached that from oil and naphtha.

In recent years, there have been natural gas curtailments resulting in losses of ammonia production. The effects of curtailments on revenue requirements are shown in Figure 4, where it can be seen that the natural

Table 6.   Naphtha–steam-reforming/ammonia production[a] (total average revenue requirements)

	No, units/t product	Unit cost ($)	Cost/t product ($)	Total annual cost ($)
Direct costs				
Materials				
Feedstock, naphtha (19,000 Btu/lb)	0.55 t	126.00/t	69.30	22,869,000
Intermediates, catalysts, and chemicals			1.00	330,000
Utilities				
Fuel, naphtha	0.34 t	126.00/t	42.84	14,137,000
Electricity	45.00 kWh	0.027/kWh	1.22	403,000
Cooling water	2.00 M gal	0.25/M gal	0.50	165,000
Boiler feedwater	0.67 M gal	1.00/M gal	0.67	221,000
Labor				
Operating labor and supervision	0.20 man-hr	8.00/man-hr	1.60	528,000
Maintenance labor and material (5% of depreciable plant investment)			12.05	3,975,000
General expense			1.00	330,000
Subtotal direct costs			130.18	42,958,000
Indirect costs				
Capital charges, 19.7% total capital investment			51.81	17,100,000
Administrative and plant overhead			5.72	1,887,000
Marketing			3.00	990,000
Subtotal indirect costs			60.53	19,977,000
Total revenue requirement			190.71	62,935,000

Abbreviations: Btu, British thermal unit; gal, gallon; kWh, kilowatt hour; lb, pound; M, thousand; man-hr, man-hour; t, ton.

[a] Basis: plant capacity 1,000 t/day, 330,000 t/year; life of plant, 15 years; depreciable plant investment, $79,500,000; total capital investment, $86,800,000.

Table 7.   Fuel oil-partial oxidation/ammonia production[a] (total average revenue requirements)

	No. units/t products	Unit cost ($)	Cost/t product ($)	Total annual cost ($)
Direct costs				
Materials				
Feedstock, crude or fuel (17,500 Btu/lb)	3.31 Bbl	13.00/Bbl	43.03	14,200,000
Intermediates, catalysts, and chemicals			0.50	165,000
Utilities				
Fuel, crude oil	2.38 Bbl	13.00/Bbl	30.94	10,210,000
Electricity	45.00 kWh	0.027/kWh	1.22	403,000
Cooling water	3.00 M gal	0.25/M gal	0.75	248,000
Boiler feedwater	0.30 M gal	1.00/M gal	0.30	99,000
Labor				
Operating labor and supervision	0.24 man-hr	8.00/man-hr	1.92	634,000
Maintenance labor and material (5% of depreciable plant investment)			18.23	6,016,000
General expense			1.00	330,000
Subtotal direct costs			97.89	32,305,000
Indirect costs				
Capital charges, 19.7% total capital investment			75.10	24,783,000
Administrative and plant overhead			8.28	2,732,000
Marketing			3.00	990,000
Subtotal indirect costs			86.38	28,505,000
Total revenue requirement			184.27	60,810,000

Abbreviations: Bbl, barrel; Btu, British thermal unit; gal, gallon; kWh, kilowatt hour; lb, pound; M, thousand; man-hr, man-hour; t, ton.

[a] Basis: plant capacity 1,000 t/day, 330,000 t/year; life of plant, 15 years; depreciable plant investment, $120,300,000; total capital investment, $125,800,000.

gas reforming plant would have to be reduced to about 60% capacity before the revenue requirements would be as high as the base case for coal.

This study indicates that:

1. If the cost for coal, naphtha, and fuel oil remain near the levels of the base cases and the cost of natural gas increases to around $3.50–$4/million Btu, ammonia from these alternative feedstocks would be competitive with ammonia produced from natural gas.
2. Coal, naphtha, and fuel oil should be considered as alternative feedstocks to meet the demand for ammonia if the supply of natural gas becomes too restricted.

Table 8.   Coal-partial oxidation/ammonia production[a] (total average revenue requirements)

	No. units/t product	Unit cost ($)	Cost/t product ($)	Total annual cost ($)
Direct costs				
Materials				
Feedstock, coal (10,800 Btu/lb)	1.27 t	27.00/t	34.29	11,316,000
Intermediates, catalysts, and chemicals			0.50	165,000
Utilities				
Fuel, coal	0.63 t	27.00/t	17.01	5,613,000
Electricity	136.00 kWh	0.027/kWh	3.67	1,211,000
Cooling water	2.80 M gal	0.25/M gal	0.70	231,000
Boiler feedwater	0.50 M gal	1.00/M gal	0.50	165,000
Labor				
Operating labor and supervision	0.48 man-hr	8.00/man-hr	3.84	1,267,000
Maintenance labor and material (5% of depreciable plant investment)			20.64	6,810,000
General expense			1.00	330,000
Subtotal direct costs			82.15	27,108,000
Indirect costs				
Capital charges, 19.7% total capital investment			84.05	27,738,000
Administrative and plant overhead			10.62	3,504,000
Marketing			3.00	990,000
Subtotal indirect costs			97.67	32,232,000
Total revenue requirement			179.82	59,340,000

Abbreviations: Btu, British thermal unit; gal, gallon; kWh, kilowatt hour; lb, pound; M, thousand; man-hr, man-hour; t, ton.

[a] Basis: plant capacity 1,000 t/day, 330,000 t/year; life of plant, 15 years; depreciable plant investment, $136,200,000; total capital investment $140,800,000.

Table 9.   Electrolytic hydrogen/ammonia production[a] (total average revenue requirements)

	No. units/t product	Unit cost ($)	Cost/t product ($)	Total annual cost ($)
Direct costs				
Materials				
Intermediates, catalysts, and chemicals			0.50	165,000
Utilities				
Electricity	12,258 kWh	0.027/kWh	330.97	109,220,000
Steam, cooling water			2.00	660,000
Labor				
Operating labor and supervision	0.12 man-hr	8.00/man-hr	0.96	317,000
Maintenance labor and material (2.5% of depreciable plant investment)			7.18	2,369,000
General expense			1.00	330,000
Subtotal direct costs			342.61	113,061,000
Indirect costs				
Capital charges, 19.7% total capital investment			67.82	22,379,000
Administrative and plant overhead			3.41	1,126,000
Marketing			3.00	990.000
Subtotal indirect costs			73.23	24,495,000
Total revenue requirement			416.84	137,556,000

Abbreviations: kWh, kilowatt hour; man-hr, man-hour.

[a] Basis: plant capacity 1,000 t/day, 330,000 t/year; life of plant, 15 years; depreciable plant investment, $94,700,000; total capital investment, $113,600,000.

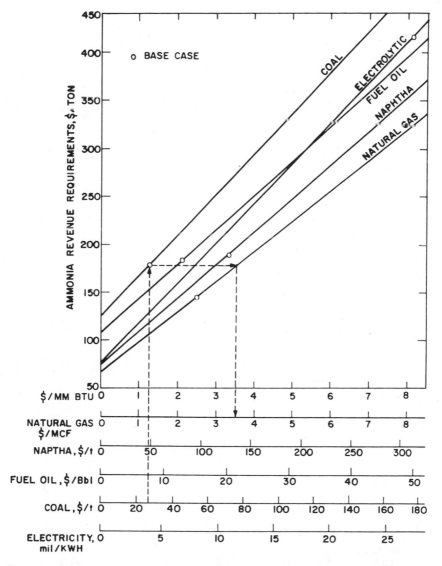

Figure 2.   Revenue requirements versus feedstock costs. Bbl, barrel; KWH, kilowatt hour; Mil, one-thousandth of a dollar; MCF, thousand cubic feet; MM Btu, million British thermal units; t, ton.

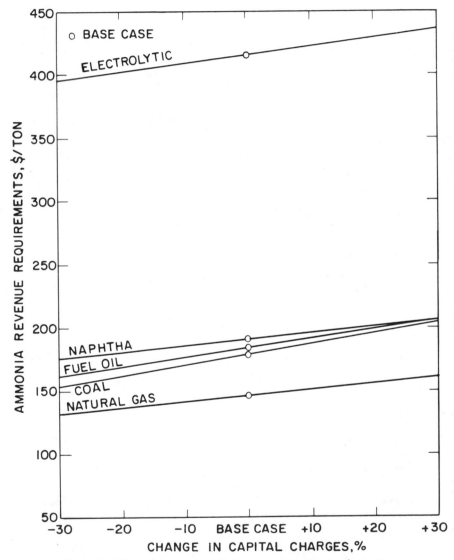

Figure 3.   Effect of changes in capital charges on revenue requirements.

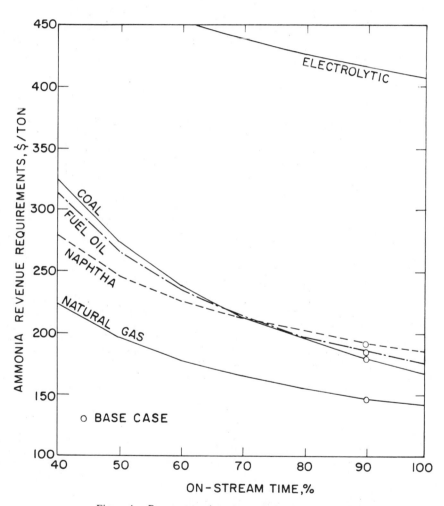

Figure 4.   Revenue requirements versus on-stream time.

3. With the assumptions of the base cases, electrolytic hydrogen is not competitive with the other processes considered. To become competitive, the cost for electricity would need to be greatly reduced and/or marketable byproducts would have to be produced.

There is always a degree of uncertainty in projected costs. Of the above costs, the degree of uncertainty is greatest in those associated with the production of ammonia using coal as the feedstock. It is expected that the TVA Ammonia from Coal Project will produce reliable cost data as well as advancing the technology of the process.

REFERENCES

American Gas Association. 1978. Committee on Natural Gas Reserves, "News Release," April 10.

Blouin, G. M., and D. E. Nichols. 1977. Economic Considerations of Chemical Nitrogen Fixation. Tennessee Valley Authority, Muscle Shoals, Alabama.

Buividas, L. J., J. A. Finneran, and O. J. Quartulli. 1974. Alternate ammonia feedstocks. Chem. Engineer. Prog., October, pp. 21–35.

Energy Research and Development Administration (ERDA). 1977. Information from ERDA—Weekly Announcements. Volume 3, July 18, No. 36, No. 77-151, and No. 77-117.

Friedman, J. 1978. U.S. chemical industry to invest in fuel change. Chem. Age, March 17, pp. 12–13, 19.

Grundt, T. 1977. Water electrolysis and its possibilities as basis for ammonia production. Paper presented at the Norsk Hydro Fertilizer Technology Seminar. January 26–27, Sao Paulo, Brazil.

Mayfield, R. R. 1977. Natural gas supply and pricing on the Gulf Coast. Chem. Engineer. Prog., October, pp. 25–31.

Meyer, R. F. 1978. A look at natural gas resources. Oil Gas J., May 8, pp. 334–346.

Quartulli, O. J., and A. Turner. 1972. British Sulfur Institute report: "Economic and Technological Factors in Giant Ammonia Plant Design, Part 1. Nitrogen No. 80:28–34.

Schanz, J. J., Jr. 1978. Oil and gas resources—Welcome to uncertainty. Resources No. 58.

Slack, A. V., and G. R. James. 1973. Ammonia—Part 1, p. 341. Marcel Dekker, Inc., New York.

Smith, A. J., and J. D. Hatfield. 1978. A New Look at the Electrolytic Production of Hydrogen for the Manufacture of Ammonia. Tennessee Valley Authority, Muscle Shoals, Alabama.

Tennessee Valley Authority (TVA). 1976. Investigative Phase Report, Ammonia from Coal Project. (Unpublished)

Waitzman, D. A. 1977. A Technical and Economic Review of Coal-Based Ammonia Production, Tennessee Valley Authority, Muscle Shoals, Alabama.

Section II
Genetics and Physiology

Nitrogen Fixation, Volume I
Edited by W. E. Newton and W. H. Orme-Johnson
Copyright 1980   University Park Press   Baltimore

Regulation and Genetics of Nitrogen Fixation in *Klebsiella pneumoniae*

T. MacNeil, G. P. Roberts, D. MacNeil, M. A. Supiano, and W. J. Brill

Klebsiella pneumoniae has been useful for studying the genetics of N_2 fixation because techniques normally used for studying *Escherichia coli* also can be used with *K. pneumoniae*. Phage P1 transduction (Streicher, Gurney, and Valentine, 1971), phage Mu mutagenesis (MacNeil, Brill, and Howe, 1978), and complementation analyses (Dixon et al., 1977) have been used to study *nif* genes. Previous mapping experiments were limited to three-factor transduction crosses with *his* as an outside marker (St. John et al., 1975; Kennedy, 1977) and to deletion mapping with a few strains containing deletions in *nif* (Bachhuber, Brill, and Howe, 1976). Most of the mutations used for these genetic studies were isolated after N-methyl-N'-nitro-N-nitrosoguanidine (NG) mutagenesis (Streicher, Gurney, and Valentine, 1971; St. John et al., 1975; Dixon et al., 1977; Kennedy, 1977). NG commonly yields multiple, closely linked mutations (Guerola, Ingraham, and Cerda-Olmedo, 1971). Mutants resulting from NG mutagenesis have been used to define *nif* genes (St. John et al., 1975; Dixon et al., 1977; Kennedy, 1977) and several of these genes have been identified on the basis of only one or two mutants. Some of these mutants, even though they revert, have been shown to contain more than one mutation (MacNeil et al., 1978).

We have isolated several hundred Nif⁻ mutants that were obtained by a variety of methods—NG, phage Mu, diethyl sulfate, hydroxylamine, and ICR191 mutagenesis (MacNeil et al., 1978). Some spontaneous Nif⁻

This research was supported by the College of Agricultural and Life Sciences, University of Wisconsin, Madison, Wisconsin, and by National Science Foundation Grants PFR77-00879 and PCM76-24271.

mutants also were isolated. Presumably this collection of mutants should contain mutations in each *nif* gene required for growth on N_2 under our conditions of growth. Deletions in the *nif* region were obtained by heat induction of a Mu phage that was inserted within or nearby the *nif* genes (Bachhuber et al., 1976; MacNeil, Brill, and Howe, 1978). The particular Mu phage used has a mutation yielding a thermolabile repressor (MacNeil, Brill, and Howe, 1978). At high temperatures the repressor is inactivated and the lytic cycle of Mu is initiated. Most of the cells are killed by phage-induced lysis. Some of the survivors of the heat induction lose the Mu prophage because of a spontaneous deletion. The deletions of Mu frequently eliminate a part of the surrounding chromosomal DNA and, therefore, a part or all of the *nif* genes (Bachhuber et al., 1976; MacNeil, Brill, and Howe, 1978). Several hundred deletions in *nif* were obtained in this way (MacNeil et al., 1978).

In order to map point mutations on the chromosome, it was necessary to transduce the mutations to the *nif*-containing plasmid, pRD1 (Dixon, Cannon, and Kondorosi, 1976), and to obtain Mu-induced deletions in that plasmid. However, pRD1 prevents the strain from being infected by phage P1 as well as by phage Mu. A derivative of pRD1, pTM4010, was obtained (MacNeil, Brill, and Howe, 1978) that did allow the bacterium to be infected by P1 or Mu. It was now possible to transduce deletion and point mutations from the chromosome to the plasmid and, in addition, plasmid-containing *nif* deletions were isolated by selection for heat-resistant strains from lysogens carrying a Mu insertion in the plasmid.

Since many recombination events were required for the construction of pRD1 (Dixon, Cannon, and Kondorosi, 1976), it was necessary to demonstrate that no rearrangement of *nif* genes occurred. Mutations in the chromosome were mapped by deletions in the plasmid and mutations in the plasmid were likewise mapped by deletions in the chromosome. With these techniques, all of the *nif* mutations were located in one of 49 deletion intervals (MacNeil et al., 1978). The arrangement of *nif* mutations on the chromosome is the same as that found on the plasmid. This fine-structure mapping indicated that many NG-induced mutants had more than one mutation. For instance, strains with the mutations *nif-4113*, *nif-4019*, and *nif-4179* contained mutations in more than one gene. These three mutations had been used to identify *nif* genes (St. John et al., 1975).

Complementation tests were performed to identify the number of *nif* cistrons. It was necessary to make the recipient strains Rec⁻ so that complementation could be distinguished from recombination in mating experiments. In these experiments, merodiploids containing a *nif* mutation in the chromosome and a *nif* mutation in the plasmid pTM4010 were constructed (MacNeil, Brill, and Howe, 1978). Transduction was used to place an *E. coli* *rec⁻* mutation into appropriate *K. pneumoniae* strains. The *rec⁻* mutation was moved to the strains by taking advantage of a strain having an insertion of the transposable element, Tn10, in *srl* that is cotransducible with *rec* (constructed by L. Csonka). In this way, tetracycline resistance was the initial selection for Rec⁻ strains. Sensitivity to ultraviolet irradiation was the final test for Rec⁻. Thirteen complementation groups were distinguished by these tests (MacNeil et al., 1978).

The polarity and number of operons can be determined by complementation patterns of strains with Mu insertion, amber, deletion, and frameshift mutations. Polar mutations prevent synthesis of products from genes operator-distal to the polar mutation. Complementation experiments were performed on polar mutations in each *nif* complementation group, resulting in the assignment of the 13 complementation groups to seven operons (MacNeil et al., 1978). The polycistronic transcripts are read from the *his*-distal to the *his*-proximal part of the particular *nif* region. Two operons seem to be monocistronic. Figure 1 presents a diagrammatic representation of the *nif* genes and their transcriptional relationships. Unlike Dixon et al. (1977), we do not find that a large fraction of our mutations have complex complementation patterns. This may be due to multiple mutations in their Nif⁻ strains. Their experiments did not utilize Rec⁻ recipients and therefore may have complexities due to the occurrence of recombination in a merodiploid.

An additional gene, *nifL*, was inferred by the finding that Mu insertions to the immediate *his*-distal end of the *nifA* complementation group were capable of reverting to Nif⁺. Mu insertions in the *his*-proximal end of the *nifA* complementation group did not revert (MacNeil et al., 1978). Although not many point mutations were detected in *nifL*, all identified point mutations were amber mutations. Most of the *nifL* mutations were caused by a Mu insertion. Thus, it seems that *nifL* is nonessential under conditions used for growing the bacteria on N₂. The Nif⁻ phenotype of all the mutants with lesions in *nifL* seems to be due to a polar effect on *nifA*.

Previous work indicated that the *nif* region was made up of two segments separated by a large region in which no *nif* mutations were mapped (Kennedy, 1977). Transduction analysis of point mutations in the 14 *nif* genes did not show that there is a large non-*nif* region (MacNeil et al., 1978). This is in agreement with the findings of Hsueh et al. (1977).

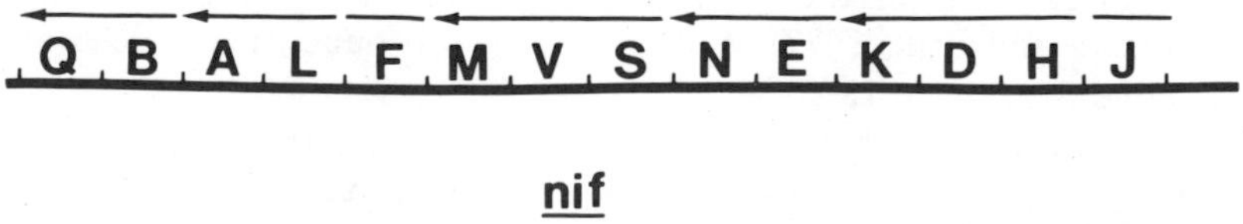

nif

Figure 1. *nif* Genes: their order, organization, and direction of transcripts. Details are discussed in MacNeil et al. (1978).

To understand the role of the 14 genes, it is important to determine the protein product of each gene. Two-dimensional polyacrylamide gel electrophoresis (O'Farrell, 1975) was performed on whole cell extracts of wild-type and several hundred Nif⁻ mutants (Roberts et al., 1978). Nine *nif*-related protein spots are readily observed by treating the gels with Coomassie Blue stain. These spots are missing in wild-type cells grown on ammonium and also in cells having a total *nif* deletion. Assignments of proteins to a gene were based on the following criteria: 1) deletions extending into the gene should eliminate both the gene product and its activity; 2) insertion mutations within a given gene should eliminate both the gene product and its activity; 3) some of the point mutations within a gene should eliminate both the protein and activity; 4) some mutations might produce an inactive protein; and 5) mutations in no other gene should eliminate only this protein. When these criteria were used, Roberts et al. (1978) were able to assign protein products to eight of the 14 genes. The results are summarized in Table 1.

It is particularly interesting that mutations in *nifA* and *nifL* prevent the synthesis of all of the detectable *nif* proteins (Roberts et al., 1978). This is further evidence that *nifA* makes a product that is required to turn on the synthesis of all of the *nif* genes (Dixon et al., 1977).

Several other *nif* genes also seem to play a role in the regulation of N_2 fixation. Many NifK⁻, NifD⁻, and NifH⁻ strains produce only low levels of all of the detectable Nif proteins (Roberts et al., 1978). This suggests that the two subunits of the component I protein and the component II protein help to regulate the synthesis of all of the Nif proteins. This could be an important regulatory function because it may be responsible for maintaining the correct ratio of the proteins for maximum N_2 fixation (Shah, Davis, and Brill, 1975). Some form of regulation may also exist by an interaction between the *nifJ* and *nifF* proteins. Most of the NifJ⁻ mutants have very low levels of a 17,000-dalton protein, which seems to be the product of *nifF* (Roberts et al., 1978).

The functions of the gene products were determined by preparing extracts from a variety of mutants and then adding pure component I, component II, or the iron-molybdenum cofactor (FeMoco) to the extract. Reconstitution of activity was determined by measuring acetylene reduction (Roberts et al., 1978). Three genes (*nifB*, *nifN*, and *nifE*) seem to be required specifically for the synthesis of FeMoco, the active site of nitrogenase (Shah and Brill, 1977). *nifM* codes for a product that is required to make active component II.

Many substrates other than N_2 are reducible by nitrogenase. Examples are acetylene (Dilworth, 1966), cyanide (Hardy and Knight, 1967), and azide (Schöllhorn and Burris, 1967). Compounds with such side groups were

Table 1.   Identification of *nif*-coded proteins[a]

Approximate molecular mass of protein (daltons)	Structural gene	Function
120,000	*nifJ*	?
60,000	*nifK*	component I subunit
56,000	*nifD*	component I subunit
50,000	*nifN*	FeMoco
46,000	*nifE*	FeMoco
35,000	*nifH*	component II
18,000	*nifS*	?
17,000	*nifF*	electron transport

[a] Details are discussed in Roberts et al. (1978).

tested for possible differential effects on wild-type versus mutant strains. One compound, 6-cyanopurine, produced white colonies when the cells were grown on ammonium, but when cells were grown on N_2 the colonies formed were a dark purple color (MacNeil and Brill, 1978). When a mutant with a total *nif* deletion was grown anaerobically on an amino acid, the colony color was white. The wild type formed purple colonies under these conditions. This color difference has been very useful as a screen for *nif* mutations.

6-Cyanopurine also was used to isolate regulatory mutants that are derepressed for nitrogenase synthesis in the presence of ammonium (MacNeil and Brill, 1978). Mutants with mutations in *glnC* yielded purple colonies on plates with excess ammonium and 6-cyanopurine; therefore, the compound may be useful for isolating strains with specific *gln* mutations. Strains of *E. coli* with the *nif*-containing plasmid pTM4010 also formed purple colonies when they fixed N_2 in the presence of 6-cyanopurine. Perhaps 6-cyanopurine will be useful for isolating derepressed mutants in other N_2-fixing bacteria as well.

The large number of genes required for N_2 fixation and the complexity of their interactions and regulation point to the difficulty expected for suitable expression of the *nif* genes in a host that is the recipient of *nif* as a product of genetic engineering. Even when the *nif* genes are introduced into the closely related organism *E. coli*, the resulting strain does not grow as well on N_2 as on wild-type *K. pneumoniae*. Besides the problems described above, an N_2-fixing organism has to keep any nitrogenase synthesized from being O_2 inactivated, and must sequester sufficient iron and molybdenum to supply the large amount required for nitrogenase and sufficient ATP and electrons to provide the great energy demand of this process. An agriculturally useful N_2-fixing maize or wheat plant, obtained by genetic manipulations of *nif* into such plants, must overcome all of these and other barriers.

REFERENCES

Bachhuber, M., W. J. Brill, and M. M. Howe. 1976. The use of bacteriophage Mu to isolate deletions in the *his-nif* region of *Klebsiella pneumoniae*. J. Bacteriol. 128:749–753.

Dilworth, M. J. 1966. Acetylene reduction by nitrogen-fixing preparations from *Clostridium pasteurianum*. Biochim. Biophys. Acta 127:285–294.

Dixon, R. A., F. C. Cannon, and A. Kondorosi. 1976. Construction of a P plasmid carrying the nitrogen fixation genes from *Klebsiella pneumoniae*. Nature 260:268–271.

Dixon, R., C. Kennedy, A. Kondorosi, V. Krishnapillai, and M. Merrick. 1977. Complementation analysis of *Klebsiella pneumoniae* mutants defective in nitrogen fixation. Molec. Gen. Genet. 157:189–198.

Guerola, N., J. L. Ingraham, and E. Cerda-Olmedo. 1971. Induction of closely linked multiple mutations by nitrosoguanidine. Nature New Biol. 230:122–125.

Hardy, R. W. F., and E. Knight, Jr. 1967. ATP-dependent reduction of azide and HCN by N_2-fixing enzymes of *Azotobacter vinelandii* and *Clostridium pasteurianum*. Biochim. Biophys. Acta 139:69–90.

Hsueh, C-T., J-C. Chin, Y-Y. Yu, H-C. Chen, W-C. Li, M-C. Shen, C-Y. Chiang, and S-C. Shen. 1977. Genetic analysis of the nitrogen fixation system in *Klebsiella pneumoniae*. Sci. Sinica 20:807–817.

Kennedy, C. 1977. Linkage map of the nitrogen fixation (*nif*) genes in *Klebsiella pneumoniae*. Molec. Gen. Genet. 157:199–204.

MacNeil, D., and W. J. Brill. 1978. 6-Cyanopurine, a color indicator useful for isolating deletion, regulatory, and other mutations in the *nif* (nitrogen fixation) genes. J. Bacteriol. 136:247–252.

MacNeil, T., W. J. Brill, and M. M. Howe. 1978. Bacteriophage Mu–induced deletions in a plasmid containing the *nif* (N_2 fixation) genes of *Klebsiella pneumoniae*. J. Bacteriol. 134:821–829.

MacNeil, T., D. MacNeil, G. P. Roberts, M. A. Supiano, and W. J. Brill. 1978. Fine-structure mapping and complementation analysis of *nif* (nitrogen fixation) genes in *Klebsiella pneumoniae*. J. Bacteriol. 136:253–266.

O'Farrell, R. H. 1975. High resolution two-dimensional electrophoresis of proteins. J. Biol. Chem. 250:4007–4021.

Roberts, G. P., T. MacNeil, D. MacNeil, and W. J. Brill. 1978. Regulation and characterization of protein products coded by the *nif* (nitrogen fixation) genes of *Klebsiella pneumoniae*. J. Bacteriol. 136:267–279.

St. John, R. T., H. M. Johnston, C. Seidman, D. Garfinkel, J. K. Gordon, V. K. Shah, and W. J. Brill. 1975. Biochemistry and genetics of *Klebsiella pneumoniae* mutant strains unable to fix N_2. J. Bacteriol. 121: 759–765.

Schöllhorn, R., and R. H. Burris. 1967. Reduction of azide by the N_2-fixing enzyme system. Proc. Natl. Acad. Sci. U.S.A. 57:1317–1323.

Shah, V. K., and W. J. Brill. 1977. Isolation of an iron-molybdenum cofactor from nitrogenase. Proc. Natl. Acad. Sci. U.S.A. 74:3249–3253.

Shah, V. K., L. C. Davis, and W. J. Brill. 1975. Nitrogenase VI. Acetylene reduction assay: dependence of nitrogen fixation estimates on component ratio and acetylene concentration. Biochim. Biophys. Acta 384:353–359.

Streicher, S., E. Gurney, and R. C. Valentine. 1971. Transduction of nitrogen fixation genes in *Klebsiella pneumoniae*. Proc. Natl. Acad. Sci. U.S.A. 68:1174–1177.

Nitrogen Fixation, Volume I
Edited by W. E. Newton and W. H. Orme-Johnson
Copyright 1980   University Park Press   Baltimore

Transcriptional Organization of the *Klebsiella pneumoniae* *nif* Gene Cluster

R. Dixon, M. Merrick, M. Filser, C. Kennedy, and J. R. Postgate

Genetic studies in our laboratory are directed toward an in-depth analysis of the cluster of genes that determines nitrogen fixation (*nif*) in *Klebsiella pneumoniae*. This paper describes some of our recent findings, with particular emphasis on our efforts to elucidate the operon structure of the nitrogen fixation genes.

KLEBSIELLA *nif* CISTRONS

It is now well established that *nif* genes are located between the histidine (*his*) operon and a gene for shikimic acid permease (*shiA*) in *K. pneumoniae*. In order to investigate the number of genes involved in nitrogen fixation, we have isolated a large number of *nif* mutants and performed an extensive complementation analysis. In our initial studies with 65 mutants, we assigned *nif* mutations to seven complementation groups: *nifB*, *nifA*, *nifF*, *nifE*, *nifK*, *nifD*, and *nifH* (Dixon et al., 1977). Transductional analysis of these mutations suggested the possibility of two gene clusters, *nifBAF* and *nifEKDH*, separated by a sequence of DNA with unknown function (Kennedy, 1977).

More recently, we have isolated a further 90 independent *nif⁻* point mutants using the mutagens nitrosoguanidine, ethyl methane sulfonate or ICR170 (which primarily induces frameshift mutations). Complementation tests with our total collection of mutants have defined 13 *nif* cistrons in the order *nifB*, *nifA*, *nifF*, *nifM*, *nifN*, *nifS*, *nifI*, *nifE*, *nifK*, *nifG*, *nifD*, *nifH*, and *nifJ*. This is a minimal estimate of the number of genes determining nitrogen fixation. We recognize that we have not yet "saturated" our complementation data and there may be additional *nif* cistrons that remain to be identified. Deletion mapping and three-factor transductional analysis with phage P1 have established that the recently discovered *nif* cistrons *nifM*, *nifN*, *nifS*, and *nifI* map in between the previously identified gene clusters *nifBAF* and *nifEKDH*, indicating that there is no major "gap" between the *nif* cistrons (Figure 1). Indeed, many of these genes are likely to be contiguous.

The existence of at least 13 *nif* genes is a somewhat surprising finding, particularly since only a small number of proteins are indicated from biochemical studies to be involved in the nitrogen fixation process. Relatively few of the *nif* gene products have so far been identified. The *nifB* product is necessary for MoFe protein activity. Extracts prepared from *nifB* mutants can be activated by addition of FeMo cofactor (FeMoco) prepared from wild-type MoFe protein (Shah and Brill, 1977), suggesting that the *nifB* product is involved in FeMoco synthesis or insertion of the cofactor into the apoprotein. The *nifA* product has a regulatory role. Mutations in this cistron have a pleiotropic effect on the synthesis of nitrogenase polypeptides, resulting in a nonderepressible phenotype. *nifA*$^+$ is dominant to *nifA*$^-$ and restores nitrogenase synthesis *in trans*, suggesting that the *nifA* gene determines a diffusable product, probably a *nif*-specific activator protein (Dixon et al., 1977). The *nifF* product is thought to be an electron-transport protein (St. John et al., 1975). *nifK*, *nifD*, and *nifH* are most probably nitrogenase structural genes. *nifH* is clearly the gene for Fe protein, because a mutation in this cistron results in a Fe protein with an altered electrophoretic mobility (Dixon et al., 1977). We have recently subdivided gene *K* into two cistrons, *nifK* and *nifG*, raising doubts as to which cistrons determine the α and β subunits of the MoFe protein. Mutations in all three genes, *nifK*, *nifG*, and *nifD*, can affect MoFe protein activity, yet clearly only two of these cistrons can code for the two subunit types. The remaining gene must determine synthesis rather than the structure of this molecule. SDS–polyacrylamide gel electrophoresis of extracts prepared from several *nifJ* mutants indicates the lack of a protein of molecular weight around 120,000. This protein is either the *nifJ* product or is subject to regulation by *nifJ*. It is also absent in *nifA* mutants, suggesting that it is under *nifA* control. Elmerich et al. (1978) have observed similar findings. The products of genes *M*, *N*, *S*, *I*, and *E* have not been identified as yet.

Physical mapping of the *nif* region has been performed by restriction endonuclease digestion of *nif* DNA fragments cloned on small amplifiable plasmids (Cannon, Riedel, and Ausubel, 1977; Cannon, 1978). The physical

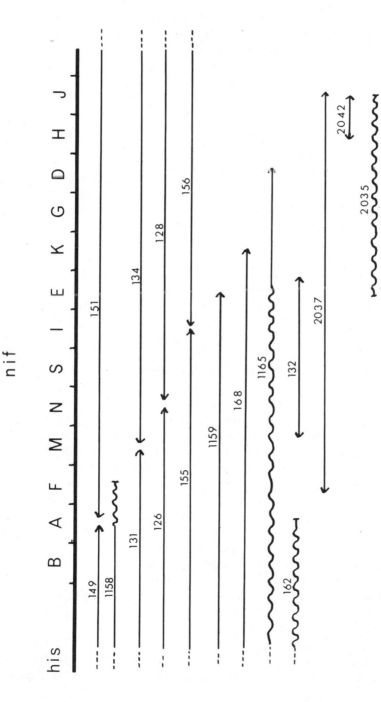

Figure 1. Map of the *Klebsiella pneumoniae nif* gene cluster showing the order of the 13 *nif* cistrons now identified and the extent of deletions and inversions derived by imprecise excision of translocatable genetic elements from the *Klebsiella* chromosome. Deletions of the *nif* region are shown as solid lines and undefined end points as dashed lines. Inversions are represented by a wavy line.

and genetic data are largely in agreement and suggest that the *nif* region is approximately 20 kb in length, which is sufficient coding capacity for about 19 average-size genes. However, the nitrogenase structural polypeptides and the presumptive *nifJ* product have a total molecular weight (as monomers) of 266,000. The four genes coding for these products must therefore be around 8 kb in length. This leaves 12 kb for the remaining nine *nif* genes and does not take into account nontranscribed sequences such as operator-promoter regions and terminators.

INSERTION MUTATIONS IN *nif* GENES

The insertion of a translocatable drug resistance element (transposon) into a bacterial gene inactivates that gene and (in the appropriate orientation) produces a strongly polar effect on the transcription of genes located promoter-distal to the insertion site in an operon (for reviews, see Kleckner, 1977; Kleckner, Roth, and Botstein, 1977). We are exploiting the known polar effects of the transposons Tn5 (determining kanamycin resistance), Tn7 (determining trimethoprim, streptomycin, and spectinomycin resistance), and Tn10 (determining tetracycline resistance) to study the transcriptional organization of the *nif* gene cluster. The majority of Tn5 insertions show strong polar effects, although some insertions show low constitutive expression of genes downstream from the insertion site (Berg, 1977). Tn10 causes strong polarity when inserted in one orientation and weak polarity when inserted in the opposite orientation (Foster and Willetts, 1977). Methods for insertion of these transposons into the *Klebsiella nif* genes are described in detail elsewhere (Merrick et al., 1978). Selection involved translocation of the element from the *E. coli* chromosome to a His⁺Nif⁺ plasmid. Translocation of Tn7 into the Nif⁺ plasmid pRD1, and Tn5 and Tn10 into plasmid pMF100, which is a drug-sensitive derivative of pRD1 (Filser and Cannon, unpublished results), occurred at frequencies similar to those reported elsewhere for these transposons. In all cases, around 2%–5% of these plasmid insertions were Nif⁻.

nif insertion mutants were tested for complementation with representatives of 11 of the 13 *nif* cistrons. Data for *nifS* and *nifG* are incomplete at present and are therefore not considered further. Many of the insertions have been mapped by P1 transduction with known point mutants or by deletion analysis. The distribution of insertion mutations within the *nif* gene cluster is shown in Figure 2. Tn5 insertions were observed in all 11 cistrons tested. Three of the Tn5 insertions were not assignable in complementation tests and probably carry mutations in additional *nif* genes. Tn7 insertions were not observed in *nifM* or *nifI* and Tn10 mutations were not found for *nifF*, *nifE*, *nifD*, and *nifH*. This distribution probably reflects both the site

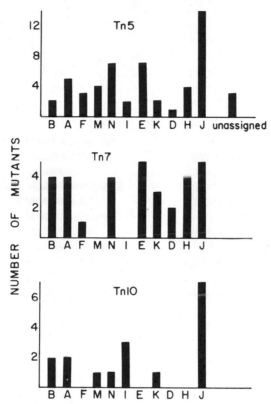

Figure 2.   Distribution of Tn5, Tn7, and Tn10 insertions in the *Klebsiella nif* region.

specificity of insertion for each transposon and their relative frequencies of translocation. The polar effects of each insertion mutation were determined by complementation analysis. Quantitative acetylene reduction assays were performed with plasmids carrying Tn7 and Tn10 (Figures 3, 4, and 5) and qualitative tests on nitrogen-free agar plates were performed with Tn5 insertions (Figure 6). The combined results from all three transposons suggest that there are several transcriptional units in the *nif* gene cluster:

1. No polarity has been detected among insertions in *nifB* and *nifA*, indicating that both of these genes are on independent operons (Figure 3).

2. Insertions in *nifF* are nonpolar, whereas Tn5 and Tn10 mutations in *nifM* show an M⁻F⁻ phenotype in plate complementation tests. However, this effect was not observed with *nifM2011*::Tn10 using acetylene reduction assays (Figure 4). Both *nifN*::Tn7 and *nifN*::Tn10 insertions showed an N⁻M⁻ phenotype in plate complementation tests,

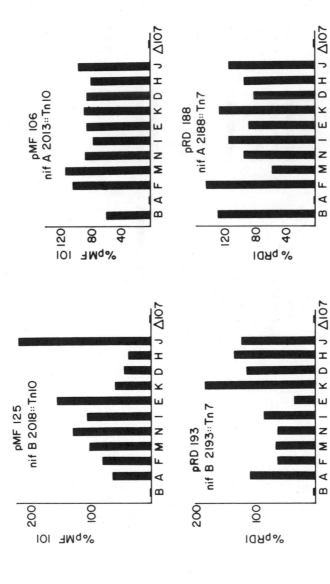

Figure 3.   Complementation profiles of *nifB* and *nifA* insertion mutants. Plasmids carrying *nifB*::Tn and *nifA*::Tn insertions were transferred by conjugation to representatives of the 11 *nif* cistrons shown on the horizontal axis. Transconjugant colonies were assayed for acetylene reduction as described by Dixon et al. (1977). Results are expressed as a percentage of the acetylene-reducing activity given by the Nif⁺ plasmids (pRD1 or pMF101) in each mutant background. Δ107 is a total *his-nif* deletion; values shown in this deletion mutant therefore indicate the background level of activity for each insertion mutation. (pMF101 is a Tn10 insertion of unknown location in pMF100.)

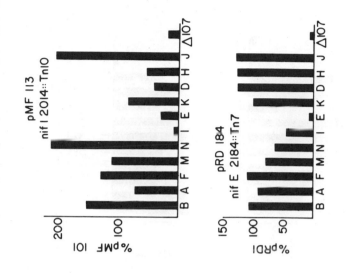

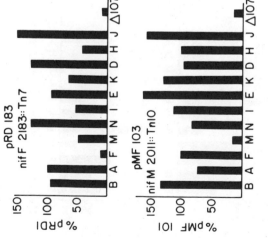

Figure 4.   Complementation profiles of *nifF*, *nifM*, *nifN*, *nifI*, and *nifE* insertion mutants. Details as for Figure 3.

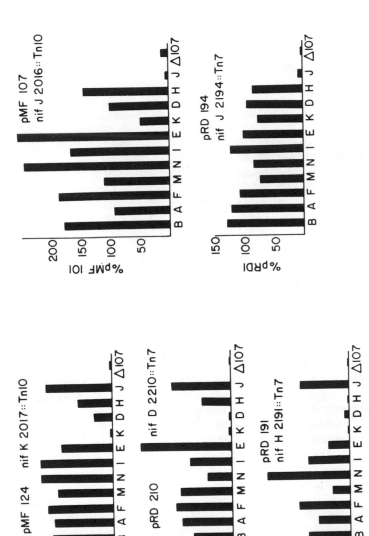

Figure 5. Complementation profiles of *nifK*, *nifD*, *nifH*, and *nifJ* insertion mutants. Details as for Figure 3.

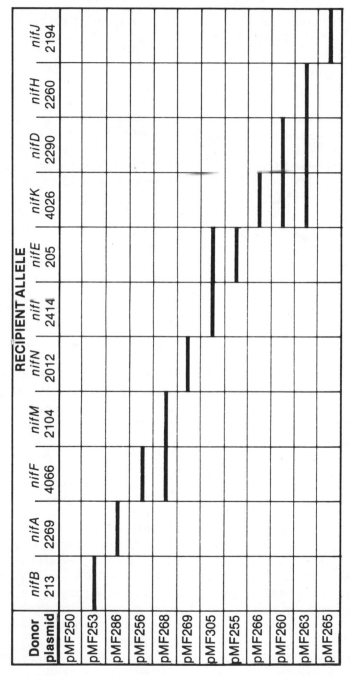

Figure 6.   Complementation tests with pMF100::Tn5 mutants. Complementation analysis was carried out on nitrogen-free agar plates as described by Merrick et al. (1978). Horizontal bars indicate that the insertion mutant failed to complement the recipient allele(s) shown. Blanks indicate that complementation was observed (pMF250 is a Tn5 insertion of unknown location in pMF100).

but again no effect was observed in assays. Although these results are equivocal, they suggest that *nifN*, *nifM*, and *nifF* are in the same transcriptional unit.

3. Several insertions showed an I⁻E⁻ phenotype in complementation tests, indicating that these two genes are in the same operon. Other insertions were solely NifE⁻ in complementation assays (Figure 4). These results could suggest that the operon is transcribed from *nifI* to *nifE*, although the results obtained with deletions are contradictory (see below).

4. Clear evidence for transcriptional polarity was obtained with insertions in the nitrogenase structural genes. Three phenotypes were obtained: a) NifH⁻NifD⁻NifK⁻ (pRD191, pMF263), b) NifD⁻NifK⁻ (pRD210, pMF260), and c) NifK⁻ (pMF124, pMF266). This confirmed that the three genes are transcribed as a single operon, from *nifH* to *nifK*, with *nifH* being the promoter-proximal gene (Figures 5 and 6). Biochemical analysis of strains carrying insertions in *nifK*, *nifD*, and *nifH* has substantiated genetic evidence for polarity within the structural *nif* gene cluster. *nifH* insertion strains do not synthesize any of the nitrogenase polypeptides, whereas strains carrying *nifD* insertions lack the α and β subunits of nitrogenase, but synthesize Fe protein. Surprisingly, *nifK* insertions strains have the same biochemical phenotype as *nifD* mutants. All three mutant classes synthesize the *nif*-specific polypeptide of molecular weight 120,000.

5. Insertions in *nifJ* were obtained with all three transposons and none of these showed polar effects. *nifJ* insertion mutants lacked the 120,000–molecular weight *nif*-specific protein, but synthesized all three nitrogenase polypeptides. This gene therefore is on a separate transcriptional unit.

IMPRECISE EXCISION OF TRANSPOSONS

Selection for drug-sensitive derivatives of Tn10 and Tn7 strains is possible using an antibiotic enrichment technique. Three major classes can be distinguished among strains that have lost the drug resistance phenotype of the insertion mutation: 1) strains that have lost the transposon by precise excision; 2) strains in which a mutation or deletion within the element itself has occurred, and 3) strains in which adjacent bacterial DNA sequences have been deleted or rearranged because of imprecise excision. We were interested in the latter class because deletion or inversion of adjacent *nif* DNA by imprecise excision could provide additional information regarding the operon structure and map location of the *nif* genes.

    We have been able to obtain deletions with each of the three transposons. By selecting for His⁻ derivatives of *nif*::Tn5 insertions, we have obtained deletions that extend from the insertion site to the nearby histidine operon. Selection for tetracycline-sensitive derivatives of Tn10 insertions has yielded deletions that extend from the site of insertion in either one direction or the other into adjacent *nif* DNA. It was essential to demonstrate that these derivatives failed to act as donor in transductional crosses with several *nif⁻* point mutations to be certain that these strains carried deletions. Another class of imprecise excision events was also detected. Drug-sensitive mutants were obtained that had a different phenotype from the original insertion strain. They were able to act as donors in transductional crosses with *nif⁻* point mutants, but they gave only a few Nif⁺ transductants when used as recipients in crosses with the same *nif⁻* point mutations. These are presumptive inversion strains (Botstein and Kleckner, 1977). Figure 1 shows some of the deletion and inversion strains derived from *nif*::Tn7 and *nif*::Tn10 insertions in the *Klebsiella* chromosome. An interesting observation was that imprecise excision of Tn7 frequently gave rise to a deletion at one side of the insertion site and an inversion at the other, e.g., UNF1165.

In general, complementation analysis of *nif* deletion and inversion strains did not conflict with the data obtained with the insertion mutants. For example, UNF2042, which has a deletion extending from *nifJ* into *nifH*, had a strong polar effect on *nifD* and *nifK*, and strain UNF162, which has an inversion extending from the histidine operon into *nifA*, had a His⁻NifB⁺NifA⁻ phenotype, confirming that *nifB* is on a separate transcriptional unit. The only exception was observed with deletions and inversions that had end points in *nifI* and *nifE*. Strain UNF155, which carries a deletion extending from the histidine operon into *nifI*, was NifE⁺ in complementation tests and UNF2035, which has an inversion extending from *nifJ* into *nifE*, had a NifJ⁻NifE⁻NifI⁻ phenotype. This suggests that *nifI* and *nifE* are transcribed from *nifE* to *nifI*, in contrast to the results obtained with transposons. Possible explanations for this contradiction are discussed below.

CONCLUSIONS

Several factors must be taken into account when considering polarity effects induced by transposons.

1. The orientation of the transposon may affect the degree of polarity observed.
2. With the exception of genes *K*, *D*, *H*, and *J*, our estimates of polarity are based solely on complementation studies with heterogenotes, since most of the *nif* gene products have yet to be identified. The existence of weak internal promoters within an operon can provide sufficient expression of distal genes to allow complementation to occur, even

though the products of these distal genes cannot be detected in haploid strains containing the polar mutation (Atkins and Loper, 1970; Kleckner et al., 1975).

3. Some insertion mutations show a gradient of antipolarity for operator proximal genes (Jordan, Saedler, and Starlinger, 1967; Besemer and Herpers, 1977). This phenomenon could explain the apparent ambiguity in the data for the *nifI* and *nifE* genes, i.e., an insertion in *nifI* could be "antipolar" on *nifE*. This problem is unlikely to occur with deletion strains. Antipolarity could also explain the absence of both the α and β subunits of MoFe protein in *nifK* insertion mutants. Indeed, an apparent antipolar effect was observed in complementation tests with *nifK* :: Tn7 and *nifK* :: Tn10 strains (Figure 5).

4. Most strains with insertions in *nifF*, *nifM*, *nifN*, *nifI*, and *nifE* were "leaky" in acetylene reduction tests. This observation is not understood, but it could indicate that these genes are not absolutely necessary for nitrogenase function. The leakiness of these mutations causes some difficulty when attempting to determine polarity by complementation analysis.

Our results are in complete agreement with those of Elmerich et al. (1978), who have used bacteriophage Mu to induce polar mutations in the *nif* gene cluster. Data from our two laboratories suggest that there are at least six transcriptional units in the *nif* region:

1. *nifB*
2. *nifA*
3. *nifF*, *nifM*, and *nifN;* possibly transcribed in the direction *nifN* to *nifF*
4. *nifI* and *nifE;* direction of transcription is uncertain
5. *nifK*, *nifD*, and *nifH;* transcribed from *nifH* to *nifK*
6. *nifJ*

The next important steps will be to identify all the *nif* gene products, to determine the location of the *nif* promoters, and to investigate how each of these transcriptional units is regulated.

REFERENCES

Atkins, J. F., and J. C. Loper. 1970. Transcription initiation in the histidine operon of *Salmonella typhimurium*. Proc. Natl. Acad. Sci. U.S.A. 65:925–932.

Berg, D. 1977. Insertion and excision of the transposable kanamycin resistance determinant Tn5. In: A. T. Bukhari, J. A. Shapiro, and S. Adhya (eds.), DNA Insertion Elements—Plasmids and Episomes, pp. 205–212. Cold Spring Harbor Laboratory, New York.

Besemer, J., and M. Herpers. 1977. Suppression of polarity of insertion mutations within the *gal* operon of *E. coli*. Molec. Gen. Genet. 151:295–304.

Botstein, D., and N. Kleckner. 1977. Translocation and illegitimate recombination by the tetracycline resistance element Tn10. In: A. T. Bukhari, J. A. Shapiro, and S. Adhya (eds.), DNA Insertion Elements—Plasmids and Episomes, pp. 185–204. Cold Spring Harbor Laboratory, New York.

Cannon, F. C. 1978. The *Klebsiella nif* genes on small amplifiable plasmids. Paper presented at the EMBO workshop on plasmids and other extrachromosomal genetic elements, April 1–5, West Berlin.

Cannon, F. C., G. E. Riedel, and F. M. Ausubel. 1977. Recombinant plasmid that carries part of the nitrogen fixation (*nif*) gene cluster of *Klebsiella pneumoniae*. Proc. Natl. Acad. Sci. U.S.A. 74:2963–2967.

Dixon, R. A., C. Kennedy, A. Kondorosi, V. Krishnapillai, and M. Merrick. 1977. Complementation analysis of *Klebsiella pneumoniae* mutants defective in nitrogen fixation. Molec. Gen. Genet. 157:189–198.

Elmerich, C., J. Houmard, L. Sibold, I. Manheimer, and N. Charpin. 1978. Genetic and biochemical analysis of mutants induced by bacteriophage Mu DNA integration into *Klebsiella pneumoniae* nitrogen fixation genes. Molec. Gen. Genet. 165:181–189.

Foster, T. J., and N. S. Willetts. 1977. Characterisation of transfer-deficient mutants of the R100-1 Tc^S plasmid pDU202 caused by insertion of Tn10. Molec. Gen. Genet. 156:107–114.

Jordan, E., H. Saedler, and P. Starlinger. 1967. Strong polar mutations in the transferase gene of the galactose operon in *E. coli*. Molec. Gen. Genet. 100:296.

Kennedy, C. 1977. Linkage map of the nitrogen fixation (*nif*) genes in *Klebsiella pneumoniae*. Molec. Gen. Genet. 157:199–204.

Kleckner, N. (1977). Translocatable elements in prokaryotes. Cell 11:11–23.

Kleckner, N., R. K. Chan, B. K. Tye, and D. Botstein. 1975. Mutagenesis by insertion of a drug-resistance element carrying an inverted repetition. J. Molec. Biol. 97:561–575.

Kleckner, N., J. Roth, and D. Botstein. 1977. Genetic engineering *in vivo* using translocatable drug-resistance elements. J. Molec. Biol. 116:125–159.

Merrick, M., M. Filser, C. Kennedy, and R. Dixon. 1978. Polarity of mutations induced by insertion of transposons Tn5, Tn7, and Tn10 into the *nif* gene cluster of *Klebsiella pneumoniae*. Molec. Gen. Genet. 165:103–111.

St. John, R. T., M. H. Johnston, C. Seidman, D. Garfinkel, J. Gordon, V. K. Shah, and W. J. Brill. 1975. Biochemistry and genetics of *Klebsiella pneumoniae* mutant strains unable to fix N_2. J. Bacteriol. 121:759–765.

Shah, V. K., and W. J. Brill. 1977. Isolation of an iron-molybdenum co-factor from nitrogenase. Proc. Natl. Acad. Sci. U.S.A. 74:3249–3253.

Nitrogen Fixation, Volume I
Edited by W. E. Newton and W. H. Orme-Johnson
Copyright 1980   University Park Press   Baltimore

Transcriptional Studies
with Cloned Nitrogen-fixing Genes

K. A. Janssen, G. E. Riedel,
F. M. Ausubel, and F. C. Cannon

We have recently described the construction, in vitro, of a small bacterial plasmid carrying some of the nitrogen-fixing (*nif*) genes from the enteric bacterium *Klebsiella pneumoniae* (Ausubel et al., 1977; Cannon, Riedel, and Ausubel, 1977). This plasmid, pCRA37, is 22.5 kilobase pairs (kb) and carries the *hisD* gene as well as *nif* genes *B* through *E* (see Figure 1). In this paper, we describe the construction of two additional plasmids, pCM1 and pSA30, that, between them, carry most if not all of the remaining *nif* genes in the *his-nif* cluster (see articles in this volume by MacNeil et al. and Dixon et al.).

Amplifiable *nif* plasmids have enabled us to assign physical locations to the *hisD* gene and to some of the *nif* genes. Using purified *nif* DNA in hybridization experiments, we have studied the control and direction of transcription of *nif* genes in vivo and in vitro. These studies suggest that all *nif* genes are repressed in the presence of ammonia and coordinately derepressed in the absence of ammonia. The direction of transcription of the entire *nif* region probably proceeds toward *his*. In DNA protection experiments, we have established conditions for binding RNA polymerase (RNAP) and glutamine synthetase (GS) to *nif* DNA. These latter studies provide a means for identifying transcription start sites (promoters) as well as protein effectors that specifically affect *nif* transcription.

This work was supported by a National Science Foundation grant (PCM75-21435-A01) to F.M.A. and by the Agricultural Research Council, Unit of Nitrogen Fixation, at the University of Sussex. K.A.J. was supported by a Cystic Fibrosis Foundation Postdoctoral Fellowship.

CONSTRUCTION OF *nif* PLASMIDS

Figure 1 depicts regions of the *nif* cluster that have been incorporated into amplifiable plasmids. Some properties of these plasmids are summarized in Table 1. The source of *nif* DNA for all primary cloning experiments was purified DNA of the stringent plasmid pRD1 (formerly termed RP41). This large 125-megadalton plasmid (A. Pühler, unpublished data) contains the entire *nif* cluster of *K. pneumoniae* (Dixon, Cannon, and Kondorosi, 1976). Purified pRD1 DNA was cleaved with the restriction endonucleases *Eco*RI or *Hind*III and the resulting fragments were litigated into an appropriate plasmid cloning vector. The resulting mixture of recombinant plasmids was transformed into various *his⁻nif⁻* derivatives of *K. pneumoniae*. Transformants carrying a genetic determinant of the cloning vector were selected and subsequently screened for a Nif⁺ phenotype. In some cases, His⁺ transformants were selected directly and were also screened for a Nif⁺ phenotype. The plasmids pCRA10, pCRA37, pGR102, and pGR103 were constructed prior to July, 1976, and the properties of pCRA10 and pCRA37 have been described previously (Cannon, Riedel, and Ausubel, 1977; Ausubel et al., 1977). Since July, 1976, when the NIH guidelines were published, all recombinant DNA work with *K. pneumoniae* has been performed at the University of Sussex.

Plasmids pGR102 and pGR103

Purified pCRA37 plasmid DNA was incubated with restriction endonucleases *Bam*I or *Hind*III under conditions that yielded partial digests of the

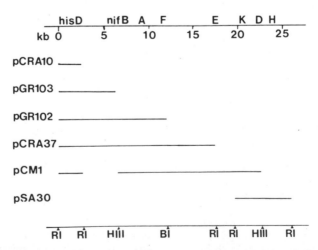

Figure 1.   Amplifiable *his-nif* plasmids constructed in vitro. The dark lines indicate regions of the *his-nif* cluster of *K. pneumoniae* that have been cloned in amplifiable plasmid cloning vectors. The vectors for each plasmid are indicated in Table 1.

Table 1.   Properties of amplifiable *his-nif* plasmids

Plasmid	Cloning vector	Size of cloning vector (kb)	Size of *his/nif* insert (kb)	Total size of plasmid (kb)	Drug resistance phenotype	*his* and *nif* genes present
pCRA10	pMB9	5.25	2.75	8.00	Tc^r	*his*D
pGR103	pMB9 (deleted)	4.92	6.45	11.37	Tc^r	*his*D
pGR102	pMB9 (deleted)	4.65	12.3	16.95	none	*his*D, *nifB-nifF*
pCRA37	pMB9	5.25	17.25	22.50	Tc^r	*his*D, *nifB-nifE*
pCM1	pCRA10	8.0	17.0	25.0	none	*his*D, *nifB-nifK* (*nifD, nifK*)^a
pSA30	pACYC184	4.1	6.9	11.0	Tc^r	*nifH*, (*nifK nifD, nifJ, shiA*)^a

^{a} Experiments to determine whether pCM1 contains *nifD* and *nifH* and experiments to determine whether pSA30 contains *nifK*, *nifD*, *nifJ*, and *ShiA* are currently in progress.

plasmid DNA. The partially digested plasmid DNA was recircularized with T4 DNA ligase and then used to transform KP5058, a *hisD⁻nifB⁻* derivative of *K. pneumoniae*. Plasmid DNA was isolated from several different His⁺Nif⁺ and His⁺Nif⁻ transformants and two plasmids, pGR102 and pGR103, were chosen for further study (see Figure 1).

Plasmid pCM1

Because the *K. pneumoniae* DNA cloned in pCRA37 contains a single cleavage site for *Hind*III (see Figure 1), pRD1 probably contains a *Hind*III restriction fragment that partially overlaps and extends beyond the end point of *nif* DNA in pCRA37. The small plasmid pCRA10 was chosen as the cloning vector for *Hind*III fragments of pRD1 DNA. pCRA10 is a 8.00-kb plasmid carrying *his*D and a single restriction site for *Hind*III within the promoter region of the tetracycline (Tc) resistance genes of the pMB9 portion of the recombinant plasmid (see Table 1). Cloning at the *Hind*III site of pMB9 frequently results in the loss of tetracycline resistance, providing a convenient indication of the presence of a fragment inserted at the HindIII site (Boyer et al., 1977). Thus, *Hind*III-digested pRD1 DNA was ligated to *Hind*III-digested pCRA10 DNA and the resulting recombinant plasmids were used to transform a *hisD⁻nifK⁻ K. pneumoniae* strain (CK317). Among 1500 His⁺Tc^s transformants, three gave rise to Nif⁺ colonies at a frequency of twenty-five to thirtyfold greater than the spontaneous reversion frequency of the *nif* mutation in CK317. The plasmids isolated from these three transformants were all 25.0 kb and one of them, pCM1, was further characterized.

Plasmid pSA30

The *Eco*RI restriction map of the *Hind*III *nif* DNA fragment from pRD1 cloned in pCM1 shows that a restriction site for *Eco*RI is close to the right hand end of the fragment (see Figure 1). pRD1 must therefore contain an *Eco*RI fragment that overlaps with this *Eco*RI-*Hind*III end fragment of pCM1 and extends beyond the *Hind*III site on pRD1. The plasmid cloning vector used to clone this *Eco*RI fragment was pACYC184, a 4.1-kb amplifiable plasmid carrying chloramphenicol (Cm) and tetracycline resistance genes (Chang and Cohen, 1978). A single recognition site for *Eco*RI maps within the Cm gene. pRD1 and pACYC184 were digested with *Eco*RI, joined with T4 ligase, and used to transform a *nifH⁻ K. pneumoniae* strain (CK260). Approximately 2% of the Tc-resistant Cm-sensitive transformants gave rise to Nif⁺ colonies at a high frequency, and plasmids isolated from five of these transformants were all 11.0 kb. One of these plasmids, pSA30, was selected for further study.

LOCATING *nif* GENES ON AMPLIFIABLE PLASMIDS

The *hisD* gene and several *nif* genes were located on amplifiable *nif* plasmids on the basis of *nif* complementation or *nif* marker rescue experiments. The entire *hisD* gene was found to be contained within the 2.75-kb *Eco*RI fragment cloned in pCRA10, since pCRA10 complements *hisD⁻* mutations and deletions in both *K. pneumoniae* and *Escherichia coli* (Cannon, Riedel, and Ausubel, 1977). The *nifB* gene was placed to the right of the *his*-proximal *Hind*III site in pCM1 because pCM1 complements *nifB⁻* mutations and because pGR103 does not complement or marker rescue *nifB⁻* mutations. The *nifF* gene was placed approximately at the location of the single *Bam*I site in pGR102 because pGR102 can marker rescue (but does not complement) *nifF* mutations. Similarly, *nifE* was placed to the left of the most *his*-distal *Eco*RI site in pCRA37, *nifK* was placed to the left of the *his*-distal *Hind*III site in pCM1, and *nifH* was placed to the right of the *his*-proximal *Eco*RI site in pSA30. Experiments designed to determine whether *nifJ* and/or *shiA* are present on pSA30 are currently in progress.

TRANSCRIPTION STUDIES

Nitrogenase is repressed in the presence of ammonia. To determine if *nif* expression is regulated at the level of transcription or translation, we have developed hybridization techniques using cloned *nif* sequences that assay directly for the presence of *nif*-specific RNA. As a positive control for these experiments, we assayed for the production of *his*-specific RNA, which is only present in cultures starved for histidine.

Coordinate Derepression of *nif* Genes

Total cellular RNA was isolated from wild-type cells grown anaerobically under nitrogen in the presence of varying amounts of ammonia (0 to 30 mM) and histidine (0 to 20 μg/ml). The 5' end of purified and partially hydrolyzed RNA was radioactively labeled in vitro with polynucleotide kinase (Maizels, 1976). Purified pCM1 DNA was digested with the restriction endonuclease *Eco*RI and subjected to electrophoresis through 1% agarose. The DNA fragments were transferred directly to nitrocellulose filters using the method described by Southern (1975), and then hybridized with the ³²P-labeled RNA. The results are shown in Figure 2. Cells grown in the presence of 30 mM ammonia and no histidine (lane A) synthesize RNA that hybridized to the *hisD*-specific fragment (band 4) but not to any of the *nif*-containing sequences (bands 1, 2, 3, and 5). Cells grown in no or low ammonia and no histidine (lanes B and C) produce RNA that hybridized to bands 1, 2, 3, and 4 (but not to band 5). In the presence of excess histidine

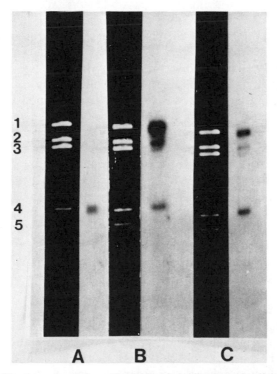

Figure 2.   Identification of cellular *nif* mRNA: hybridization of cellular [³²P]mRNA to *Eco*RI-digested pCM1 DNA. Each lane consists of a photographic composite of an ethidium bromide–stained agarose gel pattern of *Eco*RI-digested pCM1 DNA and an autoradiogram of ³²P-labeled cellular RNA hybridized to a "Southern blot" of the DNA fragments in the agarose gel.

Lane A: RNA isolated from cultures containing 30 mM ammonia and no histidine.
Lane B: RNA isolated from cultures containing 2 mM ammonia and no histidine.
Lane C: RNA isolated from cultures containing no ammonia and no histidine.

and no ammonia, there was hybridization to bands 1, 2, and 3 but no detectable hybridization to band 4 (data not shown). Thus, in the presence of ammonia, the wild-type cell represses transcription of the entire *nif* region. Band 5 is derived from the region between *nifE* and *nifK*. The fact that we observe no hybridization to this fragment in *nif*-derepressed cultures suggests either that there is no *nif* transcription in this region or that transcription occurs at such a low level that it cannot be detected by our methods.

Direction of *nif* Transcription

Experiments utilizing polar mutations, reported at this symposium by C. Elmerich, R. Dixon, and W. Brill, and their co-workers, indicate that there are several *nif* transcription units and that transcription in units containing more than one gene is in the direction toward *his*. We have developed an alternative procedure for determining the direction of *nif* transcription that involves hybridization of *nif* mRNA to separated strands of cloned *nif* DNA.

Procedures that result in physical separation of the two complementary strands of a linear DNA molecule were followed using *nif* plasmid DNA (Hayward, 1972). Linearized pCM1 DNA was denatured with alkali and subjected to electrophoresis through agarose gels. The separated pCM1 strands were transferred to nitrocellulose filters and used in hybridization experiments as described above. The results are shown in Figure 3. Labeled RNA isolated from cultures that were derepressed for *his* (lane D), for *nif* (lane E), or for both *his* and *nif* (lane F) all hybridized to the same slower-moving strand of pCM1. Thus, transcription of *his* and of the *nif* region cloned in pCM1 proceeds in the same direction. Since the direction of *his* transcription is from right to left in Figure 1, we can conclude that the direction of *nif* transcription is toward *his*.

We also assayed transcription products (copy RNA) synthesized in vitro for their ability to hybridize to pCM1 single strands (lanes A, B, and C in Figure 3). We used as templates purified *Eco*RI fragments of pCM1 containing the *hisD* gene (lane A), *nif* genes *B* through *F* (lane B), and *nif* genes *M* through *E* (lane C), which correspond to agarose gel bands 4, 2, and 3, respectively. All transcripts synthesized in vitro showed the same hybridization pattern as the cellular transcripts. This asymmetry of transcription in vitro implies that RNA polymerase is probably initiating at physiologically important promoters without specific transcriptional effectors. Our transcription mix, however, contained heparin at 100 μg/ml and glycerol at 10%. Heparin can bind weak promoters and glycerol has been shown to mimic the effect of transcriptional activator proteins such as CAP (Majors, 1975).

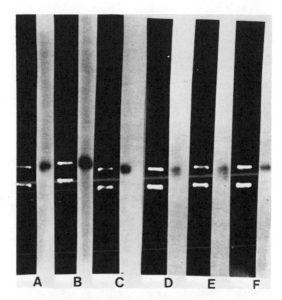

Figure 3.   Direction of *nif* transcription: hybridization of [³²P]RNA synthesized in vitro and in vivo to separated strands of linearized pCM1 DNA. Each lane consists of a photographic composite of an ethidium bromide–stained agarose gel of separated strands of alkali-denatured pCM1 DNA and an autoradiogram of ³²P-labeled RNA hybridized to a "Southern blot" of the DNA strands in the agarose gel.

Lane A: cRNA transcribed in vitro using *hisD* DNA (band 4 of Figure 2) as template.
Lane B: cRNA transcribed in vitro using *nifB-F*DNA (band 3 of Figure 2) as template.
Lane C: cRNA transcribed in vitro using *nifM-E* DNA (band 2 of Figure 2) as template.
Lane D: mRNA transcribed in vivo in the presence of 30 mM ammonia and no histidine.
Lane E: mRNA transcribed in vivo in the presence of no ammonia and 20 μg/ml histidine.
Lane F: mRNA transcribed in vivo in the presence of no ammonia and no histidine.

RNA Polymerase and Glutamine Synthetase Binding to *nif* DNA

A very sensitive method for studying specific protein-DNA interactions is to find conditions under which DNA binding protein will protect specific endonuclease recognition sites. The binding sites are detected by comparing the restriction patterns of a DNA molecule digested in the absence and the presence of the DNA binding protein (Jones et al., 1977). We have examined the ability of purified RNA polymerase (RNAP) and of purified glutamine synthetase (GS) to protect specific endonuclease recognition sites on the *his-nif* plasmids. Genetic studies have implicated GS as a positive effector regulating *nif* transcription in vivo (Streicher et al., 1974). Figure 4 shows some of the results of these binding and protection studies. Lanes A and F show the limit digestion patterns obtained with endonuclease *Hae*III for pCRA10 and pCM1, respectively, and lanes B and E show the patterns

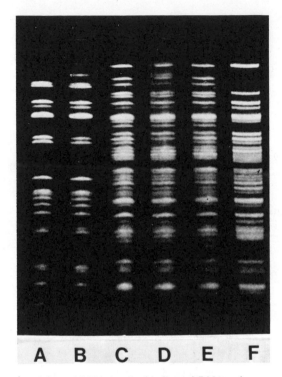

Figure 4. Protection of *his-nif* DNA by the binding of RNA polymerase (RNAP) or glutamine synthetase (GS) from digestion by *Hae*III endonuclease.

Lane A: Unprotected pCRA10 DNA
Lane B: RNAP-protected pCRA10 DNA, 200 mM NaCl
Lane C: GS-protected pCM1 DNA, 200 mM NaCl
Lane D: RNAP-protected pCM1 DNA, 100 mM NaCl
Lane E: RNAP-protected pCM1 DNA, 200 mM NaCl
Lane F: Unprotected pCM1 DNA

obtained when the DNA is protected with RNAP. Finally, lane C shows the pattern of pCM1 that has been prebound with GS. A comparison of lanes C and E indicates that GS protects most of the same sites as RNAP on pCM1. This result might be expected if GS directly affects *nif* transcription by binding at or near RNAP binding sites. A comparison of lanes D and E shows the effects of changing salt concentrations in the binding assay; as the salt concentration decreases, RNAP binds to and therefore protects more sites on the plasmid. Thus, RNAP binding sites identified in this manner may not necessarily be equivalent to biological promoters. The binding experiments with GS appear to be *nif* specific. The purified GS preparation, even at low salt concentrations, was not seen to protect sites on pCRA10, which contains only the *his* sequence.

CONCLUSION

The amplifiable *his-nif* plasmids we have described have enabled us to correlate the physical and genetic maps of much of the *his-nif* region. These plasmids have also enabled us to isolate large amounts of *nif*-specific DNA relatively easily. The techniques for hybridizing highly labeled cellular or copy RNA back to *nif* DNA provide a powerful probe for examining the nature of *nif* expression. Our results from these experiments suggest that the entire *nif* region is coordinately controlled at the level of transcription. Furthermore, transcription in vitro of pCRA10 and of pCM1 yields asymmetric (and therefore probably correct) products, and protection of *nif* and *his* DNA by RNAP from restriction endonuclease digestion in high salt conditions yields a consistent and specific pattern.

REFERENCES

Ausubel, F., G. Riedel, F. Cannon, A. Peskin, and R. Margolskee. 1977. Cloning nitrogen-fixing genes from *Klebsiella pneumoniae in vitro* and the isolation of *nif* promoter mutants. In: A. Hollaender (ed.), Genetic Engineering for Nitrogen Fixation, pp. 111–128. Plenum Publishing Corp., New York.

Boyer, H. W., M. Betlach, F. Bolivar, R. L. Rodriguez, H. L. Heyneker, J. Shine, and H. M. Goodman. 1977. The construction of molecular cloning vehicles. In: R. F. Beer and E. G. Bassett (eds.), Recombinant Molecules: Impact on Science and Society, pp. 9–20. Raven Press, New York.

Cannon, F. C., G. E. Riedel, and F. M. Ausubel. 1977. Recombinant plasmid that carries part of the nitrogen fixation (*nif*) gene cluster of *Klebsiella pneumoniae*. Proc Natl. Acad. Sci. U.S.A. 74:2963–2967.

Chang, A. C. Y., and S. N. Cohen. 1978. Construction and characterization of amplifiable multicopy DNA cloning vehicles derived from the P15A Cryptic Miniplasmid. J. Bacteriol. 134: 1141–1156.

Dixon, R., F. C. Cannon, and A. Kondorosi. 1976. Construction of a P plasmid carrying nitrogen fixation genes from *Klebsiella pneumoniae*. Nature 260:268–271.

Hayward, G. S. 1972. Gel electrophoretic separation of the complementary strands of bacteriophage DNA. Virology 49:342–344.

Jones, B. B., H. Chan, S. Rothstein, R. Wells, and W. Reznikoff. 1977. RNA polymerase binding sites in λp*lac*5 DNA. Proc. Natl. Acad. Sci. U.S.A. 74: 4914–4918.

Maizels, N. 1976. *Dictyostelium* 17 S, 25 S, and 5 S rDNAs lie within a 38,000 base pair repeated unit. Cell 9:431–438.

Majors, J. 1975. Initiation of *in vitro* mRNA synthesis from the wild type *lac* promoter. Proc. Natl. Acad. Sci. U.S.A. 72:4394–4398.

Southern, E. M. 1975. Detection of specific sequences among DNA fragments separated by gel electrophoresis. J. Mol. Biol. 98:503–517.

Streicher, S., K. Shanmugan, F. Ausubel, C. Morandi, and R. Goldberg. 1974. Regulation of nitrogen fixation in *Klebsiella pneumoniae:* Evidence for a role of glutamine synthetase as a regulator of nitrogenase synthesis. J. Bacteriol. 120:815–823.

Nitrogen Fixation, Volume I
Edited by W. E. Newton and W. H. Orme-Johnson
Copyright 1980   University Park Press   Baltimore

Hydrogenase Activity and Hydrogen Evolution by Nitrogenase in Nitrogen-fixing *Azotobacter chroococcum*

M. G. Yates and C. C. Walker

The first indications that nitrogenase evolved H_2 in vivo were obtained by Hoch, Little, and Burris (1957) Hoch, Schneider, and Burris (1960), Bergersen (1963), and others. Dixon (1967, 1968, 1972) confirmed the observations of Phelps and Wilson (1941) that the nodules of *Pisum sativum* inoculated with *Rhizobium leguminosarum* strain ONA311, unlike the free-living rhizobia, contained an uptake hydrogenase that supported oxidative phosphorylation. When the hydrogenase of these nodules was saturated with D_2, they evolved H_2, presumably by the nitrogenase. Schubert and Evans (1976, 1977) noted that several species of nodulated plants "wasted" a substantial proportion of their energy for N_2 fixation as H_2 evolution. However, Evans et al. (1977) reported smaller losses (30%) through H_2 evolution. Bulen (as quoted by Postgate, 1971) suggested that hydrogenase recycled H_2 produced by nitrogenase in aerobes, and Dixon (1972) proposed three possible roles for the hydrogenase in nodules: 1) to aid respiratory protection (Dalton and Postgate, 1969) by scavenging O_2; 2) to prevent H_2 produced by nitrogenase from inhibiting N_2 reduction; and 3) to recycle H_2 to improve metabolic efficiency.

Following Brotonegoro (1974), we observed that N_2-fixing *Azotobacter chroococcum* evolved H_2 when the uptake hydrogenase was inhibited by CO and C_2H_2 (Smith, Hill, and Yates, 1976), which led us to test Dixon's hypothesis with *Azotobacter*. We also observed that N_2-fixing *A. chroococcum* evolved H_2 in continuous culture (Walker and Yates, 1978a). We have used this technique to study the effect of different nutrient limitations on uptake hydrogenase activity and the proportion of nitrogenase activity "wasted" as H_2 evolution during growth.

MATERIALS AND METHODS

These have been described in detail by Walker and Yates (1978a). The rates of H_2 evolution or acetylene reduction by continuous cultures were measured by sampling the isolated gas phase above the culture over a short period without stopping medium flow. Hydrogenase activity was measured in washed, EDTA/NaF-treated or disrupted cells as described by Hyndman, Burris, and Wilson (1953) with methylene blue as the electron acceptor. The assay for purified nitrogenase components of A. chroococcum, the MoFe protein (Ac1) and the Fe protein (Ac2), was as described by Yates and Planqué (1975). NH_4^+ production was measured as described by Maryan and Vorley (1978).

RESULTS

H_2-Dependent Acetylene Reduction

The O_2 dependence of H_2- or mannitol-supported acetylene reduction by batch-grown A. chroococcum is shown in Figure 1. The C_2H_2 concentration was sufficiently low to prevent significant inhibition of hydrogenase activity, but is considerably higher than the apparent K_m for C_2H_2 with an equimolar ratio of Ac1 to Ac2 in vitro (0.01 atm; M. G. Yates, unpublished data). Zero O_2 concentrations were obtained by using flasks with side arms containing $Na_2S_2O_4$ solution. Clearly, H_2-supported acetylene reduction (750 nmoles C_2H_4 in 1 hr) is far in excess of control experiments without added substrates (<2.7 nmoles in 1 hr). At very low O_2 concentrations (<0.025 atm) some control preparations supported acetylene reduction after a lag phase of 30–60 min, but at a maximum rate less than 10% of that obtainable with H_2 as electron donor. H_2 also supported reduction at saturating concentrations of mannitol (Figure 2), i.e., concentrations at which extra mannitol increased neither the rate of O_2 uptake nor acetylene reduction. Under these circumstances H_2 did not increase the rate of O_2 uptake; presumably electron flow to nitrogenase limited the rate of acetylene reduction. A possible alternative effect of H_2—that it interacts with nitrogenase to enhance mannitol-supported C_2H_2 reduction—cannot be discounted, although H_2 did not enhance $Na_2S_2O_4$-dependent C_2H_2 reduction with equimolar Ac1:Ac2 in vitro (M. G. Yates, unpublished data).

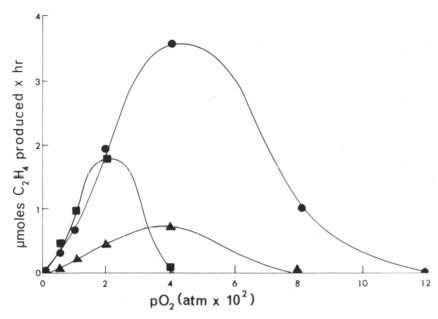

Figure 1.   H₂-supported acetylene reduction by *A. chroococcum*. Batch grown cells were washed twice in carbon-free medium and then gently shaken for 1 hr to deplete carbon reserves further. The atmosphere contained 8% acetylene, O₂ as indicated, 70% H₂ when included, and argon as the normalizing gas control. ▲——▲, H₂-dependent C₂H₂ reduction; ■——■, mannitol (0.06 mg/ml)–dependent acetylene reduction; ●——●, H₂ + mannitol-dependent acetylene reduction. The control without added substrate produced less than 3 nmoles C₂H₄ in 1 hr.

H₂-Supported Respiratory Protection of Nitrogenase

H₂ increased both the rate and the optimum pO₂ for acetylene reduction at mannitol concentrations below 1 g/liter (Figures 1 and 3). Above this value H₂ did not affect the optimum pO₂ for nitrogenase and, presumably, did not afford respiratory protection. It should be emphasized, therefore, that the influence of H₂ both as a support for acetylene reduction and as a means of aiding respiratory protection is paramount when mannitol is limiting. H₂ is apparently a relatively poor electron donor for nitrogenase compared with carbon substrates: at the same optimum pO₂, H₂ supports only one-third of the rate of acetylene reduction supported by low concentrations of mannitol.

H₂ Evolution by Nitrogenase

Table 1 and Figures 4 and 5 show that N₂-, O₂-, and mannitol-limited continuous cultures of *A. chroococcum* evolved H₂ under air. This rate of evolution was linear and increased significantly when air was replaced by an

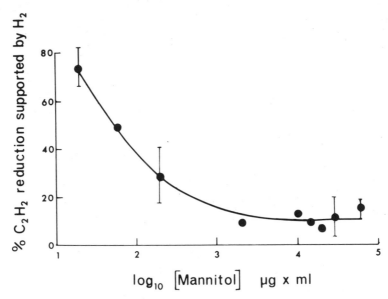

Figure 2. H_2-supported acetylene reduction by *A. chroococcum*. The experimental methods are as described for Figure 1. The percentages (ordinate) were derived from the maximum rates of acetylene reduction obtained at optimum pO_2 values. Reprinted from Walker and Yates 1978a).

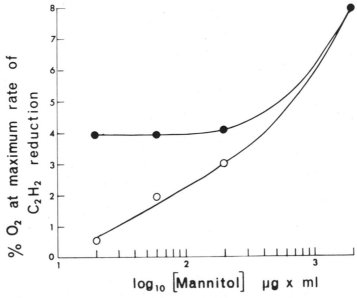

Figure 3.   Respiratory protection of nitrogenase by hydrogenase activity. The experimental methods are as described for Figure 1. ●——●, with H_2; O——O, without H_2. From data of Walker and Yates (1978a).

Table 1. H₂ evolution in continuous and batch cultures of *Azotobacter chroococcum*[a]

State of culture	Optical density	H₂ evolution under air (nmol·min⁻¹·mg of protein⁻¹)	H₂ evolution under Ar:O₂:CO₂ (nmol·min⁻¹·mg of protein⁻¹)	Percent nitrogen activity expended as H₂ evolution $\left(\dfrac{\text{column 3}}{\text{column 4}} \times 100\right)$	H₂:N₂ $\left(\text{Col. 3: } \dfrac{\text{Col. 4} - \text{Col. 3}}{3}\right)$
O₂ limited	0.33	12.50	32.50	38	1.87
O₂ limited	0.17	5.69	11.73	49	2.83
N₂ limited	0.33	65.00	149.50	43	2.31
Carbon limited	0.05	8.80	65.00	14[b] (30)	0.47[b] (1.31)
Carbon limited	0.10	16.00	134.20	12[b] (36)	0.41[b] (1.70)
Batch culture	0.40	0.003	0.016		

[a] Experimental techniques as described in "Methods."

[b] Substantial uptake hydrogenase activity present; therefore, some of the H₂ evolved by nitrogenase would be recycled and these figures do not represent the correct percentage of nitrogenase activity devoted to H₂ formation (Column 5) or the correct H₂:N₂ ratio (Column 6). The figures in parentheses were obtained after acetylene treatment to inhibit hydrogenase.

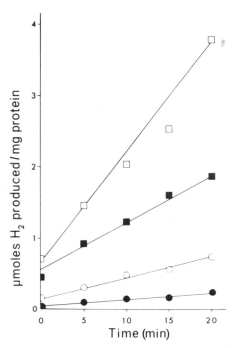

Figure 4. H_2 evolution by O_2- and N_2-limited continuous cultures of *A. chroococcum*. Cells were grown on nitrogen-free Burk's medium containing 10 g/liter mannitol at D = 0.1. $Ar:O_2:CO_2$ (79:21:0.3) mixture was flushed through for 25 min to replace air. O_2-limited cultures: ●——●, H_2 evolution under air; ○——○, H_2 evolution under $Ar:O_2:CO_2$. N_2-limited cultures: ■——■, H_2 evolution under air; □——□, H_2 evolution under $Ar:O_2:CO_2$.

$Ar:O_2:CO_2$ (79:21:0.03) mixture. H_2 was not evolved in the absence of O_2. Because *Azotobacter* hydrogenase catalyzes H_2 uptake in vivo (see Mortenson and Chen, 1974), we suggest that the H_2 evolved was ATP-dependent H_2 evolution by nitrogenase. In O_2- and N_2-limited cultures, the uptake hydrogenase activity was almost undetectable. It was significantly more active in mannitol-limited cultures, where a considerable proportion of the H_2 produced by nitrogenase was recycled. When the hydrogenase was inhibited (83%) by pre-incubation of the culture with 40% acetylene, the rate of H_2 evolution increased under air or $Ar:O_2:CO_2$ mixture and remained linear (Figure 5). Some of this increase may be caused by increased nitrogenase activity because double exposures of nitrogenase to C_2H_2 increased the rate of acetylene reduction in *Anabaena cylindrica* (David and Fay, 1977) and *A. chroococcum* (Walker and Yates, 1978b).

Effect of Component Ratio and
ATP Concentration on the $H_2:N_2$ Ratio in vitro

Silverstein and Bulen (1970) showed that low ATP concentrations favored H_2 evolution over N_2 reduction by crude *Azotobacter* nitrogenase in vitro.

A similar effect occurs in vitro when the Fe protein of *Klebsiella* nitrogenase is limiting (R. R. Eady, unpublished data). Figures 6 and 7 show that a combination of these restrictions increases the $H_2:N_2$ ratio dramatically in vitro. In these experiments, the $H_2:N_2$ ratio was calculated as:

$$\left[H_2 \text{ evolved under } N_2 \right] \Big/ \left[\frac{(H_2 \text{ evolved under Ar}) - (H_2 \text{ evolved under } N_2)}{3} \right] \quad (1)$$

This formula was adopted because the NH_4^+ assay (Maryan and Vorley, 1978) was not sufficiently sensitive to measure the low rate of N_2 reduction at low ATP concentrations and limiting Fe protein. When sufficient N_2 was reduced to render the NH_4^+ assay reliable, the results agreed within 10% with values obtained from applying equation 1. In no instance did the $H_2:N_2$ ratio drop below 1. The value was closest to 1 at optimum ATP and creatine phosphokinase concentrations and a component ratio Ac2:Ac1 of 46.

H₂:N₂ Ratios in vivo

Table 1 shows that the $H_2:N_2$ ratios in vivo were always greater than 1 except when uptake hydrogenase activity was substantial.

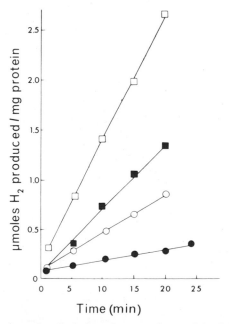

Figure 5.   H₂ evolution by carbon-limited continuous cultures of *A. chroococcum*. The experimental techniques are as described for Figure 4. Pretreatment with 40% acetylene was for 20 min and the culture was flushed for 1 hr before measuring H₂ evolution. ●———●, H₂ evolution under air; ■———■, H₂ evolution under Ar:O₂:CO₂; ○———○, H₂ evolution under air after exposure to C₂H₂; □———□, H₂ evolution under Ar:O₂:CO₂ after exposure to C₂H₂. Reprinted from Walker and Yates (1978a).

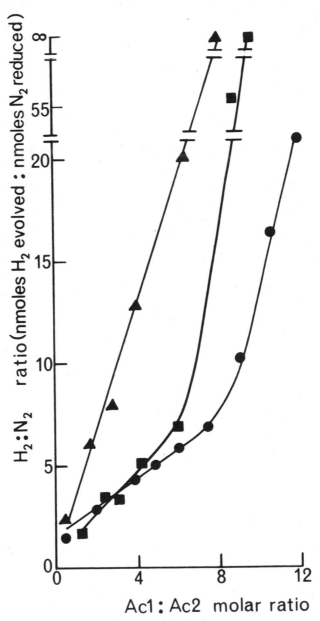

Figure 6.   Effect of component ratio and ATP concentration on the $H_2:N_2$ ratio in vitro. The molar ratio was calculated on the basis of estimated molecular weights of Ac1 and Ac2 (Yates and Planqué, 1975); Ac2 concentration was constant at 9.2 μg/ml, 0.14 μM. ATP concentration: ●——●, 5 mM; ■——■, 1 mM; ▲——▲, 0.1 mM.

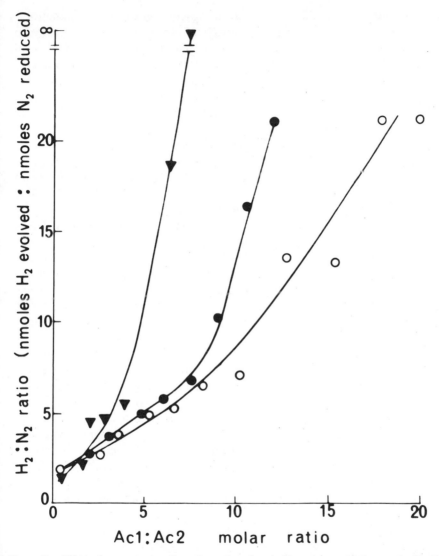

Figure 7.   Effect of component ratio and creatine phosphokinase concentration on the H₂:N₂ ratio in vitro. Conditions as described for Figure 6 with initial MgATP concentration at 5 mM. Creatine phosphokinase concentrations: ▼——▼, 5 μg; ●——●, 100 μg; O——O, 300 μg.

DISCUSSION

The evidence in this paper and that of Walker and Yates (1978a) shows that H₂ will support acetylene reduction by nitrogenase and, by inference, nitrogen fixation by *A. chroococcum* in vivo if a sufficiently active uptake hydrogenase is present. The maximum activity obtained with H₂ as donor was 15% of that obtained with saturating concentrations of mannitol in these experiments. The possible pathways of electron transfer from hydrogenase are: 1) to the respiratory chain, and 2) to nitrogenase. With regard to the first pathway, H_2 uptake was O_2 dependent in our experiments, and H_2 is known to support oxidative phosphorylation in *Azotobacter vinelandii* (Hyndman, Burris, and Wilson, 1953; Dixon, 1972). With regard to the second pathway, a direct electron transfer may occur from hydrogenase through unknown carriers to nitrogenase, but a possible sparing effect on the utilization of carbon reserves cannot be excluded.

The data in Figure 2 show that H_2 supported additional acetylene reduction at saturating levels of mannitol when it did not enhance respiratory protection (Figure 3) and did not increase the rate of O_2 uptake. Since the equivalent rates of acetylene reduction under O_2-limited conditions with mannitol and mannitol plus H_2 (Figure 1) imply that H_2-supported ATP synthesis is not better "coupled" than mannitol-supported ATP synthesis, it follows that the rate of ATP synthesis is not increased by H_2 at high levels of mannitol. Therefore, it is likely that at saturating mannitol concentrations H_2 contributes reducing equivalents to reductant-limited nitrogenase activity. However, as mentioned earlier, the possibility that H_2 interacts with nitrogenase to enhance mannitol-dependent acetylene reduction cannot be ruled out.

Yakovlev and Levchenko (1966) showed that H_2 supported nitrogenase activity in azotobacter. Our experiments show that hydrogenase is a relatively poor electron donor to nitrogen as compared with carbon substrates. The difficulty of ascribing to the unidirectional hydrogenase a role of electron donor to nitrogenase is that the enzyme will not reduce methyl viologen ($E_0 = -446$ mV) and will only reduce benzyl viologen ($E_0 = -359$ mV) very slowly in vitro (Van der Werf and Yates, unpublished data), nor will it reduce *Azotobacter* ferredoxin or flavodoxin in vitro (Yates, unpublished data). On the other hand, the reversible hydrogenase of *Clostridium pasteurianum* will reduce ferredoxin (Mortenson and Chen, 1974). Eisbrenner et al. (1978) showed that the hydrogenase of blue-green algae would only support acetylene reduction in the light, whereas $Na_2S_2O_4$ was effective in the light or the dark. This observation suggested that the hydrogenase could not transfer electrons directly to nitrogenase. However, requirement for an energized membrane (Haaker, de Kok, and Veeger, 1974) for electron transfer to nitrogenase in azotobacter may make the situation in this organism very different from that in the blue-green algae.

The presence of H_2 increased the optimal pO_2 for maximum rate of acetylene reduction significantly at mannitol concentrations below 1 g/liter (Figures 1 and 3), indicating that hydrogenase activity effectively aided respiratory protection of nitrogenase under carbon-limited conditions. Hydrogenase had no apparent effect on respiratory protection at mannitol concentrations above 1 g/liter. These observations, together with the data presented in Figure 2 (that H_2 supported a high percentage of nitrogenase activity at low mannitol concentrations), emphasize the importance of hydrogenase for nitrogen fixation under carbon-limited conditions. Such a conclusion may also apply to the blue-green algae (Benemann and Weare, 1974a, 1974b; Bothe, Tennigkeit, and Eisbrenner, 1977) and *Rhodopseudomonas capsulata* (Kelly et al., 1977).

It is quite clear that H_2 production represents a considerable proportion of nitrogenase activity in all of the continuous cultures studied here (see also Walker and Yates, 1978a, 1978b). The question arises as to whether or not the amount of energy saved by recycling this H_2 is significant. When carbon substrate is not limiting, less than 1% of the total electron flow to O_2 is diverted to nitrogen fixation (Yates, 1970); recycling H_2 evolved by nitrogenase would be a trivial contribution energetically, particularly under oxygen-limited conditions when nitrogenase activity is limited by ATP synthesis. When carbon substrate was limiting (1.5 g mannitol/liter), one continuous culture took up 2.7 μmole O_2/hr while H_2 evolution by nitrogenase was 42 nmol/hr under air. Assuming that all of this could be recycled to O_2, it would enhance the rate of O_2 uptake by only 0.5%. However, when the carbon supply was very low (0.01 g/liter), nitrogenase activity was also very low and was more than doubled when H_2 was in the atmosphere. Assuming that nitrogenase used 40% of its energy to produce H_2 (average from Table 1) at an ATP:$2e^-$ ratio of 5 and this was completely recycled to produce ATP at a P:O ratio of 3 and a 50% efficiency of energy conversion, then the organism could regain up to 30% of the energy lost as H_2 evolution or 12% of its overall energy for nitrogen fixation. It is possible, therefore, that circumstances exist in which recycled H_2, produced by nitrogenase, could be utilized to provide a significant proportion of energy for nitrogenase function or, alternatively, electrons for N_2 reduction.

The reasons for H_2 evolution by nitrogenase are not understood. It is clear that the enzyme evolves H_2 in vivo in the absence of N_2, but the literature contains no suggestion as to why this should happen. HD exchange occurs during nitrogen reduction (see Newton et al., 1977), but this should yield no net gain of H_2. The experiments of Schrauzer and his colleagues (see Schrauzer, 1977) and of Nikonova and Shilov (1977) give rise to the suggestion that a stoichiometry exists between H_2 production and N_2 reduction according to equation 2:

$$N_2 + 8H^+ + 8e^- \rightarrow 2NH_3 + H_2 \qquad (2)$$

which gives a theoretical $H_2:N_2$ ratio of 1. The synthesis of some N_2 complexes yields stoichiometric amounts of H_2 (Chatt and Leigh, 1972). However, Schrauzer suggests that the cause of H_2 evolution is a disproportionation reaction of a diazene-type intermediate to yield a hydrazine-type intermediate together with traces of the elements:

$$-N_2H_2 \begin{cases} \longrightarrow N_2 + H_2 \\ \longrightarrow N_2H_4 \end{cases}$$

An alternative possibility is:

$$M{=}N{-}NH_2 \rightarrow [N{-}NH_2] \begin{cases} \longrightarrow N_2 + H_2 \\ \longrightarrow \frac{2}{3}NH_3 + \frac{2}{3}N_2 \end{cases}$$

(Chatt and Richards, 1977; Wiberg, Fischer, and Bachhuber, 1977). Both these mechanisms must yield an $H_2:N_2$ ratio of less than 1. Rivera-Ortiz and Burris (1975) predicted that this $H_2:N_2$ ratio in the enzymic reduction of N_2 would be below 1 at infinite (hyperbaric) N_2 concentration.

The $H_2:N_2$ ratios obtained with O_2-, N_2-, or carbon-limited continuous cultures of *A. chroococcum* were always greater than 1 (Table 1). The apparent low ratios obtained with carbon-limited cultures (<1) were due to higher uptake hydrogenase activity in these cultures that recycled some of the H_2 produced by nitrogenase. When the hydrogenase was irreversibly inhibited by pretreatment with acetylene, the $H_2:N_2$ ratio increased to >1. O_2-limited cultures grown at a low oxygen-solution rate yielded the highest $H_2:N_2$ ratios. Such high ratios may be due to a high ratio of MoFe protein (Ac1) to the Fe protein (Ac2), to a low ATP concentration, or to a low ATP:ADP ratio—or, indeed, to a combination of all three (Figures 6 and 7). Even if H_2 evolution and N_2 reduction are stoichiometric (because no observed $H_2:N_2$ ratio was less than 1), this hypothesis is not disproved. A large percentage of the H_2 produced under air must arise by a mechanism intrinsic to the enzyme rather than to N_2 reduction. It is therefore possible that during N_2 reduction nitrogenase has two mechanisms for H_2 evolution: one that is intrinsically associated with nitrogenase activity and a second that is determined by the stoichiometry of the N_2 reduction process. This suggestion means that, under some conditions in vitro (saturating Fe protein and optimum ATP concentration), N_2 completely inhibits intrinsic H_2 evolution by nitrogenase. Genetic manipulation to obtain multiple copies of the Fe protein in vivo might be one way of decreasing H_2 evolution by nitrogenase to a minimum.

Dixon (1972) postulated that one function of hydrogenase was to prevent H_2 from accumulating in the vicinity of the nitrogenase to inhibit N_2 reduction. However, the linear rate of H_2 evolution by carbon-limited *A. chroococcum* under air, after the hydrogenase was inhibited by previous treatment with acetylene (Figure 5), suggests either that insufficient H_2 was produced to inhibit nitrogenase or that the gas was dispersed rapidly into the medium. If this were not so, the rate of H_2 evolution would have increased with time.

Our results can be summarized as follows:

1. Nitrogenase produced significant quantities of H_2 in O_2-, N_2-, or mannitol-limited continuous cultures of *A. chroococcum* in vivo. Most of this H_2 was recycled in mannitol-limited cultures where the hydrogenase was most active. O_2- or N_2-limited cultures had very low hydrogenase activity.
2. The $H_2 : N_2$ ratios in continuous cultures were always greater than 1, and were particularly high in O_2-limited cultures with low oxygen-solution rates. $H_2 : N_2$ ratios in vitro were ≥ 1.
3. Hydrogenase activity can support nitrogenase activity by aiding respiratory protection and as a source of energy and reducing power. These functions are most important at low carbon and energy substrate concentrations.
4. H_2 produced by nitrogenase apparently does not accumulate to inhibit N_2 reduction in vivo.

ACKNOWLEDGMENTS

We thank Professor J. R. Postgate and Dr. R. L. Richards for useful discussions.

REFERENCES

Benemann, J. R., and N. M. Weare. 1974a. Nitrogen fixation by *Anabaena cylindrica*, III. Hydrogen-supported nitrogenase activity. Arch. Microbiol. 101:401–408.

Benemann, J. R., and N. M. Weare. 1974b. H₂ evolution by N₂-fixing *Anabaena cylindrica*. Science 184:174.

Bergersen, F. J. 1963. The relationship between hydrogen evolution, hydrogen exchange, nitrogen fixation and applied O₂ tension in soybean root nodules. Aust. J. Biol. Sci. 16:669–680.

Bothe, H., J. Tennigkeit, and G. Eisbrenner. 1977. The utilization of molecular hydrogen by the blue-green algae *Anabaena cylindrica*. Arch. Microbiol. 114:43–49.

Bothe, H., E. Distler, and G. Eisbrenner. 1978. Hydrogen metabolism in blue-green algae. Biochimie. In press.

Brotonegoro, S. 1974. Nitrogen fixation and nitrogenase activity of *Azotobacter chroococcum*. Ph.D. Thesis, Agricultural University, Wageningen, The Netherlands.

Chatt, J., and G. J. Leigh. 1972. Nitrogen fixation. Chem. Soc. Rev. 1:121–144.

Chatt, J., and R. L. Richards. 1977. The binding of dinitrogen and dinitrogen hydrides to molybdenum. J. Less-Common Metals. 54:477–484.

Dalton, H., and J. R. Postgate. 1969. Effect of oxygen on growth of *Azotobacter chroococcum* in batch and continuous cultures. J. Gen. Microbiol. 54:463–473.

David, K. A., and P. Fay. 1977. Effects of long-term treatment with acetylene on nitrogen-fixing microorganism. Appl. Environ. Microbiol. 34:640–646.

Dixon, R. O. D. 1967. Hydrogen uptake and exchange by pea root nodules. Ann. Bot., N.S. 31:179–188.

Dixon, R. O. D. 1968. Hydrogenase in pea root nodule bacteroids. Arch. Mikrobiol. 62:272–283.

Dixon, R. O. D. 1972. Hydrogenase in legume root nodule bacteroids: Occurrence and properties. Arch. Mikrobiol. 85:193–201.

Eisbrenner, G., E. Distler, L. Floener, and H. Bothe. 1978. The occurrence of hydrogenase in blue-green algae. Arch. Microbiol. Submitted.

Evans, H. J., T. Ruiz-Argüeso, N. T. Jennings, and J. Hanus. 1977. Energy coupling efficiency of symbiotic nitrogen fixation. In: A. Hollaender (ed.), Genetic Engineering for Nitrogen Fixation, pp. 333–354. Plenum Publishing Corp., New York.

Haaker, H., A. de Kok, and C. Veeger. 1974. Regulation of dinitrogen fixation in intact *Azotobacter vinelandii*. Biochim. Biophys. Acta 357:344–357.

Hoch, G. E., Little, and R. H. Burris. 1957. Hydrogen evolution from soybean root nodules. Nature 179:430–431.

Hoch, G. E., K. C. Schneider, and R. H. Burris. 1960. Hydrogen evolution and exchange, and conversion of N_2O to N_2 by soybean root nodules. Biochim. Biophys. Acta. 37:273–279.

Hyndman, L. A., R. H. Burris, and P. W. Wilson. 1953. Properties of hydrogenase from *Azotobacter vinelandii*. J. Bacteriol. 65:522–531.

Kelly, B. C., C. M. Meyer, C. Gandy, and P. M. Vignaid. 1977. Hydrogen recycling by *Rhodopseudomonas capsulata*. FEBS Lett. 81:281–285.

Maryan, P. S., and W. T. Vorley. 1978. An improved spectrophotometric method for the determination of ammonia as a measure of *in vitro* nitrogenase activity. Anal. Biochem. Submitted.

Mortenson, L. E., and J.-S. Chen. 1974. Hydrogenase. In: J. B. Newlands (ed.), Microbial Iron Metabolism, pp. 232–282. Academic Press, Inc., New York.

Newton, W. E., W. A. Bulen, K. A. Hadfield, E. I. Stiefel, and G. D. Watt. 1977. HD formation as a probe for intermediates in N_2 reduction. In: W. E. Newton, J. R. Postgate, and C. Rodriguez-Barrueco (eds.), Recent Developments in Nitrogen Fixation, pp. 119–130. Academic Press, London.

Nikonova, L. A., and A. E. Shilov. 1977. Dinitrogen fixation in homogenous protic media. In: W. E. Newton, J. R. Postgate, and C. Rodriguez-Barrueco (eds.), Recent Developments in Nitrogen Fixation, pp. 41–52. Academic Press, London.

Phelps, A. G., and P. W. Wilson. 1941. The occurrence of hydrogenase in nitrogen fixing organisms. Proc. Soc. Exp. Biol. Med. 47:473–476.

Postgate, J. R. 1971. Fixation by free-living microbes: Physiology. In: J. R. Postgate (ed.), The Chemistry and Biochemistry of Nitrogen Fixation, p. 186. Plenum Press, London.

Rivera-Ortiz, J. M., and R. H. Burris. 1975. Interactions among substrates and inhibitors of nitrogenase. J. Bacteriol. 123:537–545.

Schrauzer, G. N. 1977. Nitrogenase model systems and the mechanism of biological nitrogen reduction: Advances since 1974. In: W. E. Newton, J. R. Postgate, and C. Rodriguez-Barrueco (eds.), Recent Developments in Nitrogen Fixation, pp. 109–118. Academic Press, London.

Schubert, K. R., and H. J. Evans. 1976. Hydrogen evolution: A major factor affecting the efficiency of nitrogen fixation in nodulated symbionts. Proc. Natl. Acad. Sci. U.S.A. 73:1207–1211.

Schubert, K. R., and H. J. Evans. 1977. The relationship of hydrogen reactions to nitrogen fixation in nodulated symbionts. In: W. E. Newton, J. R. Postgate, and C. Rodriguez-Barrueco (eds.), Recent Developments in Nitrogen Fixation, pp. 469–487. Academic Press, London.

Silverstein, R., and W. A. Bulen. 1970. Kinetic studies of the nitrogenase-catalyzed hydrogen evolution and nitrogen reduction reactions. Biochemistry 9:3809–3815.

Smith, L. A., S. Hill, and M. G. Yates. 1976. Inhibition by acetylene of conventional hydrogenase in nitrogen-fixing bacteria. Nature 262:209–210.

Walker, C. C., and M. G. Yates. 1978a. The hydrogen cycle in nitrogen fixing *Azotobacter chroococcum*. Biochimie. In press.

Walker, C. C., and M. G. Yates. 1978b. H₂ evolution and acetylene effects in N₂-fixing *Azotobacter chroococcum* under different nutrient limitations. Proceedings of the Steenbock-Kettering Symposium on Nitrogen Fixation, June 12–16, 1976, Madison, Wisconsin. Abstract A-2, p. 1.

Wiberg, N., G. Fischer, and H. Bachhuber. 1977. *cis* and *trans* Diazene (diimine). Angew. Chem. Int. Ed. Engl. 16:780–781.

Yakovlev, V. A., and L. A. Levchenko. 1966. The hydrogenase and succinic dehydrogenase of *Azotobacter vinelandii* and their relation to nitrogen fixation. Dokl. Akad. Nauk. S.S.S.R. 171:1224–1226.

Yates, M. G. 1970. Control of respiration and nitrogen fixation by oxygen and adenine nucleotides in N₂-grown *Azotobacter chroococcum*. J. Gen. Microbiol. 60:393–401.

Yates, M. G., and K. Planqué. 1975. Nitrogenase from *Azotobacter chroococcum:* Purification and properties of component proteins. Eur. J. Biochem. 60:467–476.

Nitrogen Fixation, Volume I
Edited by W. E. Newton and W. H. Orme-Johnson
Copyright 1980   University Park Press   Baltimore

Membrane Energization and Nitrogen Fixation in *Azotobacter vinelandii* and *Rhizobium leguminosarum*

C. Veeger, C. Laane, G. Scherings,
L. Matz, H. Haaker, and L. Van Zeeland-Wolbers

Abbreviations: TTFB, 4,5,6,7-tetrachloro-2-trifluoromethylbenzimidazol; CCCP, carbonyl cyanide *m*-chlorophenylhydrazon; N_2ase, nitrogenase; BSA, bovine serum albumin; ACMA, 9-amino-6-chloro-2-methoxyacridine; H_2ase, hydrogenase; FeS II, *A. vinelandii* iron-sulfur protein II; Av1, *A. vinelandii* N_2ase protein I; Av2, *A. vinelandii* N_2ase protein II; Ac2, *A. chroococcum* N_2ase protein II.

During the last few years, it has become clear that, in the aerobe *Azotobacter vinelandii*, a number of factors contribute to the complicated process of nitrogen fixation (Haaker, de Kok, and Veeger, 1974; Haaker, Scherings, and Veeger, 1977; Haaker and Veeger, 1977; Scherings, Haaker, and Veeger, 1977; Veeger, Haaker, and Scherings, 1977). It was shown that membrane energization rather than a high ATP:ADP ratio is the determining factor in aerobic nitrogen fixation. Furthermore, we showed that flavodoxin hydroquinone is the major electron donor to N_2ase, although the presence of a NAD(P)H flavodoxin oxidoreductase was also reported. Autooxidation of flavodoxin hydroquinone by O_2 is the major factor that "switches off" (cf. Dalton and Postgate, 1969) nitrogen fixation, whereas N_2ase itself was stabilized toward O_2 inactivation and regulated in activity by the presence of a stoichiometric amount of the pink 2Fe-2S protein

This investigation was supported by the Netherlands Foundation for Chemical Research (S.O.N.), with financial support from the Netherlands Organization for the Advancement of Pure Research (Z.W.O.).

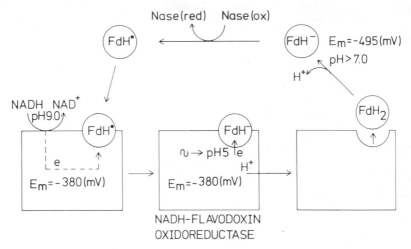

Figure 1.   Proposed scheme for electron transport from NADH to N_2ase in *A. vinelandii*.

isolated and characterized earlier (Shethna, DerVartanian, and Beinert, 1968; DerVartanian, Shethna, and Beinert, 1969). In fact, an O_2-stable complex had already been described by Bulen and Le Comte (1972). In the scheme of Figure 1 (cf. Haaker and Veeger, 1977), it was visualized that the proton-driving force over the membrane, as coupled to respiration, lowers the local pH in the vicinity of the binding site of flavodoxin semiquinone to the inner membrane via membrane energization. Under the conditions of a local pH of 4.5–5, the potential of the couple flavodoxin semiquinone/flavodoxin hydroquinone is around −380 mV, low enough to allow reduction by NAD(P)H via a site with pH 8–9. Dissociation of the hydroquinone and subsequent deprotonation in the cytoplasm at pH > 7 decrease the potential (~ −460 mV). The present paper deals with an extended study of the factors mentioned above, both in *A. vinelandii* and *Rhizobium leguminosarum*.

MEMBRANE ENERGIZATION
IN RELATION TO NITROGEN FIXATION

Little is known about the supply of reducing equivalents to N_2ase in bacteroids. Appleby, Turner, and MacNicol (1975) noted that the correlation between the ATP:ADP ratio and N_2ase activity in soybean bacteroids was poor when CCCP was used as an uncoupling agent. Although those authors favor a different interpretation, their results suggest that, in addition to the ATP:ADP ratio, the method of generating reducing equivalents is important in supporting N_2ase activity in vivo.

Several techniques have been developed to follow respiration and N_2ase activity in aerobic nitrogen-fixing bacteria (Bergersen, Turner, and Appleby, 1973; Haaker, de Kok, and Veeger, 1974; Wittenberg et al., 1974; Bergersen and Turner, 1975a). The usefulness of the "shaking assay" method for bacteroid suspensions was questioned by Stokes (1975), who deduced mathematically that with hemoprotein present effects other than facilitated O_2 diffusion might occur. Furthermore, the free O_2 concentration cannot be measured and therefore it is not certain whether or not equilibrium conditions exist. Bergersen and Turner (1975a) devised a "no-gas-phase" system in which the free O_2 concentration, the deoxygenation of O_2-binding protein, and acetylene reduction could all be recorded. It is clear that this system cannot achieve the steady-state conditions likely to occur in vivo. We devised a compromise technique that assays simultaneously the steady-state free O_2 concentration and N_2ase activity under conditions of steady-rate respiration (Figure 2) and that takes at least 2 min to reach steady-state conditions (Laane, Haaker, and Veeger, 1978).

The system consists of an 8-ml magnetically stirred, gas-tight vessel with an O_2 electrode located at the bottom (Rank Brothers, Bottisham, Cambridge). Reactants are added by syringe through a rubber stopper. An amplifier is used to detect low O_2 concentrations. The electrode response was calibrated by adding small amounts of air-saturated water to the anaerobic reaction mixture. This system was also used to determine ATP made: O_2 consumed (P:O) ratios in vesicles of R. leguminosarum bacteroids and of A. vinelandii under conditions of steady-state respiration. The steady-state concentration of free O_2 in the solution is determined by the partial pressure of O_2 in the gas phase, the rate of stirring, and the rate at which O_2 is consumed by the cells or vesicles. The stirring speed and the amount of cells or vesicles were usually held constant and the concentration of free O_2 (c.q. respiration rate) was therefore determined by varying the partial pressure of O_2 in the gas phase. The O_2 input into the solution at the standard stirring speed was calculated by adding known amounts of O_2 to the gas phase and measuring the initial rate of increase of the O_2 tension in the anaerobic liquid phase (Figure 2). Anomalies introduced by the gas-liquid interface were minimized, if necessary, by increasing the amount of liquid so that the surface-to-volume ratio decreased.

This system was used to study the effect of O_2 on the N_2ase activity of R. leguminosarum bacteroids under controlled steady-state conditions (Figure 3). Figures 3A and 3B show that the nitrogen-fixing efficiency of freshly prepared R. leguminosarum bacteroids can be considerably enhanced by the addition of fatty acid–free BSA. BSA stimulates steady-state N_2ase activity, O_2 consumption, and oxidative phosphorylation without affecting the free O_2 concentration at which maximum acetylene

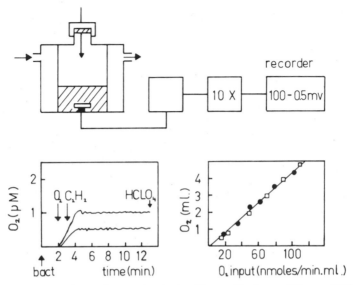

Figure 2.   Experimental system (see text and Laane, Haaker, and Veeger, 1978).

formation occurs. At high O_2 concentrations, the ATP:ADP ratio remains fairly constant and the observed decline in nitrogen fixation is therefore not due to a decreased ATP supply as proposed by Bergersen and Turner (1975b) for soybean bacteroids, but rather to inhibition by excess O_2, the so-called switch off state (cf. Haaker, Scherings, and Veeger, 1977; Haaker and Veeger, 1977). This also occurs in the presence of myoglobin and leghemoglobin as O_2 carriers, but even under these conditions fat-free BSA stimulates N_2ase activity about twofold without influencing the ATP:ADP ratio. We thus conclude that nitrogen fixation in *Rhizobium* bacteroids, as in *A. vinelandii* cells, is dependent mainly on the state of energization of its membranes and to a lesser extent on the ATP:ADP ratio. During the preparation of bacteroids, the cell membrane is exposed to the uncoupling effect of free fatty acids and to plant phospholipase D activity (Laane, Haaker, and Veeger 1978). Both effects can be counteracted by BSA provided that the fatty acid–free form is used (Table 1).

Figure 4 demonstrates the effect of the uncoupler CCCP on the N_2ase activity and the ATP:ADP ratio of *R. leguminosarum* bacteroids. CCCP strongly inhibits N_2ase activity of bacteroids with little or no effect on the ATP:ADP ratio (Figure 4A). At concentrations higher than 1 μM, an approximately linear relationship exists between ATP:ADP ratio and N_2ase activity (Figure 4B). Without uncouplers, however, the relation, obtained under conditions at which O_2 concentrations were lower than necessary to obtain maximum N_2ase activity, is also linear but with a different slope.

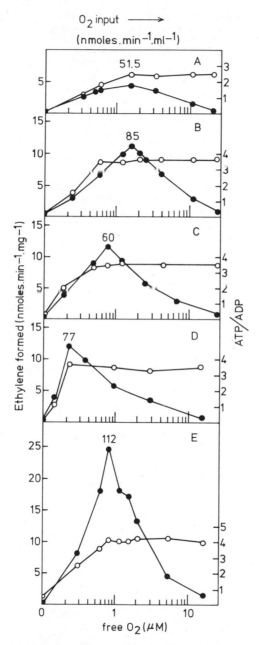

Figure 3.   The relationship between free dissolved O_2 concentration, O_2 consumption, N_2ase activity and ATP:ADP ratio of bacteroids as influenced by different effectors (cf. Laane, Haaker, and Veeger, 1978). ●———●, N_2ase; ○———○, ATP:ADP. A, without addition; B, 3.1% BSA; C, 130 μM myoglobin; D, 110 μM leghemoglobin; E, 130 μM myoglobin plus 3.1% BSA.

Table 1.   Effect of linoleic acid on N_2ase activity and ATP:ADP ratio of *R. leguminosarum* bacteroids (see Laane, Haaker, and Veeger, 1978)

Linoleic acid (μg)	Addition	C_2H_4 evolution (nmol min^{-1} mg^{-1})	ATP:ADP
—	—	12	3.5
—	C_2H_5OH	12	3.5
18	—	11.3	3.2
36	—	7.7	1.6
54	—	3.5	1.0
36	3% BSA	11.6	3.4

When ACMA is used as a fluorescent probe, a lower energized state of the membrane is observed at all CCCP concentrations used (cf. Haaker, de Kok, and Veeger, 1974). Because the only known effect of CCCP is to lower the energized state of the membrane, these results also show that, as well as being controlled by the ATP:ADP ratio, N_2ase activity is determined by the supply of reducing equivalents, which itself is related to the state of membrane energization.

Studies with oligomycin excluded the possibility that the decline of N_2ase activity was due to decreased ATP synthesis and a lowered utilization of ATP by N_2ase (Veeger, Haaker, and Scherings, 1977). Oligomycin lowers the rate of ATP synthesis but does not affect N_2ase activity. Appleby, Turner, and MacNicol (1975) also noted that the relation between the ATP:ADP ratio and N_2ase activity was poor when CCCP was used. Their explanation that separate domains of ATP formation and accumula-

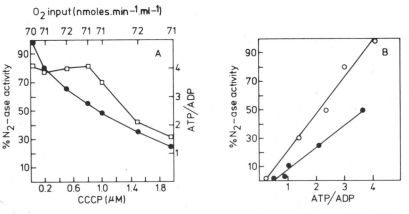

Figure 4.   A) Influence of CCCP on O_2 consumption, N_2ase activity and ATP:ADP in *R. leguminosarum* bacteroids. (●——●), N_2ase; (□——□), ATP:ADP. B) Relationship between bacteroid N_2ase activity and ATP:ADP in the presence (●——●) and absence (○——○) of CCCP (cf. Laane, Haaker, and Veeger, 1978).

tion exist within the bacteroid and that one of these domains is more sensitive to uncoupling seems very unlikely. Our results show that their data can also be explained in terms of limitation in electron supply to N_2ase.

We have tried to induce N_2ase activity in an in vitro system in which the reducing equivalents are generated by membrane vesicles from either *A. vinelandii* cells or *R. leguminosarum* bacteroids. For this purpose, flavodoxin, which is a good electron donor for N_2ase in the hydroquinone state (Scherings, Haaker, and Veeger, 1977), and N_2ase complex (Scherings, Haaker, and Veeger, 1977), both purified from *A. vinelandii*, were added to the vesicle suspension. Although we have been successful in preparing vesicles that can be highly energized by respiration (Figure 5), attempts to reduce flavodoxin to the hydroquinone state by an energized membrane were unsuccessful. One major problem could be the failure to energize these vesicles directly by ATP, because flavodoxin hydroquinone is readily oxidized to the semiquinone state by O_2. Treatment of the vesi-

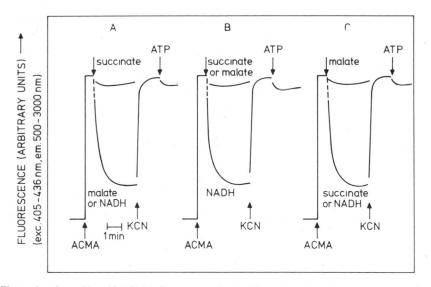

Figure 5.   Quenching of ACMA fluorescence by $NaCl^+$ and $NaCl^-$ membrane vesicles of *A. vinelandii* and *R. leguminosarum* bacteroids on energization. Vesicles of *A. vinelandii* were isolated anaerobically from cells harvested at the turnover from logarithmic into nonlogarithmic growth (A_{680} = 0.9–1.0) by sonication of lysozyme-EDTA–treated cells. The membranes were sedimented by centrifugation between 18,000 and 100,000 × *g*. Bacteroids from pea root nodules were isolated as described by Bergersen (1971) and its vesicles as for *A. vinelandii*. Vesicles were washed and suspended in a medium containing 50 mM Tes/KOH, 4 mM $MgCl_2$, and 0.33 M NaCl where indicated (pH 7.6). ACMA (2 μM), succinate (10 mM), malate (10 mM), NADH (0.25 mM), KCN (5 mM), and ATP (1 mM) were added at the times indicated. Membrane vesicles were added to a final concentration of 0.02 mg of protein/ml. Data are corrected for NADH fluorescence. (A), $NaCl^-$ and $NaCl^+$ membrane vesicles of *A. vinelandii;* (B), $NaCl^-$ membrane vesicles of *R. leguminosarum* bacteroids; (C), $NaCl^+$ membrane vesicles of *R. leguminosarum* bacteroids.

cles with trypsin (Bhattacharyya and Barnes, 1976) or by column chromatography (Racker and Horstman, 1967) to remove the ATPase inhibitor is ineffective. Furthermore, Figure 5 shows that there are major differences between the two types of vesicles. Vesicles from *A. vinelandii* cells can easily be energized by oxidation of malate and NADH, but barely with succinate both in the presence and absence of NaCl. On the other hand, in the absence of NaCl, vesicles from *R. leguminosarum* bacteroids can only be energized by NADH oxidation and not by oxidation of succinate or malate. However, when the vesicles were made in the presence of NaCl (NaCl$^+$), succinate oxidation could induce energization to the same extent as NADH oxidation, whereas malate still was hardly active.

It should be noted that soybean bacteroids also contain considerable amounts of flavodoxin (Phillips et al., 1973). Thus, in bacteroids reducing equivalents are likely to be donated by flavodoxin hydroquinone to the N_2ase, a postulate made probable by the high activities observed with bacteroidal N_2ase in the presence of *A. vinelandii* flavodoxin hydroquinone (Table 2). Furthermore, we could demonstrate the presence of a NADH-flavodoxin reductase in extracts of bacteroids.

The presence of a unidirectional, H_2-oxidizing H_2ase in a number of nitrogen-fixing organisms and its relation with nitrogen fixation have been well established (Wilson and Burris, 1947; Lindström, Lewis, and Pinsky, 1951; Dixon, 1976; Schubert and Evans, 1976; Smith, Hill, and Yates, 1976; Kelley et al., 1977). This H_2ase enables these organisms to recover some of the energy lost during wasteful production of H_2 by N_2ase, resulting in a higher nitrogen-fixing efficiency. We have studied membrane energization and oxidative phosphorylation connected with H_2 oxidation in membrane vesicles of *A. vinelandii* and *R. leguminosarum*. Figure 6A shows that membrane vesicles of *A. vinelandii* can be energized by H_2 as an electron donor. Figure 6B clearly shows no flavoprotein involvement in the respira-

Table 2. Comparison of the efficiency of several electron donors for N_2ase activity in a crude extract of *R. leguminosarum* bacteroids (strain PRE)[a]

Donor(s)	%
Dithionite	100
Flavodoxin	75
Ferredoxin I	58
Flavodoxin + ferredoxin I	127

[a] Bacteroids were isolated as described in Laane, Haaker, and Veeger (1978). A crude extract was prepared by sonicating the bacteroids suspension for 2 min and centrifuging the ruptured cells for 1 hr at 38,000 × *g*. Flavodoxin and ferredoxin I purified from *A. vinelandii* and reduced by photoreduction with deazaflavin according to Scherings, Haaker, and Veeger (1977) were added to a final concentration of 10 μM. In a standard assay 0.4 mg of crude extract protein was added to the incubation mixture. Data are expressed as percentage of acetylene reduction rate with dithionite as electron donor.

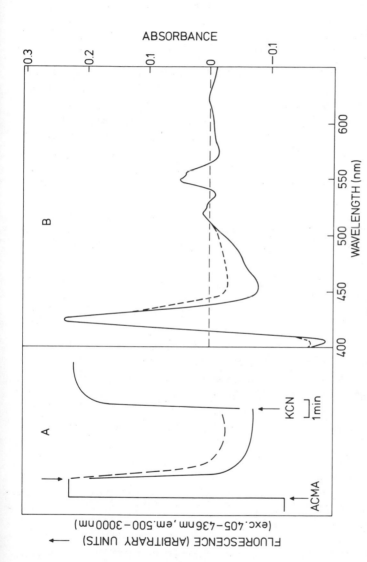

Figure 6.   Oxidation of H₂ by membranes of *A. vinelandii*. (A) Quenching of ACMA fluorescence by membrane vesicles with NADH and H₂. Vesicles were isolated as in Figure 5 and suspended in a medium containing 50 mM Tes/KOH and 4 mM MgCl₂ (pH 7.4). ACMA, NADH, and KCN were added to a final concentration of 2 μM, 0.3 mM, and 5 mM, respectively, at the times indicated by arrows. In a separate experiment (----) hydrogen (50%) was added to the gas phase of a closed cuvette that was shaken at the time indicated. 0.02 mg of vesicle protein was used for each assay. (B) Reduced minus oxidized difference spectra of membrane vesicles. Vesicles were suspended in a medium containing 50 mM Tes/KOH and 2.5 mM MgCl₂ (pH 7.0) to a final concentration of 3.4 mg of protein per ml. NADH (——) was added to a final concentration of 0.6 mM; hydrogen (----) was added as described in A.

tory chain of *A. vinelandii* when H_2 is used as substrate. Vesicles of *R. leguminosarum* (strain PRE) did not oxidize H_2 and therefore failed to induce membrane energization.

P:O RATIOS IN MEMBRANE VESICLES OF
A. VINELANDII AND *R. LEGUMINOSARUM* BACTEROIDS

Assessment of the efficiency of oxidative phosphorylation in intact cells and isolated respiratory membranes of *A. vinelandii* has been attempted (Ackrell and Jones, 1971; Baak and Postma, 1971; Eilermann et al., 1971; Haaker and Veeger, 1976). The P:O values reported for intact cells of *A. vinelandii* (Baak and Postma, 1971) suggest a complete coupling between oxidation and phosphorylation, whereas isolated respiratory membranes are much less efficient (Ackrell and Jones, 1971; Haaker and Veeger, 1976). No P:O values are reported for *R. leguminosarum* bacteroids. We decided to study the oxidative phosphorylation efficiency in membrane vesicles of *A. vinelandii* and *R. leguminosarum* in the assay system described above, which allows investigation of this parameter under controlled steady-state respiration rates. With a suspension of *A. vinelandii* vesicles and NADH as substrate, Figure 7 shows that when the O_2 supply is increased the P:O ratio rises to a maximum, but declines when O_2 becomes detectable in the medium. At the same time, the ATP concentration rises to a fairly constant level as the O_2 supply is increased. The shape of the curve clearly demonstrates that a certain amount of energization of the membrane is necessary for full ATP synthesis. Similar phenomena are observed with malate or H_2 as substrates, except that at high O_2 input the ATP concentration declines rapidly as H_2 is used as a substrate. This decline appears not to be caused by a limited amount of H_2, but to inhibition of H_2ase by excess O_2. The maximum oxidation capacities are not completely reached at the highest P:O ratio. The calculated P:O values are based on real oxidation rates and are therefore corrected for the appearance of O_2 in the medium. The maximum P:O ratio obtained with H_2 is comparable with that obtained with NADH. Furthermore, addition of acetylene to the gas phase (final concentration 20%) does not inhibit H_2 respiration. Exactly the same curves are obtained as those presented in Figure 7C, which contrasts with the previous results obtained with *Azotobacter chroococcum* (Smith, Hill, and Yates, 1976).

Figure 8 shows the results obtained with vesicles of *R. leguminosarum* bacteroids. Although higher P:O ratios are observed with NADH as substrate, the general concept is similar to that of *A. vinelandii*. Also shown is the effect of fatty acid–free BSA on the efficiency of the oxidative phosphorylation, which supports the conclusion that the cytoplasmic membrane of *R. leguminosarum* bacteroids is partially uncoupled by free fatty acids.

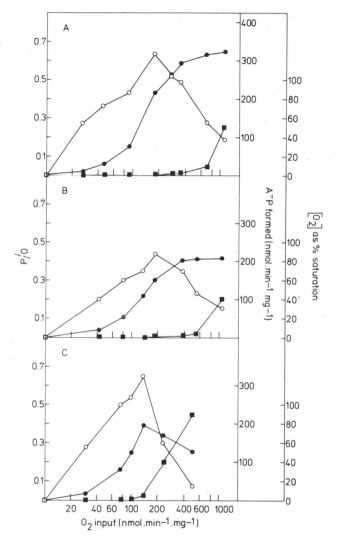

Figure 7.   Effect of O_2 on P:O ratio and ATP synthesis in membrane vesicles of *A. vinelandii*. Vesicles were isolated as in Figure 5 and suspended and stored at 0°C in a medium containing 50 mM Tes/KOH, 2.5 mM $MgCl_2$, and 1.2% fatty acid–free BSA (pH 7.0). Experiments were performed with the system described in the text. The standard incubation mixture contained at 30°C: 50 mM Tes/KOH, 2.5 mM $MgCl_2$, 10 mM glucose, and 5 mM phosphate (final pH 7.0). The incubation mixture, total volume 2.5 ml, and the reaction vessel were made anaerobic by flushing with argon. Three minutes after addition of 15 U of a dialyzed hexokinase solution, vesicles (0.35 mg protein), substrate, and variable amounts of O_2 were added to the gas phase. After 3 min stirring, when the system was found to be in equilibrium, an anaerobic solution of ADP (final concentration 0.5 mM) was added. After 2.5 min incubation with ADP, the vesicles were rapidly fixed with $HClO_4$ up to a final concentration of 4% (w/v) and the level of glucose 6-phosphate was determined (Haaker, de Kok, and Veeger, 1974). A logarithmic scale indicates the O_2 input. It is plotted against: (O——O), P:O ratio; (●——●), ATP synthesis; (■——■), O_2 concentration expressed as percentage saturation. (A), NADH, final concentration 0.6 mM; (B), malate, final concentration 5 mM; (C), H_2, 50% in the gas phase.

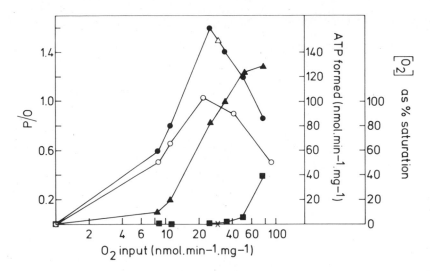

Figure 8.   Effect of O_2 on P:O ratio and ATP synthesis in membrane vesicles of *R. leguminosarum* bacteroids. Bacteroids from pea root nodules were isolated as earlier described (Bergersen, 1971). Vesicles were prepared as described in Figure 5 for *A. vinelandii*, except that 0.3 M sucrose was present during the isolation procedure, and that BSA was omitted in the final suspension. When indicated fatty acid-free BSA (final concentration 2.5%) was added to the incubation mixture. Assay conditions as in Figure 7, except that 2.4 mg of vesicle protein were added to the incubation mixture; final volume 5 ml, temperature 25°. O_2 input is plotted against: (O——O), P:O ratio with NADH; (●——●), P:O ratio with NADH and BSA; (△), P:O ratio with NADH and BSA in an incubation volume of 4 ml; (X), P:O ratio with succinate (final concentration 5mM); (▲——▲) ATP synthesis with NADH; (■——■), O_2 concentration expressed as percentage saturation.

Furthermore, the respiration rate of *R. leguminosarum* with NADH is about eighteen times less than that of *A. vinelandii* vesicles. Although succinate is rapidly oxidized by vesicles of *R. leguminosarum* bacteroids, no significant ATP synthesis could be detected, which means that succinate oxidation is uncoupled in these NaCl⁻ vesicles (cf. Figure 5B). Energization of NaCl⁺ membranes, as induced by succinate oxidation (cf. Figure 5), is due to coupled oxidative phosphorylation (ATP synthesis) under these conditions (not shown in Figure 8). The reason for this difference is not known yet, but it is possibly due to extraction of an essential component, because we observed some ATP formation in poorly washed NaCl⁻ vesicles. H_2 is not oxidized at all and O_2 appears immediately in the medium with both NaCl⁻ and NaCl⁺ vesicles. We therefore conclude that there is no active H_2ase present in both types of vesicles. The P:O ratio with NADH, determined in a 4-ml or 5-ml assay mixture, is exactly the same, which excludes the possibility that gas-liquid phase interactions interfere with this determination.

NAD(P)H-FLAVODOXIN OXIDOREDUCTASE

A reductase that reacts with flavodoxin from *A. vinelandii* in the presence of NAD(P)H was purified (200 times) until it gave one band on SDS polyacrylamide gel electrophoresis (Table 3). Yates (1971) has described a NADH dehydrogenase that was purified only 10–15 times with respect to the benzyl viologen reductase activity. We made a similar observation with our enzyme (purification 25–30 times). The major difference is that our enzyme catalyzes flavodoxin reduction by NAD(P)H and that this activity is purified about 200 times. Nevertheless, the specific activity of our purest preparations is not very high, and varies between 6 and 30 nmoles·min^{-1}·mg^{-1}. We believe it to be a flavodoxin reductase because of the spectrophotometric differences occurring at 450 nm and 615 nm on reduction of flavodoxin by NADH. If the amount of flavodoxin semiquinone (as measured by the increase in absorbance at 615 nm) is calculated and compared with the value calculated from the decrease in absorbance at 450 nm, the latter value is much larger. This difference is explained by formation of hydroquinone (no 615-nm absorption) in addition to semiquinone. After proper correction, we calculate that the ratio of hydroquinone formed to semiquinone formed varies between 0.5 and 1.5 in different preparations. From the point of view of flavoprotein catalysis (see below) two-electron transfer (Blankenhorn, 1977; Hemmerich, 1977), and

Table 3. Purification of flavodoxin reductase[a]

Preparation	Total volume (ml)	Total protein (mg)	Specific activity	Purifi- cation	Yield
Crude supernatant	212	4092	0.029	1.0	100
Protamine sulfate supernatant	262	2019	0.049	1.8	88.7
DEAE cellulose pool	250	500	0.23	82	105.9
Hydroxylapatite pool	6.6	12.2	2.1	75.0	23.2
Sucrose density gradient centrifugation	23.1	5.9	6.1	218.0	32.4

[a] *A. vinelandii* cells (114 g) that had been frozen and stored at $-20°C$ were broken in a french pressure cell by two passes through the cell in 50 mM Tes buffer (pH 7.4). After removal of the cellular debris by centrifugation at $15,000 \times g$ for 0.5 hr, the membranes were removed by ultracentrifugation at $250,000 \times g$ for 4 hr. The crude supernatant was treated with 0.1 mg protamine sulfate/mg of protein. The protamine sulfate supernatant was dialyzed against 5 mM Tes and applied to DEAE cellulose. The column was washed with 5 mM Tes (pH 7.4). After elution with 0.1–0.5 M NaCl in 5 mM Tes, the enzyme was applied to hydroxylapatite and then was eluted from it by between 0.1 and 0.3 M potassium phosphate buffer (pH 7.4). This pool was then layered on a 0%–20% sucrose (w/w) gradient and centrifuged at $100,000 \times g$ for 4 hr. The initial absorbance increase at 615 nm was registered ($\epsilon = 5.3$ mM cm^{-1}); specific activity expressed as nmoles mg^{-1} min^{-1}.

thus hydroquinone formation, is more likely to occur. Assuming two-electron reduction of the flavodoxin (the E_m for the quinone/hydroquinone couple of -270 mV does not exclude this), the formation of semiquinone thus seems to be caused by comproportionation of the hydroquinone formed with unreacted quinone.

The system of flavodoxin reduction by NAD(P)H is complicated by two phenomena:

1. The rate of reduction is only linear in a very limited range of enzyme concentrations. A tenfold increase in enzyme concentration leads to strong (50%–60%) inhibition. We have no explanation to offer yet.
2. The tracings of flavodoxin reduction at either 615 nm or 450 nm are nonlinear. With time, an activation of the rate of reduction at both wavelengths is observed. This is not due to substrate inhibition by flavodoxin since, for practical reasons, we are assaying at flavodoxin concentrations of 20–30 μM, which is below the K_m value of A. $vinelandii$ flavodoxin (~ 100 μM). However, the reaction with benzyl viologen as acceptor is linear with enzyme concentration. The pH profile of the flavodoxin reductase shows a maximum around 9 (Tris and glycine buffer), whereas the benzyl viologen reductase shows an increase in activity until pH 11 and in this respect resembles the pH profile of NADH dehydrogenase from A. $chroococcum$ (Yates, 1971).

In addition to reacting with benzyl viologen, the enzyme reacts with flavodoxins from A. $vinelandii$, $Megasphaera$ $elsdenii$, and $Desulfovibrio$ $vulgaris$ under anaerobic condition. The flavodoxins vary in reactivity—M. $elsdenii$ > D. $vulgaris$ > A. $vinelandii$, an order that corresponds to the redox potentials of the semiquinone-hydroquinone couple. The rates of the M. $elsdenii$ and D. $vulgaris$ flavodoxins also increase with time. The K_m values of these flavodoxins are lower, however (10–20 μM).

The absorption spectrum and the fluorescence excitation spectrum of this enzyme show maxima at 380 nm and 450 nm (shoulder at 480 nm). The enzyme contains FAD, which is fluorescent. The lifetime of the fluorescence emission of the oxidized flavin, excited at 450 nm, is 1.0 nsec and the degree of polarization of the emission on excitation at the same wavelength is 0.39. Titration of the reductase with NAD(P)H decreases the relative fluorescence of the enzyme while increasing the lifetime of the flavin. The enzyme-NADH complex regains its relative fluorescence on the addition of benzyl viologen. The flavin is quenched by irradiation with deazaflavin-EDTA. Addition of flavodoxin quickly restores the fluorescence.

The enzyme that we isolated resembles in many respects the NADH dehydrogenase isolated by Yates (1971). It is very easy, because of the low turnover under practical assay conditions (20–30 μM flavodoxin), to overlook the flavodoxin reductase activity. Flavodoxin hydroquinone could be produced at a slow rate without the need for membrane energy. It is known to be rapidly oxidized by N_2ase to the semiquinone (Haaker, Scherings, and Veeger, 1977; Haaker and Veeger, 1977; Scherings, Haaker, and Veeger, 1977; Veeger, Haaker, and Scherings, 1977). Assuming that these processes occur, the problem arises as to how to convert the semiquinone into the hydroquinone, a process that cannot be achieved by NAD(P)H alone. In view of the role of membrane energization in N_2ase activity of both *A. vinelandii* and *R. leguminosarum*, as well as the role of flavodoxin as electron donor in this process, our proposal that the membrane energy is the driving force for flavodoxin hydroquinone formation still appears very attractive and acceptable.

PROTECTION OF *A. VINELANDII* N_2ASE AGAINST DAMAGE BY O_2

Various aspects of O_2 tolerance in nitrogen-fixing organisms have been reviewed (Yates, 1977). Two hypotheses for O_2 tolerance of *Azotobacter* cell-free extracts have been put forward in the past: 1) N_2ase is located in a specialized particle or "azotophore," which provides a physical barrier to O_2 diffusion (Reed, Toia, and Raveed, 1974); and 2) association of N_2ase with membranes or "factors" induces an O_2-tolerant conformation of N_2ase.

The evidence for the first theory is very scarce and Haaker and Veeger (1977) could not find any evidence for localization of N_2ase in a subcellular particle. The second theory is more than an idea because removal of a factor during purification resulted in the transition from O_2 tolerant to O_2 sensitive (Haaker, Scherings, and Veeger, 1977). Haaker and Veeger (1977) subsequently did the appropriate isolation and reconstitution experiment that showed that the isolated factor gave protection toward O_2 on addition to the O_2-sensitive N_2ase components. Scherings, Haaker, and Veeger (1977) identified the factor as FeS II, previously purified and partly characterized by Beinert and co-workers (Shethna, DerVartanian, and Beinert, 1968; DerVartanian, Shethna, and Beinert, 1969). FeS II is a 2Fe-2S protein with a native molecular weight of 24,000 and a midpoint potential of -225 mV (G. Scherings, unpublished data) to -230 mV (Ke et al., 1974). The protein seems to contain more than one peptide chain (G. Scherings, unpublished data). Bulen and Le Comte (1972) reported that FeS II was a small contamination in their preparations of *Azotobacter* N_2ase. We now wish to report in some detail work carried out in part in collaboration with Dr. M. G. Yates on the protection characteristics using purified proteins.

Figure 9 shows the protection of Av1 + Av2 (incubated simultaneously) by FeS II. The activity that remains after 15 and 45 min incubation in the presence of 340 μM O_2 at room temperature is plotted versus the molar ratio of FeS II to Av2. Saturation behavior is observed. A phenomenon for

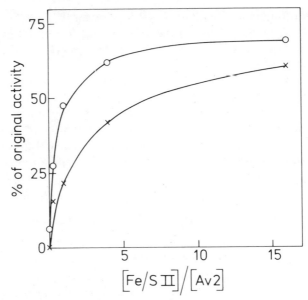

Figure 9.   Protection of Av1 + Av2 by FeS II. Anaerobic mixtures of dithionite-free Av1 and Av2 in a molar ratio of 1:2 were prepared in 25 mM Hepes-KOH buffer (pH 7.2). O_2-free FeS II was added in the specific molar ratios to Av2 as indicated. Av2 concentration in these mixtures was 5μM. Inactivation by O_2 was carried out in 1-ml Steriseal syringes. Subsequently, 0.5 ml of the protein mixtures and 0.5 ml of 25 mM Hepes-KOH buffer (pH 7.2) sparged with an argon-oxygen mixture to 0.68 mM O_2 were drawn into the syringe. Mixing was obtained by the movement of a stainless steel bead inside the syringe. The dead volume of the syringe was removed. The syringe was then sealed by plunging the needle into a rubber bung and left at room temperature for the indicated periods: (O——O), 15 min; (X——X), 45 min. Controls for 100% activity contained 1 mM $Na_2S_2O_4$ in the syringe. Remaining activity was assayed by injecting 0.2 ml into 7-ml vaccine bottles containing the complete C_2H_2 reduction assay mixture. Final concentrations were: $Na_2S_2O_4$, 20 mM; creatine phosphate, 10 mM; ATP, 1 mM; $MgCl_2$, 2 mM; creatinekinase, 22 U; Hepes-KOH, 12.5 mM; final pH 7.2; final volume 0.5 ml. The gas phase (6.5 ml) was 80% argon and 20% C_2H_2 by volume at 1 atm. Acetylene reduction was allowed to proceed for 15 min at 30°C.

which we have no explanation is that, at high molar ratios of FeS II to Av2, inactivation appears to have a rapid initial phase followed by a slower decay.

In our earlier work with the *A. vinelandii* N_2ase prepared according to Bulen and Le Comte (1972), we observed that this preparation resolved into two bands on Bio-Gel A-50m when $MgCl_2$ was present, but not when $S_2O_4^{2-}$ was also present or when $MgCl_2$ was absent. In cell-free extracts on Sepharose and in the presence of $MgCl_2$, N_2ase has a molecular weight of $1-2 \times 10^6$, but in the absence of $MgCl_2$ it is present as a 300,000-dalton species (Figure 10). In addition to its polymerizing effect, which induces multiple species observable in the analytical ultracentrifuge (Bulen and Le Comte, 1972), $MgCl_2$ appears to have a stabilizing effect on the low molecular weight form of N_2ase in the presence of $S_2O_4^{2-}$. In fact, FeS II co-elutes from Bio-Gel A-50m with N$_2$ase activity in the presence of MgCl$_2$, but lags behind in its absence. In addition, we observed that chromatography in the presence of 1 mM MgADP gives, as in the case of buffer alone, a complex with a molecular weight of about 300,000.

These pronounced effects of MgCl$_2$ on the N$_2$ase complex led to tests of the effect of MgCl$_2$ on the oxygen protection endowed by FeS II on the Av1 + Av2 system. Figure 11 shows that there is, in fact, quite a pronounced dependence on the MgCl$_2$ concentration, especially at low [FeS II]:[Av2] ratios. At still-physiological concentrations of Mg^{2+}, the stability toward exposure to 340 μM O$_2$ for 45 min can be increased such that more than 90% of the original activity is retained by adding FeS II in an equimolar ratio with respect to Av2 at a ratio [Av1]:[Av2] = 1:2. This experiment nicely confirms the stoichiometry of 1 FeS II per mole of purified Bulen–Le Comte complex reported earlier (Scherings, Haaker, and

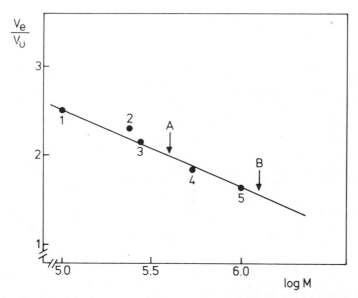

Figure 10.   Analytical Sepharose-4B column chromatography of *A. vinelandii* nitrogenase complex in cell-free extracts. The column (26 × 2.5 cm) was equilibrated with argon-saturated 25mM Tris-MES buffer (pH 7.2), containing either 10 mM MgCl$_2$ or 1 mM MgADP in the appropriate experiments, at room temperature. Standardization was with the following proteins: 1, glucose-6-P-dehydrogenase; 2, catalase; 3, xanthine oxidase; 4, ferritin; 5, pyruvate dehydrogenase complex. Cell-free extracts were prepared from frozen cells by passage through a Manton-Gaulin mill and subsequent centrifugation at 50,000 × g. The arrow at A corresponds to extract dialyzed versus elution buffer and that at B to extract dialyzed versus buffer with MgCl$_2$. Data in this figure have been obtained from large-zone experiments; peak elutions, however, gave the same results. Chromatography in the presence of MgADP has only been performed as a peak elution experiment and therefore is not included in the figure. The molecular weight was not different from that obtained in the presence of buffer alone.

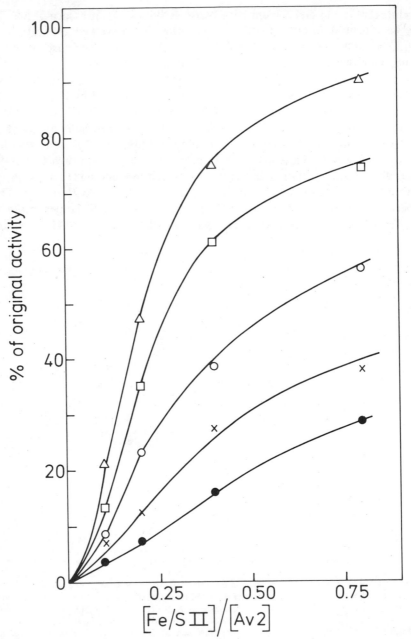

Figure 11.   Effect of MgCl₂ on the protection of Av1 + Av2 by FeS II. The percentage of original activity remaining after 45 min reaction with 0.34 mM O₂ is shown. The incubations were performed as described in Figure 9. MgCl₂ concentrations: (●——●), none; (X——X), 0.5 mM; (O——O), 1.0 mM; (□——□), 2,5 mM; (△——△), 5.0 mM. Experimental points represent the average of two activity measurements.

Veeger, 1977). The effect of variations in the [Av1]:[Av2] ratio on the degree of protection given by FeS II plus $MgCl_2$ is as yet undetermined.

It was important to investigate whether or not the pattern of Figure 11 results from additive effects of FeS II + $MgCl_2$ on Av1 and Av2 separately. Figure 12 shows that there is a slight protective effect by FeS II and by $MgCl_2$ on the O_2 sensitivity of Av1. Av1 is relatively stable toward O_2 inactivation, i.e., a 170-min incubation in buffer containing 400 μM O_2 decreased the activity to 7% of its original value. $MgCl_2$ also protects only slightly in the absence of FeS II, which is in contrast to the observation with the Av1 + Av2 system (Figure 11). FeS II alone has an effect that is, at saturation, about equal to the maximal effect of $MgCl_2$ in the absence of this protein. Added together, FeS II and $MgCl_2$ seem to protect not more than additively, but maximum protection is not very impressive. It can be calculated that the (pseudo-first order) rate constant for the inactivation of Av1 by O_2 declines to about 50% of its original value by the combined presence of $MgCl_2$ and FeS II. On the other hand, protection of Av2 seems to be much more pronounced (Figure 13). Although kinetic details have yet to be worked out, it is clear that the protection of Av2 by FeS II is still much less than that of the Av1 + Av2 system, as shown by the inactivation conditions, i.e., 45 min and 340 μM O_2 in the case of the Av1 + Av2 system as compared with 5 min and 75 μM O_2 for the Av2 system. Surprisingly, $MgCl_2$ seems to have a destabilizing effect on Av2.

In the cell, another divalent cation might participate in the stabilization of the Av1-Av2–FeS II system because the important form of Mg^{2+} there, especially at high oxygen supply, is MgATP. ATP is known to induce hypersensitivity of Ac2 toward O_2 (Yates, 1972). Similarly, we observed that equimolar concentrations of $MgCl_2$ and ATP (5–10 mM) highly destabilize the Av1-Av2–FeS II system. Because Ca^{2+} limitation induces hypersensitivity toward O_2 in *A. chroococcum* cells (J. R. Postgate, unpublished data), we tested the effect of Ca^{2+} on the Av2 system. It has an even more destabilizing action than Mg^{2+}. The effect of Ca^{2+} on the Av1 + Av2 system has not been tested yet.

The effect of FeS II on the Av2 system could be dependent on the concentration or on a fixed stoichiometry. On repeating the inactivation experiments at a 2.5-fold higher Av2 concentration, the initial slopes of the plots at the two Av2 concentrations (5.4 and 13.5 μM) were not significantly different. Thus, probably no fixed stoichiometry exists in this system. However, we expect that such a stoichiometry does exist in the Av1 + Av2 system.

The results presented here and previously (Haaker, Scherings, and Veeger, 1977; Haaker and Veeger, 1977; Scherings, Haaker, and Veeger, 1977) show that the O_2-tolerant form of *A. vinelandii* N$_2$ase consists of a stoichiometric complex of at least three different proteins, e.g., Av1, Av2,

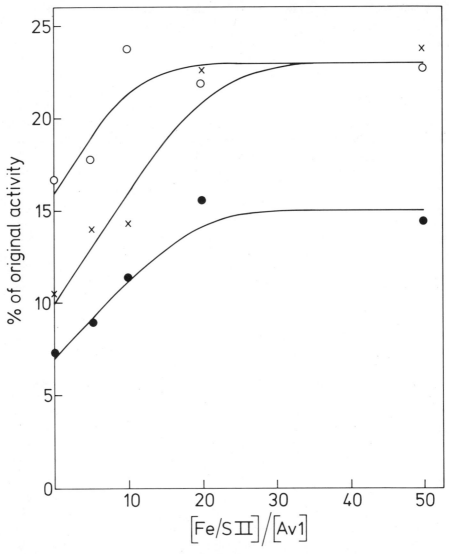

Figure 12.   Protection of Av1 by feS II and MgCl$_2$. Inactivation was carried out as described in Figure 9 except that no Av2 was present. The final Av1 concentration was 1.5 μM. Reaction was for 170 min with 400 μM O$_2$. MgCl$_2$ concentrations: (●——●), none; (X——X), 3 mM; (O——O), 9 mM. Assay for remaining activity was started by injecting 300 μl into the assay mixture described in Figure 9, plus Av2 (twentyfold excess over total Av1). Experimental points represent the average of two activity measurements.

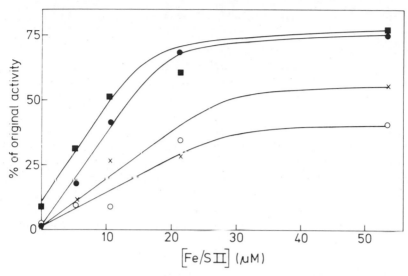

Figure 13.   Protection of Av2 by FeS II. Anaerobic mixtures of $Av2_{ox}$ and FeS II_{ox} were prepared in 25 mM Hepes-KOH (pH 7.2) and then diluted twofold in all-glass Chance syringes with 25 mM Hepes-KOH (pH 7.2) that had been sparged with an argon-oxygen mixture to $[O_2]$ = 150 μM. Mixing was done as described in Figure 9. Final concentration of $Av2_{ox}$ was 5.3 μM ($\bullet$——$\bullet$) and 13.5 μM ($\blacksquare$——$\blacksquare$). Final concentrations of FeS II as indicated. Inactivation time was 5 min. The two lower lines were obtained in the presence of either 5 mM $MgCl_2$ (X——X) or 5 mM $CaCl_2$ (O——O) at an Av2 concentration of 5.3 μM. Assay for remaining activity was started by injecting 0.25 (5.3 μM) or 0.10 (13.3 μM) ml into the assay mixture (see Figure 9) plus a fixed amount of Av1 (1.5 times Av2 concentration). Note: The often quoted optimum in activity that is observed when fixed amounts of Av2 are mixed with varying amounts of Av1 might indicate a need for titration in order to estimate correctly the remaining concentration of active Av2. However, in our conditions, this does not appear to be neccessary since the initial linear dependence of C_2H_2-reducing activity on the [Av2] in the presence of a fixed amount of Av1 (linear up to nearly equimolar concentrations) can actually be used as a "standard curve."

and FeS II, stabilized by the divalent cation Mg^{2+}. As judged from the inception of nitrogenase activity of this three-component complex in its reaction with flavodoxin hydroquinone (cf. Scherings, Haaker, and Veeger, 1977), this O_2-stable complex is probably inactive, or at least less active. "Switch on" to the active form is accomplished by an adequate supply of reducing equivalents via flavodoxin hydroquinone. Reduction leads to dissociation of the complex and thus to activation. Such a process can only offer a partial explanation for the rapid "switch off/switch on," O_2-dependent phenomenon observed in several aerobes (Hill, Drozd, and Postgate, 1972). The rapid autooxidation of flavodoxin hydroquinone, which cuts off electron supply, is at least an equally contributing factor in the deactivation of nitrogen fixation in our opinion. However, under such conditions, the O_2-stable oxidized Av1-Av2–FeS II complex is rapidly formed.

This view is supported by experiments carried out with Dr. R. N. F. Thorneley that indicate that no complex exists between Av2 and FeS II. In studying the effect of Av2 on the reduction of FeS II by $S_2O_4{}^{2-}$ by stopped-flow spectrophotometry, we observed that reduced Av2 could, at low dithionite concentrations, compete quite effectively with SO_2^- in the reduction of FeS II. Reduction of FeS II by dithionite in the absence of Av2 follows the rate expression $v = 1.6 \times 10^5$ $[SO_2^-][FeS\ II]$ M·sec^{-1} (Figure 14). The second-order rate constant is similar to those for other FeS proteins (Lambeth and Palmer, 1973). However, in the presence of Av2 a rate expression of the form $k_{obs} = k_2 + k_3$ $[S_2O_4{}^{2-}]^{1/2}$ is obtained. Since Av2 is reduced rapidly by SO_2^- ($k > 10^8$ M^{-1}·sec^{-1}), like Ac2$_{ox}$ (Thorneley, Yates, and Lowe, 1976), k_2 is interpreted as the rate constant associated with the reduction of FeS II via Av2.

We sought to determine a stability constant for the presumed Av2-FeS II complex involved in the electron transfer reaction. Because the reduction of Av2$_{ox}$ by SO_2^- is much faster than the reduction of FeS II$_{ox}$ at low dithionite concentrations, we rapidly mixed Av2 in the presence of a low di-

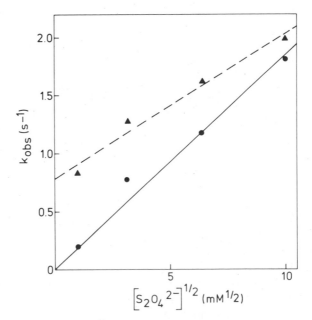

Figure 14.   Dependence of the observed rate constant for FeS II reduction on dithionite concentration in the absence and presence of Av2. Final concentrations were FeS II, 44 μM, and Av2, 22 μM. Plotted are FeS II versus $S_2O_4^-$ (●——●) and FeS II + Av2 versus $S_2O_4^-$ (▲——▲).

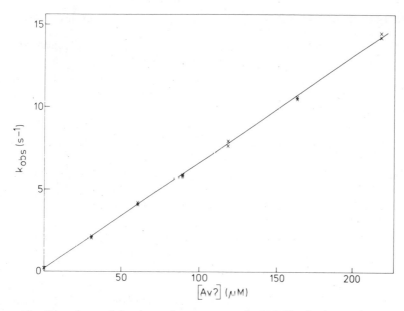

Figure 15.   Dependence of the observed rate constant for FeS II reduction on the concentration of Av2 in the presence of a low dithionite concentration. Av2 plus $S_2O_4^-$ was mixed with FeS II. Final Av2 concentrations as indicated; final FeS II concentration was 18 μM, final dithionite concentration 1 mM.

thionite concentration with FeS II_{ox} in the stopped-flow apparatus and observed a second-order rate of reduction of FeS II_{ox} by $Av2_{red}$. Figure 15 shows the dependence of k_{obs} on Av2 concentration at constant dithionite concentration. The plot is not detectably curved up to the highest concentration of Av2 (218 μM) employed. Thus, there is no evidence for complex formation under these conditions. An estimate of the minimum curvature that would have been detected by the experimental system indicates a lower limit of 1 mM for the dissociation constant of the $Av2_{red}$ plus FeS II_{ox} complex. The rate constant for electron transfer between these two iron-sulfur proteins is calculated to be 6.5×10^4 $M^{-1} \cdot sec^{-1}$.

There is good evidence, then, to show that O_2 protection in whole cells and in cell-free and purified extracts of *A. vinelandii* is brought about by the redox-dependent complex formation of Av1, Av2, and FeS II. "Switch on" and "switch off" in whole cells can be explained by our hypothesis; specifically, rapid "switch on" can occur by the rapid reduction of FeS II by Av2. FeS II also protects *A. chroococcum* N_2ase, but not cell-free extracts of *R. leguminosarum* N_2ase, against O_2 inactivation. Thus, *R. leguminosarum* N_2ase probably does not contain a FeS II–type of protein, but may be well protected against inactivation by free O_2 by the high concentration of the leghemoglobin around the bacteroid.

CONCLUSIONS

The results presented here show that, as in *A. vinelandii*, nitrogen fixation in aerobic *R. leguminosarum* bacteroids can only be achieved by energization of the cytoplasmic membrane by means of oxidation energy. The substrates are slightly different, e.g., NADH and malate in *A. vinelandii* and NADH in *R. leguminosarum*. Succinate oxidation energizes the membrane of *A. vinelandii*, and can also energize the membrane of *R. leguminosarum* when these membranes are prepared at high salt concentration. Because succinate oxidase activity is also present at low NaCl concentration, we conclude that, in *R. leguminosarum* membranes, this process proceeds in a non–energy-linked way under these conditions.

Hydrogen oxidation coupled to oxidative phosphorylation occurs in *A. vinelandii* membranes, but oxidation of hydrogen is totally absent in membranes of *R. leguminosarum*. Hydrogen oxidation in *A. vinelandii* is sensitive to increasing O_2 tensions; both respiration rate and P:O ratio decline. This result contrasts with the NADH oxidase activity in *A. vinelandii*, where, on appearance of O_2 in the medium, a decline is found in the P:O ratio, but not in ATP concentration. This decrease in P:O ratio is thus due to non–energy-linked oxidation of this substrate by O_2.

Flavodoxin hydroquinone is the electron donor for N_2ase in *A. vinelandii* and *R. leguminosarum*. NAD(P)H flavodoxin oxidoreductase is present in both organisms. This purified FAD-containing protein is able to reduce flavodoxin to both the hydroquinone and the semiquinone, but this property is insufficient for electron donation to N_2ase in the absence of energized membranes.

The O_2 stability of the *A. vinelandii* complex is achieved by formation of a tight high molecular weight ($1–2 \times 10^6$) stoichiometric complex of Av1 and Av2 with FeS II. In this complex, whose formation is promoted by Mg^{2+}, the three proteins are in the oxidized state. Reduction of the complex leads to N_2ase activity and a lowering of the molecular weight to 300,000, possibly by dissociation of this three-component complex into the usual two-component N_2ase. The protection of nitrogenase activity toward O_2 inactivation by FeS II occurs only with the intact Av1 plus Av2 complex and not with the individual component proteins.

ACKNOWLEDGMENTS

We wish to thank Mr. Berry Sachteleben for drawing the figures. We thank the European Molecular Biology Organization for the fellowship to G. S. that made it possible to carry out part of this study in cooperation with Drs. G. M. Yates and R. N. F. Thorneley of the A.R.C. Unit of Nitrogen Fixation, University of Sussex, Brighton, England. L. M. gratefully acknowledges the fellowship from the Agricultural University, Wageningen.

REFERENCES

Ackrell, B. A. C., and C. W. Jones. 1971. The respiratory system of *Azotobacter vinelandii*. 1. Properties of phosphorylating respiratory membranes. Eur. J. Biochem. 20:22–28.

Appleby, C. A., G. L. Turner, and P. K. MacNicol. 1975. Involvement of oxyleg-haemoglobin and cytochrome P-450 in an efficient oxidative phosphorylation pathway which supports nitrogen fixation in *Rhizobium*. Biochim. Biophys. Acta 387:461–474.

Baak, J. M., and P. W. Postma. 1971. Oxidative phosphorylation in intact *Azotobacter vinelandii*. FEBS Lett. 19:189–192.

Bergersen, F. J. 1971. Biochemistry of symbiotic nitrogen fixation in legumes Annu. Rev Plant Physiol. 22:121–140.

Bergersen, F. J., and G. L. Turner. 1975a. Leghaemoglobin and the supply of O_2 to nitrogen-fixing root nodule bacteroids: Studies of an experimental system with no gas phase. J. Gen. Microbiol. 89:31–47.

Bergersen, F. J., and G. L. Turner. 1975b. Leghaemoglobin and the supply of O_2 to nitrogen-fixing root nodule bacteroids: Presence of two oxidase systems and ATP production at low free O_2 concentration. J. Gen. Microbiol. 91:345–354.

Bergersen, F. J., G. L. Turner, and C. A. Appleby. 1973. Studies of the physiological role of leghaemoglobin in soybean root nodules. Biochim. Biophys. Acta 292:271–282.

Bhattacharyya, P., and E. M. Barnes, Jr. 1976. ATP-dependent calcium transport in isolates membrane vesicles from *Azotobacter vinelandii*. J. Biol. Chem. 251:5614–5619.

Blankenhorn, G. 1977. On the mode of hydrogen transfer and catalysis in nicotinamide-dependent oxido reduction. In: H. Sund (ed.), Pyridine Nucleotide Dependent Dehydrogenases, pp. 185–198. Walter de Gruyter, Berlin.

Bulen, W. A., and J. R. Le Comte. 1972. Nitrogenase complex and its components. Methods Enzymol. 24:456–470.

Dalton, H., and J. R. Postgate. 1969. Effect of oxygen on growth of *Azotobacter chroococcum* in batch and continuous cultures. J. Gen. Microbiol. 54:463–473.

DerVartanian, D. V., Y. I. Shethna, and H. Beinert. 1969. Purification and properties of two iron-sulfur proteins from *Azotobacter vinelandii*. Biochim. Biophys. Acta 194:548–563.

Dixon, R. O. D. 1976. Hydrogenases and efficiency of nitrogen fixation in aerobes. Nature 262:173.

Eilermann, L. J. M., H. G. Pandit-Hovenkamp, M. van der Meer-van Buren, A. H. J. Kolk, and M. Feenstra. 1971. Oxidative phosphorylation in *Azotobacter vinelandii*: Effect of inhibitors and uncouplers on P/O ratio, trysin-induced ATPase and ADP stimulated respiration. Biochim. Biophys. Acta 245:305–312.

Haaker, H., A. de Kok, and C. Veeger. 1974. Regulation of dinitrogen fixation in intact *Azotobacter vinelandii*. Biochim. Biophys. Acta 357:344–357.

Haaker, H., G. Scherings, and C. Veeger. 1977. Aerobic nitrogen fixation in *A. vinelandii*. In: W. Newton, J. R. Postgate, and C. Rodriguez-Barrueco (eds.), Recent developments in nitrogen fixation, pp. 271–285. Academic Press, Inc., New York.

Haaker, H., and C. Veeger. 1976. Regulation of respiration and nitrogen fixation in different types of *Azotobacter vinelandii*. Eur. J. Biochem. 63:499–507.

Haaker, H., and C. Veeger. 1977. Involvement of the cytoplasmic membrane in nitrogen fixation by *Azotobacter vinelandii*. Eur. J. Biochem. 77:1–10.

Hemmerich, P. 1977. Discussions on mechanism of hydrogen transfer. In: H. Sund (ed.), Pyridine Nucleotide Dependent Hydrogenases, pp. 203–205. Walter de Gruyter, Berlin.

Hill, S., J. W. Drozd, and J. R. Postgate. 1972. Environmental effects on the growth of nitrogen fixation bacteria. J. Appl. Chem. Biotechnol. 22:541–558.

Ke, B., W. A. Bulen, E. R. Shaw, and R. H. Breeze. 1974. Determination of oxidation reduction potentials by spectropolarimetric titration: application to several iron-sulfur proteins. Arch. Biochem. Biophys. 162:301–309.

Kelley, B. C., C. M. Meijer, C. Gandy, and R. M. Vignais. 1977. Hydrogen recycling by *Rhodopseudomonas capsulata*. FEBS Lett. 81:281–285.

Laane, C., H. Haaker, and C. Veeger. 1978. Involvement of the cytoplasmic membrane in nitrogen fixation by *Rhizobium leguminosarum* bacteroids. Eur. J. Biochem. 87:147–153.

Lambeth, D. O., and G. Palmer. 1973. The kinetics and mechanism of reduction of electron transfer proteins and other compounds of biological interest by dithionite. J. Biol. Chem. 248:6095–6103.

Lindström, E. S., S. M. Lewis, and M. J. Pinsky. 1951. Nitrogen fixation and hydrogenase in various bacterial species. J. Bacteriol. 61:481–487.

Phillips, D. A., R. M. Daniel, C. A. Appleby, and H. J. Evans. 1973. Isolation from *Rhizobium* of factors which transfer electrons to soybean nitrogenase. Plant Physiol. 51:136–138.

Racker, E., and L. L. Horstman. 1967. Partial resolution of the enzymes catalyzing oxidative phosphorylation. J. Biol. Chem. 242:2547–2551.

Reed, D. W., R. E. Toia, and D. Raveed. 1974. Purification of azotophore membranes containing the nitrogenase from *Azotobacter vinelandii*. Biochem. Biophys. Res. Commun. 58:20–26.

Scherings, G., H. Haaker, and C. Veeger. 1977. Regulation of nitrogen fixation by Fe-S protein II in *Azotobacter vinelandii*. Eur. J. Biochem. 77:621–630.

Schubert, K., and H. J. Evans. 1976. Hydrogen evolution: A major factor affecting the efficiency of nitrogen fixation in nodulated symbionts. Proc. Natl. Acad. Sci. U.S.A. 73:1207–1211.

Shethna, Y. I., D. V. DerVartanian, and H. Beinert. 1968. Non-heme (iron-sulfur) proteins of *Azotobacter vinelandii*. Biochem. Biophys. Res. Commun. 31:862–868.

Smith, L. A., S. Hill, and M. G. Yates. 1976. Inhibition by acetylene of conventional hydrogenase in nitrogen-fixing bacteria. Nature 262:209–210.

Stokes, A. N. 1975. Facilitated diffusion; the elasticity of oxygen supply. J. Theor. Biol. 52:285–297.

Thorneley, R. N. F., M. G. Yates, and D. J. Lowe. 1976. Nitrogenase of *Azotobacter chroococcum*. Kinetics of the reduction of oxidized iron-protein by sodium dithionite. Biochem. J. 155:137–144.

Veeger, C., H. Haaker, and G. Scherings. 1977. Energy transduction and nitrogen fixation in *Azotobacter vinelandii*. In: K. Van Dam and B. F. Van Gelder (eds.), Structure and Function of Energy-transducing Membranes, pp. 81–93. Elsevier–North Holland, Amsterdam.

Wilson, P. W., and R. H. Burris. 1947. The mechanism of biological nitrogen fixation. Bacteriol. Rev. 11:41–73.

Wittenberg, J. B., F. J. Bergersen, C. A. Appleby, and G. L. Turner. 1974. Facilitated oxygen diffusion; the role of leghemoglobin in nitrogen fixation by bacteroids isolated from soybean root nodules. J. Biol. Chem. 249:4057–4066.

Yates, M. G. 1971. Electron transport to nitrogenase in *A. chroococcum*. Purification and some properties of NADH dehydrogenase. Eur. J. Biochem. 24:347–357.

Yates, M. G. 1972. The effect of ATP upon the oxygen sensitivity of nitrogenase from *Azotobacter chroococcum*. Eur. J. Biochem. 29:386–392.

Yates, M. G. 1977. Physical aspects of nitrogen fixation. In: W. Newton, J. R. Postgate, and C. Rodriguez-Barrueco (eds.), Recent Developments in Nitrogen Fixation, pp. 233–253. Academic Press, Inc., New York.

Nitrogen Fixation, Volume I
Edited by W. E. Newton and W. H. Orme-Johnson
Copyright 1980   University Park Press   Baltimore

Nitrogen Fixation by Photosynthetic Bacteria: Properties and Regulation of the Enzyme System from *Rhodospirillum rubrum*

P. W. Ludden

The photosynthetic bacterium *Rhodospirillum rubrum* was first shown to metabolize gaseous N_2 by Gest and Kamen (1949). Lindström, Burris, and Wilson (1949) quickly confirmed their results by showing that $^{15}N_2$ was taken up into cellular material by illuminated bacteria. The studies of Gest, Kamen, and Bergoff (1950) on gas exchange reactions by the photosynthetic bacteria provided a number of essential observations about the physiology of N_2 fixation by *R. rubrum*. They found that *R. rubrum* would assimilate N_2 when grown either in the absence of fixed nitrogen or with glutamate as the sole nitrogen source. Growth on N_2 also caused the cells to produce large amounts of H_2. This H_2 evolution could be inhibited by either N_2 or ammonia and has since been shown to be evolved by nitrogenase, not hydrogenase (Wall, Weaver, and Gest, 1975). Note, however, that the experiments of Wall, Weaver, and Gest (1975) were conducted with *Rhodopseudomonas capsulata*, not *R. rubrum*. The N_2 and H_2 metabolism of the two organisms is quite similar and the generalization of their conclusion to include *R. rubrum* seems reasonable. The fact that nitrogenase-mediated H_2 evolution is inhibited in vivo by fixed nitrogen (NH_4^+) is important, since nitrogenase activity in other bacteria proceeds unabated after the addition of NH_4^+ to the medium (Daesch and Mortenson, 1972; Tubb and Postgate, 1973). Gest and Kamen also showed that NH_4^+ inhibited N_2 uptake. Schick (1971) demonstrated that NH_4^+ reversibly inhibited N_2 uptake by nitrogenase in whole cells of *R. rubrum* and that the duration of inhibition of nitrogenase activity was proportional to the amount of NH_4^+ added. Neilson and Nordlund (1975) confirmed these observations using acetylene reduction. They also showed that asparagine, glutamine, and low concentrations of O_2 could cause a similar temporary inhibition of acetylene reduction. The regulation of *R. rubrum* nitrogenase in whole cells does show similarities to other nitrogen-fixing organisms in that NH_4^+ represses the synthesis of the enzyme and methionine sulfoximine derepresses the enzyme in the presence of NH_4^+ (Weare and Shanmugam, 1976) as it does in *Azotobacter vinelandii* (Gordon and Brill, 1974).

Studies on nitrogenase activity in extracts of *R. rubrum* began almost immediately after cell-free extracts of *Clostridium pasteurianum* were obtained (Carnahan et al., 1960). Schnieder et al. (1961) provided evidence for activity in extracts of *R. rubrum* and several other free-living nitrogen fixers. The system was not further characterized until Burns and Bulen (1966) made the interesting observations that a high Mg^{2+}:ATP ratio (25 mM:5 mM) was needed for maximal activity of N_2 uptake or H_2 evolution and that the photosynthetic membranes (chromatophores) of *R. rubrum* inhibited activity greatly. It is interesting to speculate on what might be considered a "normal" system of regulation for nitrogenase if Burns and Bulen had pursued the *R. rubrum* nitrogenase rather than the *A. vinelandii* nitrogenase.

Biggins, Kelly, and Postgate (1971) were able to partially purify the MoFe protein from *R. rubrum* and to demonstrate cross-reactions with several Fe proteins from other organisms. Emerich and Burris (1978) have studied the cross-reactions of both components of *R. rubrum* nitrogenase extensively. Evans' group has partially purified the nitrogenase components from a green photosynthetic bacterium and the Fe protein of *Chromatium* strain D. (Smith and Evans, 1971). They have also purified and characterized the MoFe protein of *Chromatium* strain D and found it to be very similar to other MoFe proteins (Evans, Telfer, and Smith, 1973).

Munson and Burris (1969) studied the nitrogenase of *R. rubrum* cells grown in continuous culture and found that the time course of substrate reduction by crude extracts was nonlinear. They also observed the inhibition by chromatophores and the requirement for high Mg^{2+} concentration.

Thus, there are four observations that set *R. rubrum* nitrogenase apart from other nitrogenase systems: 1) the rapid "turning off" of the enzyme activity by NH_4^+ in vivo; 2) high Mg^{2+} required for nitrogenase activity in crude extracts; 3) inhibition of nitrogenase activity by chromatophores; and 4) nonlinearity of the time course for activity in crude extracts.

PROPERTIES OF CRUDE EXTRACTS

We repeated the results of Burns and Bulen and of Munson and Burris and also noted that the addition of low levels of Mn^{2+} (0.5 mM) greatly stimulated activity in crude extracts (Ludden and Burris, 1976). Figure 1 shows the stimulatory effect of high Mg^{2+} concentration and of Mn^{2+}, and also shows the nonlinear nature of the time course of acetylene reduction by a crude extract of *R. rubrum* nitrogenase.

We also found the inhibition by chromatophore membranes to be a problem until the cells were collected anaerobically. The cells were, of

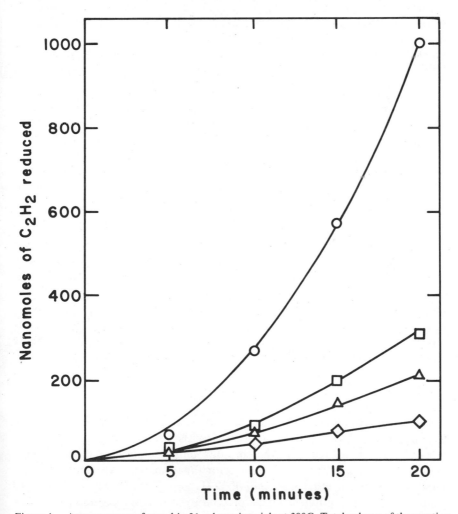

Figure 1.   Assays were performed in 21-ml vaccine vials at 30°C. Total volume of the reaction mixture was 1 ml; it contained 5 mM ATP, 30 mM creatine phosphate, 0.05 mg creatine phosphokinase, 0.4 ml of crude extract that had a protein concentration of 8.4 mg/ml, 10 mM $Na_2S_2O_4$, and 40 mM triethanolamine-acetate buffer. The gas phase was 90% H_2 and 10% C_2H_2. C_2H_2 was measured by gas chromatography. O——O, 25 mM Mg^{2+} and 0.5 mM Mn^{2+}; □——□, 10 mM Mg^{2+} and 0.5 mM Mn^{2+}; △——△, 25 m*M* Mg^{2+}; ——— , 10 mM Mg^{2+}. (Reprinted from Ludden and Burris, 1976. Copyright 1976 by the American Association for the Advancement of Science.)

Table 1.   Separation of activating factor from anaerobically collected chromatophores[a]

Components of reaction mixture	C_2H_2 reduced (nmole) per 20 min
Crude extract	588
Crude extract plus resuspended chromatophores	2100
Crude extract plus buffer wash from chromatophores	1008
Crude extract plus salt wash from chromatophores	2436
Crude extract plus final chromatophores	798
Resuspended chromatophores	210
Final chromatophores	0
Buffer wash	54
Salt wash	0

[a] Assays were performed as described for Figure 1 with 25 mM $MgCl_2$ and 0.5 mM $MnCl_2$ added. Chromatophores were sedimented by centrifugation at 45,000 × g for 2 hr; supernatant fluid was the buffer wash. Chromatophores were resuspended in 30 ml of 0.5 M NaCl in buffer and again were centrifuged at 45,000 × g for 1 hr to yield the salt wash and final chromatophores; 0.2 ml of crude extract or other fraction was added where indicated. Reprinted from Ludden and Burris (1976). Copyright 1976 by the American Association for the Advancement of Science.

course, grown and lysed anaerobically, but most methods of cell harvesting result in exposure of the cells to air at some point. Only when the cells were collected by siphoning into centrifuge tubes that were being sparged with N_2 did the chromatophores cease to inhibit. Table 1 shows that chromatophores from anaerobically collected cells not only do not inhibit, they greatly stimulate activity. Table 1 also demonstrates that 0.5 M NaCl can be used to elute the activating factor (AF) from the chromatophores. Nordlund, Erickson, and Baltescheffsky (1977) have repeated some of these observations and refer to AF as membrane component, or MC.

PURIFICATION AND PROPERTIES OF AF

The known properties of AF are, at this writing, few. It is O_2 labile and trypsin sensitive. AF is stabilized greatly by the presence of dithiothreitol (2 mm). AF can be eluted from the chromatophores with 0.5 M NaCl and the membranes can then be removed by centrifugation. The AF can be precipitated with 30% polyethylene glycol 4000 (Union Carbide) when 0.5 M NaCl is present; 15% polyethylene glycol precipitates very little of the activity. AF can be further purified by DEAE cellulose chromatography. Blue Sepharose[R] will bind AF, but recovery of AF from blue Sepharose columns is low and the eluting agent (10 mM NAD) is expensive.

AF does not display NADase, ATPase, or phosphodiesterase (bis-*p*-NO_2-phenylphosphate as substrate) activity. The only known activity of AF is the activation of the Fe protein from *R. rubrum* or *Azospirillum lipoferum*. In contrast to the Fe protein and MoFe protein of *R. rubrum* nitrogenase, the synthesis of AF is not repressed by ammonia. Therefore, cells grown on ammonia are a good source of AF with no contaminating Fe or MoFe proteins.

PURIFICATION AND PROPERTIES
OF NITROGENASE FROM *R. rubrum*

Once the necessity of AF for activity was understood, the MoFe and Fe proteins could be purified. The MoFe protein had been partially purified previously (Biggens, Kelly, and Postgate, 1971). The MoFe protein has properties very similar to other MoFe proteins and does not require activation. Shah and Brill (1977) have purified the iron-molybdenum cofactor (FeMoco) from *R. rubrum* MoFe protein and found no differences between *R. rubrum* FeMoco and the cofactor from other organisms. Table 2 shows some of the properties of the MoFe (Rr1) and Fe (Rr2) proteins. The measurements on the Fe protein were made on the protein as isolated, i.e., inactive. The inactivity of the protein is not due to a lack of an iron-sulfur center because the inactive Fe protein has its full complement of iron. The EPR signal of the inactive Fe protein is similar to that of the active protein, so inactivity is not a result of an unusual environment for the iron-sulfur cluster. The Fe protein from *R. rubrum* is unusual in a number of ways. It appears to have two different subunit types, as demonstrated on SDS gels, but still has an amino acid composition very similar to that of other Fe proteins (Ludden and Burris, 1978a). An even more interesting difference is the presence of phosphate, ribose, and an adenine-like compound covalently attached to the protein.

Evidence for PO_4^{3-} on the protein is shown in Figure 2. For this experiment, cells were derepressed for nitrogenase in the presence of $^{32}PO_4^{3-}$, harvested, extracted, and the extract applied to a small DEAE cellulose column. The Fe protein was eluted with 400 mM NaCl and all of

Table 2.   Properties of nitrogenase components from *R. rubrum*[a]

	Molecular weight	Subunit	Fe	Mo	Pentose	PO_4^{3-}	Ad[b]	Specific activity
Rr1	234,000	4 (58,500)	20	1.7	0.5	0.5	0.5	1700
Rr2	61,500	2 (30,000)	3.5	N.D.[c]	2–3	2	2	800–1200
(Inactive)		(31,500)						

[a] Data from Ludden and Burris (1978a).

[b] Ad = adenine-like compound.

[c] N.D. = not determined.

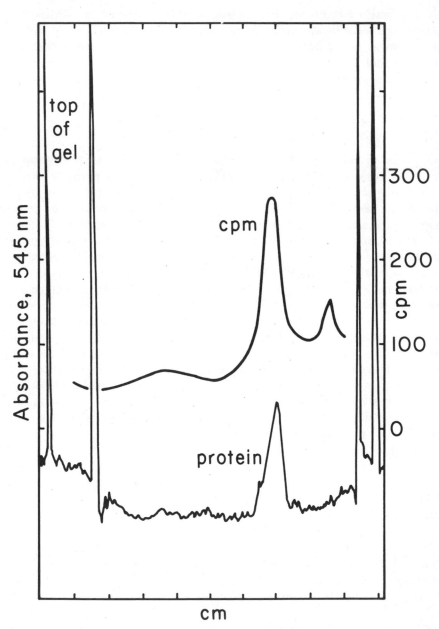

Figure 2.   10% polyacrylamide, 0.1% SDS gel of [^{32}P]Fe protein. (Reprinted from Ludden and Burris, 1978a, by permission of the *Biochemical Journal*.)

the Fe protein fraction was put on an anaerobic acrylamide gel. After electrophoresis for 2 hr, the brown band containing Fe protein was cut out and the Fe protein extracted by mashing the gel into buffer containing SDS and dithiothreitol. The extracted protein was precipitated by trichloroacetic acid and resuspended in SDS buffer before being applied to an SDS gel for analysis as described by Laemmli (1970). The gel was stained for protein, scanned, and then sliced into 2-mm pieces that were assayed for the presence of ^{32}P by scintillation counting. Figure 3 shows that the protein peak and the peak of radioactivity coincide. Phosphate can also be shown to be present in H_2O_2 digests of purified Fe protein using a colorimetric assay (Chen, Toribara, and Warner, 1956) which shows 2 PO_4^{3-} per Rr2.

The presence of ribose was less rigorously established, but it is always associated with purified Rr2 at a level of 2–3 moles per mole of protein. The colorimetric assay of Dische (1962) was used, which involves the production of furfural in the presence of concentrated HCl. Since it is not specific for ribose, however, the actual 5-carbon sugar present is therefore not necessarily ribose.

Figure 3 shows the fluorescence spectrum (emission) of purified Fe protein treated with glyoxal hydrate and acetic acid. The spectrum of AMP is shown for comparison (Yuki et al., 1972). Av2, which has a very similar amino acid composition, shows a very low level of fluorescence in this assay; likewise, Rr2 does not give a positive result if the assay mixture is not heated. Therefore, the fluorescence spectrum cannot be attributed to protein. Yuki et al. (1972) claim that this assay is specific to adenine to the extent that no other major purine or pyrimidine gave a positive test. Adenine was rendered inactive in this assay if the 6-amino group was blocked.

Figure 4 shows the absorption spectrum of Rr2, Av2, and Cp2 as well as the spectrum of material obtained after hydrolysis of Rr2 for 1 at 100°C in 1 N HCl. The spectra of the proteins were determined after precipitation in $HClO_4$ to remove iron-sulfur centers and dithionite. The Fe protein from *R. rubrum* has an extra shoulder in its spectrum at 268 nm that is not seen in the others. Hydrolysis of Rr2 followed by centrifugation to remove unhydrolyzed material enriched the 268-nm absorbing material. A positive test in the fluorescent assay for adenine was also found for the hydrolysate material. Neither the fluorescent spectrum of glyoxal hydrate/acetic acid-treated protein nor the UV spectrum of the modifying group are identical to those for adenine.

Other nitrogenase proteins available in Burris' lab have been investigated for the presence of ribose or phosphate. Cp2, Av2, and Bp2 showed only low levels (<0.5 nmoles/nmole of protein) of ribose or phosphate. Av2 showed only traces of adenine.

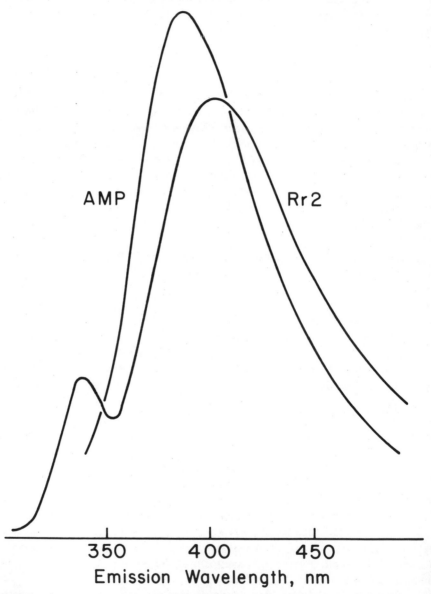

Figure 3.   Fluorescence spectra of AMP and *R. rubrum* Fe protein that had been treated with glyoxal hydrate. (Reprinted from Ludden and Burris, 1978a, by permission of the *Biochemical Journal.*)

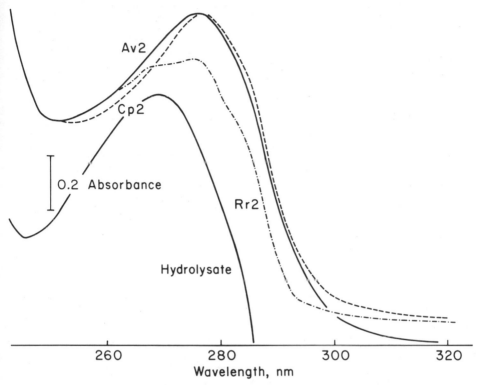

Figure 4.   Ultraviolet spectra of Fe proteins from *A. vinelandii*, *C. pasteurianum* and the inactive form from *R. rubrum*. Also shown is the top half of the UV spectrum of the soluble material from 1 N HCl hydrolysis (100°C) of *R. rubrum* Fe protein. (Reprinted from Ludden and Burris, 1978a, by permission of the *Biochemical Journal*.)

ACTIVATION OF THE Fe PROTEIN

The mechanism of activation of the Fe protein was studied in detail. Several possibilities existed for the activation, and the ones most seriously considered are: 1) AF is an enzyme that activates the Fe protein by covalent modification, probably involving the PO_4^{3-}, ribose, and adenine-like molecule; 2) AF carries or accepts a small molecule involved in activation and thus "charges" the protein, with one AF activating one Fe protein; and 3) the AF and the Fe protein form a stable complex required for catalytic activity of nitrogenase. We have concluded that the first model is correct (Ludden and Burris, 1978b). Nordlund, Erickson, and Baltescheffsky (1977) have suggested the stable complex model.

Activation of the reduced Fe protein is a process that requires ATP and M^{2+}. The metals that have been found to work are Mn^{2+}, Mg^{2+}, and Fe^{2+}. MoFe protein is not required for activation. Figure 5 shows a time course

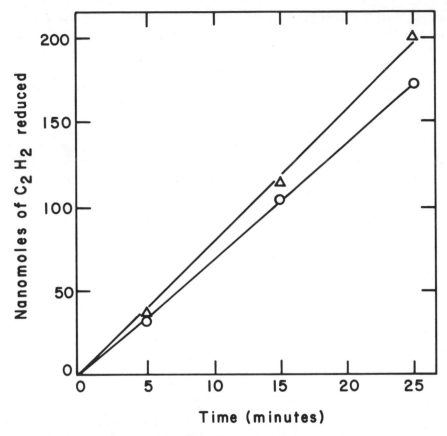

Figure 5.   Activity of activated *R. rubrum* Fe protein in the absence of AF. △——△, 10 mM Mg^{2+}; O——O, 25 mM Mg^{2+} and 0.5 mM Mn^{2+}. (Reprinted from Ludden and Burris, 1976. Copyright 1976 by the American Association for the Advancement of Science.)

and metal requirements for acetylene reduction in the usual assay mixture by Fe protein that has been activated in an M^{2+} ATP + Fe protein + AF mixture. The activated Fe protein was separated from AF on a small DEAE column. The time course is linear and Mn^{2+} no longer stimulates activity. The AF that is separated from activated Fe protein retains its ability to activate Fe protein. Native gels have been run on Fe protein in various stages of purification and on both active and inactive pure Fe protein (Ludden and Burris, 1978a). There does not appear to be any difference in migration, nor does the activated protein show the additional band that would be expected if AF and Fe protein formed a relatively stable complex. From these results, and because no additional band that might be AF appears in the activated Fe protein fraction (either in native or O_2-inacti-vated Fe protein fractions), we conclude that the activated Fe protein is a stable, modified form that, once activated, retains its activity in the absence of AF. In fact, the search for conditions that allow the activated protein to return to the inactive state in vitro has been completely unsuccessful to date.

ROLE OF THE COVALENTLY ATTACHED GROUP IN ACTIVATION

The difference spectrum in Figure 6 for inactive minus active Fe protein shows a peak similar to the peak seen for the hydrolysate of Rr2 (Figure 4). From the amount of Rr2 present, an E_{mm} at 268 nm can be calculated for the material that is lost. Because it is not known to what extent the activation is complete, this extinction coefficient (3.5) is regarded as a lower limit for the material removed. It is the inactive protein that shows the greater absorbance, and since it has not been exposed to ATP the ΔA cannot be accounted for by incomplete removal of ATP. The difference spectrum clearly supports the idea of activation of Rr2 by covalent modification.

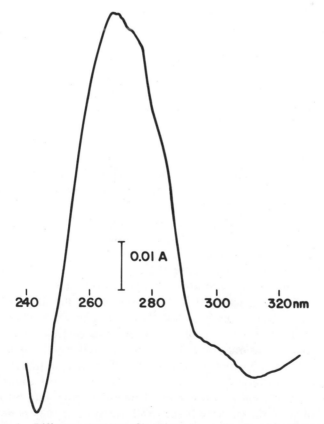

Figure 6.   Difference spectrum of inactive minus active *R. rubrum* Fe protein.

Table 3.   Effects of AF, alkaline phosphatase (AP), and snake venom diesterase (SVD) on Rr2[a]

| | Percent change | | |
	AF	AP	SVD
Inactive Enzyme			
Ribose	−16.6	0	+12
Phosphate	−7.6	−6	0
Adenine-like compound	−45.3	N.D.[b]	0
Activity[c]	+(400-fold)	0	0(±ATP)
Active Enzyme			
Phosphate		−50	
Activity[c]		0	

[a] Samples were treated as described by Ludden and Burris (1978a, 1978b) before determination of ribose, phosphate, and adenine-like compound.

[b] N.D. = not determined.

[c] Refers to activity of the Fe protein in an acetylene reduction assay mixture containing MgATP, an ATP-generating system, dithionite, and MoFe protein but no Mn^{2+} or AF.

The data in Table 3 are the result of studies on the effect of various enzymes on the modifying group of Rr2. Alkaline phosphatase from *Escherichia coli* cleaves phosphomonoesters but not phosphodiesters. Snake venom diesterase (SVD) cleaves nucleic acid diesters and requires a free 3′ hydroxyl group on the substrate for activity. SVD will cleave AMP off of glutamine synthetase (Stadtman et al., 1972) and ADPR off of RNA polymerase (Goff, 1974). SVD had no effect on inactive Rr2; no part of the covalently attached group was removed and no increase in Fe protein activity was seen on incubation with SVD ± MgATP. Alkaline phosphatase had no effect on the phosphate content of the inactive protein, but removed up to 50% of the PO_4^{3-} from the activated Fe protein. This indicates that, although PO_4^{3-} is not removed during activation, it becomes accessible to alkaline phosphatase. The last column of Table 3 shows that AF removes up to 50% of the adenine-like compound, but relatively small amounts of PO_4^{3-} and ribose are lost. Because the adenine-like compound is lost to a much greater degree than the other components of the prosthetic group, it seems reasonable to conclude that it is outermost of the three parts. Ribose and phosphate remain bound, but of these two phosphate is thought to be outermost from the protein because it becomes susceptible to alkaline phosphatase (which cleaves only phosphomonoesters) after activation. The above reasoning rests on the assumption that the ribose, phosphate, and adenine-like compound are linked together in a single unit. Furthermore, we have never been able to remove all of the adenine-like compound or make all of the PO_4^{3-} alkaline phosphatase labile; 50% seems to be the upper limit.

ROLE OF ATP IN ACTIVATION

ATP is required for activation of the Fe protein. Because ATP has been shown to bind to the Fe protein (Tso and Burris, 1973), the ATP requirement could be due to ATP binding to the Fe protein or to additional binding of ATP to AF with or without ATP hydrolysis. When AF, ATP, inactive Fe protein, and metals are mixed, only low levels of PO_4^{3-} (less than 0.1 ATP hydrolyzed per Fe protein present), and no pyrophosphate are released. Therefore, ATP is not hydrolyzed during activation. Figure 7 shows that both the active and inactive forms of Rr2 bind ATP. The assay of Walker and Mortenson (1974) as modified by Ljones and Burris (personal communication) was used to demonstrate the ATP-dependent release of iron-sulfur centers to dipyridyl dyes. There may be quantitative differences in MgATP binding that we have not detected, but, qualitatively, MgATP has an effect on both forms of Rr2.

Activation of the protein in the absence of ATP was achieved when phenozine methosulfate was used to oxidize the inactive protein in the

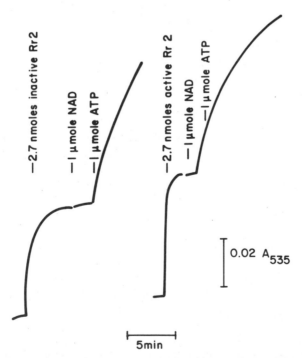

Figure 7.   Binding of ATP to both active and inactive Fe protein demonstrated by release of iron-sulfur centers to 1 mM bathophenanthroline sulfonate in the presence of ATP. (Reprinted from Ludden and Burris, 1978a, by permission of the *Biochemical Journal*.)

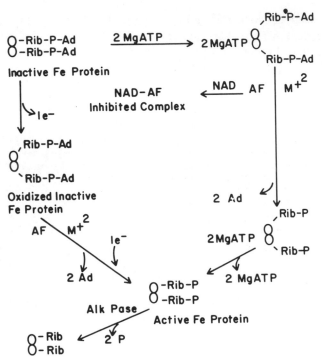

Figure 8.   Model for the activation of *R. rubrum* Fe protein. Rib = ribose, P = phosphate, and Ad = adenine-like compound.

absence of O_2 (Table 4). Although much of the Fe protein was lost under the oxidizing conditions, some activated protein was obtained when AF, inactive Rr2, and metals were mixed. A divalent metal was still required for the activation in the absence of ATP.

Figure 8 shows the present model for the activation of Rr2 by AF (Ludden and Burris, 1978b). Either the binding of MgATP or the loss of an

Table 4.   Activation of oxidized Rr2 in the absence of ATP

Incubation[a]			C_2H_2 assays[b]	
			$-AF$	$+AF$
Rr2	ATP	AF	(nmoles/25 min)	(nmoles/10 min)
100 μl	0	0.1 ml	20	36
100 μl	5 mM	0.1 ml	26	22
100 μl	0	0	0	16

[a] The incubations were carried out for 20 min under H_2 in the presence of 1 mM PMS. The Fe protein used was from the second DEAE column. All incubations contained 25 mM Mg^{2+} and 0.5 mM Mn^{2+}.

[b] AF C_2H_2 assays contained 25 mM Mg^{2+} and 0.5 mM Mn^{2+}, and $-AF$ assays contained 10 mM Mg^{2+} and no Mn^{2+}.

electron is seen as causing a conformational change in Rr2 that allows the metal-dependent AF to cleave the adenine-like molecule off of the protein. The ribose and PO_4^{3-} remain. The protein can be further modified by alkaline phosphatase, although loss of the PO_4^{3-} does not affect the activity of the enzyme.

BLOCK IN THE CATALYTIC FUNCTION OF Rr2

When mixed with MoFe protein, the inactive Fe protein from *R. rubrum* is equally incapable of carrying out either the reductions of N_2, C_2H_2, and H^+ or the hydrolysis of ATP. Therefore, inactivation of Rr2 does not cause an uncoupling of electron transfer from ATP hydrolysis, nor does it mimic CO inhibition of the enzyme, which inhibits electron transfer to all substrates except H^+. It was shown in Figure 7 that the inactive Fe protein would bind ATP. Figure 9 shows that the inactive Fe protein can be reversibly oxidized and reduced. In these experiments, the change in A_{430} (Ljones, 1973) was measured in an assay mixture in which the reductant (dithionite) limited the reaction. A catalytically active mixture of Av2 and Rr1 was used to exhaust the reductant in the presence of excess MgATP. Trace A in Figure 9 shows the ΔA_{430} due to oxidation of Av2. When inactive Rr2 is placed in the mixture in addition to Av2 and Rr1 (Rr2 and Av2 at equal concentrations), the

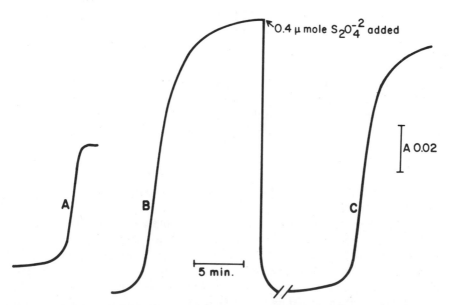

Figure 9.   Oxidation (ΔA_{430}) of inactive *R. rubrum* Fe protein in the presence of *A. vinelandii* Fe protein, *R. rubrum* MoFe protein, and nitrogenase reaction mixture. No AF or Mn^{2+} are present, and dithionite limits the reaction.

ΔA_{430} is doubled, indicating that the inactivve Rr2 has been oxidized. Addition of more dithionite to the cuvette reduces both Av2 and Rr2. In an experiment done in collaboration with Orme-Johnson (Ludden et al., 1978), inactive Rr2 was found to be incapable of reducing the MoFe protein (reduction of MoFe protein was measured as loss of EPR signal at g = 3.65). Recent evidence of Hageman and Burris (1978) that strongly suggests that ATP hydrolysis occurs during the transfer of an electron from the Fe protein to the MoFe protein can be invoked to argue against transfer of electrons from Rr2 to EPR-silent electron sinks in the MoFe protein, since ATP is not hydrolyzed by MoFe protein plus inactive Rr2.

Inactive Fe protein from *R. rubrum* is capable of binding ATP and undergoing oxidation and rereduction, but is incapable of transferring electrons to MoFe protein or substrate and does not hydrolyze ATP, either alone or in conjunction with MoFe protein. The simplest explanation is that the inactive Fe protein is simply incapable of binding to the MoFe protein. The inactive Fe protein does not inhibit active nitrogenase, so it seems that tight binding complexes that are seen when *C. pasteurianum* and *A. vinelandii* components are mixed are not formed (Emerich and Burris, 1976).

SUMMARY

The Fe protein of nitrogenase from *R. rubrum* is isolated in an inactive form. The enzyme can be activated by an activating factor (AF) that requires a divalent metal for activity. Either ATP or the oxidation of the Fe protein is also required for activation. ATP is not hydrolyzed during activation. The inactive protein has two phosphate, two ribose (or other pentose), and two adenine-like molecules per protein dimer. Activation involves the removal of the adenine-like molecule. The inactive Fe protein can bind MgATP and be reversibly oxidized and reduced, but does not reduce the MoFe protein. The inactive protein does not reduce any of the substrates tested (N_2, C_2H_2, and H^+), nor does it hydrolyze ATP in the presence of MoFe protein. Inactive Rr2 does not inhibit the activity of nitrogenase, and the probable lesion in the sequence of catalytic events is in the binding of Rr2 to MoFe protein.

It has long been known that NH_4^+ will rapidly turn off nitrogenase in vivo in *R. rubrum*. This is in contrast to the lack of inhibition of nitrogenase by NH_4^+ in other free-living organisms that have nitrogenase. The activation system described here may be an integral part of the regulatory system involved in turning off nitrogenase in *R. rubrum*. While one side of a potential regulatory system has been described here in some detail (the turning-on side), the factors that allow *R. rubrum* Fe protein to be turned off in vitro remain to be discovered.

ACKNOWLEDGMENTS

I wish to thank my colleagues in Dr. R. H. Burris' lab, D. W. Emerich, R. V. Hageman, S. L. Albrecht, and T. Ljones for suggestions and nitrogenase components. I thank Dr. Burris, in whose laboratory these experiments were conducted, for encouragement and support. I also wish to thank W. J. Brill, W. H. Orme-Johnson, and V. K. Shah for numerous helpful discussions.

REFERENCES

Biggens, D. R., M. Kelly, and J. R. Postgate. 1971. Resolution of nitrogenase of *Mycobacterium flavum* 301 into two components and cross reaction with nitrogenase components from other bacteria. Eur. J. Biochem. 20:140–143.

Burns, R. C., and W. A. Bulen. 1966. A procedure for the preparation of extracts from *Rhodospirillum rubrum* catalyzing N_2 reduction and ATP-dependent H_2 evolution. Arch. Biochem. Biophys. 113:461–463.

Carnahan, J. E., L. E. Mortenson, H. F. Mower, and J. E. Castle. 1960. Nitrogen fixation in cell-free extracts of *Clostridium pasteurianum*. Biochim. Biophys. Acta 38:188–189.

Chen, P. S., T. Y. Toribara, and H. Warner. 1956. Microdetermination of phosphorus. Anal. Chem. 28:1756–1758.

Daesch, G., and L. E. Mortenson. 1972. Effect of ammonia on the synthesis and function of the N_2-fixing system in *Clostridium pasteurianum*. J. Bacteriol. 110:103–109.

Dische, Z. 1962. Colorimetric assay of pentoses. In: Methods in Carbohydrate Chemistry, Vol. 1, pp. 485–486. R. L. Whistler and M. L. Wolfram (eds.), Academic Press, Inc., New York.

Emerich, D. W., and R. H. Burris. 1976. Interactions of heterologous nitrogenase components that generate catalytically inactive complexes. Proc. Natl. Acad. Sci. U.S.A. 73:4369–4373.

Emerich, D. W., and R. H. Burris. 1978. Cross reactions between nitrogenase components from microorganisms. J. Bacteriol. 134:936–943.

Evans, M. C. W., A. Telfer, and R. V. Smith. 1973. The purification and some properties of the molybdenum iron protein of *Chromatium* nitrogenase. Biochim. Biophys. Acta 310:344–352.

Gest, H., and M. D. Kamen. 1949. Evidence for a nitrogenase system in the photosynthetic bacterium *Rhodospirillum rubrum*. Science 109:560.

Gest, H., M. D. Kamen, and H. M. Bergoff. 1950. Studies on the metabolism of photosynthetic bacteria. V. Photoproduction of hydrogen and nitrogen fixation by *Rhodospirillum rubrum*. J. Biol. Chem. 182:153–170.

Goff, C. G. 1974. Chemical structure of a modification of the *Escherichia coli* ribonucleic acid polymerase polypeptides induced by bacteriophage T_4 infection. J. Biol. Chem. 249:6181–6190.

Gordon, J. K., and W. J. Brill. 1974. Depression of nitrogenase synthesis in the presence of excess NH_4^+. Biochem. Biophys. Res. Commun. 59:967–971.

Hageman, R. V., and R. H. Burris. 1978. Nitrogenase and nitrogenase reductase associate and dissociate with each catalytic cycle. Proc. Natl. Acad. Sci. U.S.A. 75:2699–2702.

Laemmli, U. K. 1970. Cleavage of structural proteins during the assembly of the head of bacteriophage T_4. Nature 227:680–682.

Lindström, E. S., R. H. Burris, and P. W. Wilson. 1949. Nitrogen fixation by photosynthetic bacteria. J. Bacteriol. 58:313–316.

Ljones, T. 1973. Nitrogenase from *Clostridium pasteurianum*. Changes in optical spectrum during electron transfer and effects of ATP, inhibitors and alternative substrates. Biochim. Biophys. Acta 321:103–113.

Ludden, P. W., and R. H. Burris. 1976. Activating factor for the iron protein of nitrogenase from *Rhodospirillum rubrum*. Science 194:424–426.

Ludden, P. W., and R. H. Burris. 1978a. Purification and properties of nitrogenase from *Rhodospirillum rubrum*: Evidence for the presence of phosphate, ribose and an adenine-like unit covalently bound to the Fe protein. Biochem J. 175:251–259.

Ludden, P. W., and R. H. Burris. 1978b. The mechanism of activation of the Fe protein of nitrogenase from *Rhodospirillum rubrum*. Submitted for publication.

Ludden, P. W., R. V. Hageman, W. H. Orme-Johnson, and R. H. Burris. 1978. Activities of the 'inactive' Fe protein of nitrogenase from *Rhodospirillum rubrum*. In preparation.

Munson, T. O., and R. H. Burris. 1969. Nitrogen fixation by *Rhodospirillum rubrum* grown in nitrogen limited continuous culture. J. Bacteriol. 97:1093–1097.

Neilson, A. H., and S. Nordlund. 1975. Regulation of nitrogenase synthesis in intact cells of *Rhodospirillum rubrum*: Inactivation of nitrogen fixation by ammonia, L-glutamine and L-asparagine. J. Gen. Microbiol. 91:53–62.

Nordlund, S., J. Erickson, and H. Baltescheffsky. 1977. Necessity of a membrane component for nitrogenase activity in *Rhodospirillum rubrum*. Biochim. Biophys. Acta 462:187–195.

Schick, H. J. 1971. Substrate and light dependent fixation of molecular nitrogen in *Rhodospirillum rubrum*. Arch. Microbiol. 75:89–101.

Schnieder, K. C., C. Bradbeer, R. N. Singh, L. C. Wang, P. W. Wilson, and R. H. Burris. 1961. Nitrogen fixation by cell-free preparations from microorganisms. Proc. Natl. Acad. Sci. U.S.A. 46:726–733.

Shah, V. K., and W. J. Brill. 1977. Isolation of an iron-molybdenum cofactor from nitrogenase. Proc. Natl. Acad. Sci. U.S.A. 74:3249–2353.

Smith, R. V., and M. C. W. Evans. 1971. Nitrogenase activity in cell-free extracts of the blue-green alga, *Anabaena cylindrica*. J. Bacteriol. 105:913–917.

Stadtman, E. R., M. Brown, A. Segal, W. A. Anderson, B. Hennig, A. Ginsburg, and J. H. Mangum. 1972. Regulation of glutamine synthetase. In: O. Wieland (ed.), Metabolic Interconversion of Enzymes, pp. 231–244. Springer-Verlag, New York.

Tso, M. Y., and R. H. Burris. 1973. The binding of ATP and ADP by nitrogenase components from *Clostridium pasteurianum*. Biochim. Biophys. Acta 309:263–270.

Tubb, R. S., and J. R. Postgate. 1973. Control of nitrogenase synthesis in *Klebsiella pneumoniae*. J. Gen. Microbiol. 79:103–117.

Wall, J. D., Weaver, P. F., and H. Gest. 1975. Genetic transfer of nitrogenase-hydrogenase activity in *Rhodopseudomonas capsulata*. Nature 258:630–631.

Walker, G. A., and L. E. Mortenson. 1974. Effect of magnesium ATP on the accessibility of the iron of clostridial azoferredoxin, a component of nitrogenase. Biochemistry 13:2382–2388.

Weare, N. M., and K. T. Shanmugam. 1976. Photoproduction of ammonium ion from N_2 in *Rhodospirillum rubrum*. Arch. Microbiol. 110:207–213.

Yuki, H., C. Sempuka, M. Park, and K. Takiura. 1972. Fluorometric determination of adenine and its derivatives by reaction with glyoxal hydrate trimer. Anal. Biochem. 46:123–126.

Section III
Nitrogenase and Cofactors

Nitrogen Fixation, Volume I
Edited by W. E. Newton and W. H. Orme-Johnson
Copyright 1980   University Park Press   Baltimore

Redox Properties of
Oxidized MoFe Protein

G. D. Watt, A. Burns, and S. Lough

The MoFe protein isolated from a number of nitrogen-fixing organisms (Zumft and Mortenson, 1975) in the presence of excess $S_2O_4^{2-}$ yields a low temperature EPR spectrum with *g* values near 4.3, 3.6, and 2.01. However, under nitrogen-reducing conditions in the presence of MgATP and Fe protein, the MoFe protein undergoes a transition to an EPR-silent state and is thought to have undergone further reduction to become "super-reduced" (Orme-Johnson et al., 1972; Smith, Lowe, and Bray, 1972; Zumft, Palmer, and Mortenson, 1973). An EPR-silent state of MoFe has also been observed (Palmer et al., 1972) by oxidizing the MoFe protein with various dye oxidants. Orme-Johnson et al. (1977) have discussed both the super-reduced and dye-oxidized forms of the MoFe protein and have summarized the available information concerning these two redox states. Only the oxidized form of the MoFe protein can be treated thermodynamically, because the super-reduced protein is presumably the substrate-reducing species and once formed should spontaneously undergo oxidation by substrate. So far, it has only been identified as an intermediate during steady-state turnover of the MoFe protein.

Two redox regions of the dye-oxidized MoFe protein are indicated as present (Orme-Johnson et al., 1977), but the reported $E^{\circ\prime}$ values and the number of electrons involved in both regions are quite variable. $E^{\circ\prime}$ values as positive as -33 mV and as negative as -450 mV and reduction equivalents ranging from two to 12 electrons per mole have been reported. We sought to determine what redox states exist in the dye-oxidized MoFe protein, their spectroscopic properties, their redox potentials, and the number of electrons involved. This information should give insight into the identity and properties of the metal cluster types in the MoFe protein.

In this article, we describe several redox states of the MoFe protein produced by dye oxidation under various conditions. Also included are some of the spectral properties that have been measured.

EXPERIMENTAL PROCEDURES

Nitrogenase Complex

Nitrogenase complex from *Azotobacter vinelandii* was prepared by the Bulen and Lecomte (1972) method. The dye-oxidized form was prepared by a procedure similar to that described below for dye-oxidation of the MoFe protein.

MoFe Protein

MoFe protein in the presence of excess $Na_2S_2O_4$ was prepared by a minor modification of the Shah and Brill (1973) method. Oxidized MoFe protein was prepared by adding degassed oxidant to the MoFe protein until a detectable excess of oxidant persisted for about 10–30 min. The oxidants used were: methylene blue, thionine, 2,6-dichlorophenolindophenol, and for some experiments air. After the protein had been oxidized by the dyes, it was placed on an anaerobic P-2 gel column and eluted with anaerobic buffer. The excess dye and other oxidation products remained near the top of the column and the oxidized protein was collected anaerobically as it emerged from the bottom of the column. Specific activities were measured and protein and metal analyses were performed on the dye-oxidized, gel-filtered MoFe protein.

Instrumental Methods

The redox state of the dye-oxidized MoFe protein was determined by two methods. The first was a polarographic procedure (Watt and Burns, 1977) that measured the total number of electrons transferred to the MoFe protein by measuring the decrease in concentration of an amperometrically standardized $S_2O_4^{2-}$ solution after reaction with a known amount of oxidized MoFe protein. For this measurement, 2 ml of thoroughly degassed 0.05 M Tris (pH 8.0) containing 0.1 M NaCl was transferred by gas-tight syringe to one compartment of an argon-flushed double compartment polarographic H-Cell (the other compartment contained a saturated calomel electrode (SCE). The dropping mercury electrode (DME) was then placed in the buffer and small crystals of solid $Na_2S_2O_4$ were added (with stirring) to the buffer until the desired concentration ($\sim 1 \times 10^{-4}$ M) of $S_2O_4^{2-}$ was attained as indicated by the DME. The stirrer was turned off and the precise concentration of the $S_2O_4^{2-}$ was recorded. At this point, 0.1-ml samples of the oxidized MoFe protein were added, the solution was stirred briefly, and the $S_2O_4^{2-}$ concentration was remeasured. The change in $S_2O_4^{2-}$ concentration measured the number of electrons transferred from $S_2O_4^{2-}$ to MoFe protein. $S_2O_4^{2-}$ is a two-electron reductant, as shown by

$$S_2O_4^{2-} + 2H_2O = 2SO_3^= + 4H^+ + 2e$$

at pH 8 and above.

The second technique of characterizing the redox state of the oxidized MoFe protein is a controlled potential electrolysis method (Watt and Bulen, 1976). This method is quite selective and measures the number of electrons transferred to the MoFe protein at various applied potentials. The controlled potential reduction experiments consist of bringing a 2-ml solution of buffer and mediator to the desired potential using a three-electrode potentiostat. When the desired potential is reached as evidenced by a steady current flow, a sample of oxidized MoFe protein is added. If electron transfer occurs from the electrode through the mediator into the MoFe protein, a current flow is observed as the protein undergoes reduction. The integration of this current flow over the time of reduction gives the total number of electrons transferred to the MoFe protein at the applied potential. To completely define the redox state of the protein, a number of such reduction measurements are made as a function of applied potential.

Both the polarographic cell and the controlled potential reduction cell are equipped with gas lines that allow them to be flushed with various gases and run under any desired gaseous atmosphere. The reduction experiments reported here were run under argon, nitrogen, hydrogen, 10% carbon monoxide in argon, and acetylene atmospheres.

RESULTS

Oxidation Conditions

The MoFe protein as isolated contains excess $S_2O_4^{2-}$ to protect it during purification. This excess was removed and the protein oxidized rapidly by adding anaerobic oxidants. The excess oxidant and other small products of $S_2O_4^{2-}$ oxidation were then removed by anaerobic chromatography on P-2 or Sephadex G-25 gel columns. By investigating the conditions under which dye oxidation of the MoFe protein occurs, we have identified several variables that influence the outcome of the oxidations. The variables so far identified are: 1) the redox potential of the oxidant; 2) the time interval that excess oxidant and protein are incubated; 3) the NaCl concentration of the mixture during oxidation; and 4) the pH and temperature. By manipulation of these variables we have obtained several redox states of the MoFe protein, and describe below their electrochemical characterization and some of their spectral properties.

MoFe Protein Oxidations

When MoFe protein at pH 8 in 0.1 M NaCl is reacted with excess oxidant, allowed to stand at 25°C under argon for 30 min or longer, and then passed through an anaerobic gel column, oxidized protein results that will accept 12 electrons when reduced by $S_2O_4^{2-}$ or methyl viologen. Figure 1 is a polarographic titration of the oxidized MoFe protein as it undergoes reduction by $S_2O_4^{2-}$.

Controlled potential reduction of this oxidized protein has been previously reported (Watt and Bulen, 1976). Two redox regions were found, one at −530 mV and another at −690 mV (versus SCE), each accepting six electrons. Thus, a total of 12 electrons is required for complete reduction of the dye-oxidized MoFe protein. Oxidation of the MoFe protein by brief (0.5- to 2-min) exposure to air produces oxidized, but essentially inactive, protein that also accepts 12 electrons on reduction. However, the shape of the reduction curve (Figure 2) of this air-inactivated sample is different from that of active oxidized protein, indicating that exposure to air has produced other alterations in the protein besides the removal of 12 electrons.

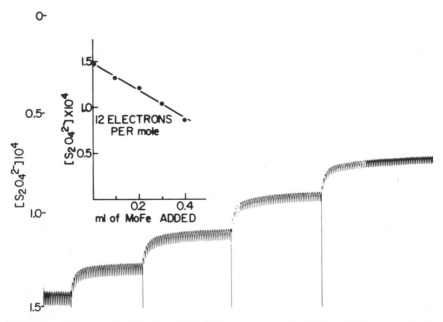

Figure 1.   Polarographic titration of the MoFe protein. The initial $S_2O_4^{2-}$ concentration (lower left of curve) was prepared by adding solid $Na_2S_2O_4$ to 2 ml of anaerobic buffer in the polarographic H-Cell. Then 0.1-ml additions of oxidized MoFe protein were made, followed by brief stirring, and the resulting $S_2O_4^{2-}$ concentration was remeasured. The insert is a plot of $S_2O_4^{2-}$ concentration against ml of MoFe protein added.

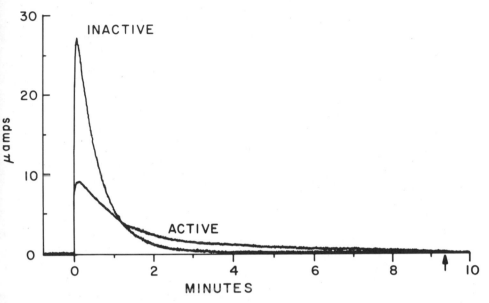

Figure 2.   Controlled potential reduction curves of air-oxidized (inactive) and dye-oxidized (active) MoFe protein. Both samples required 12 electrons for complete reduction, although the rate of reduction was quite different. The curves correspond to ~20 nmoles of reducing equivalents being transferred to the protein samples.

A shorter exposure of the MoFe protein to excess dye oxidant and an increase of NaCl concentration to 0.25 M cause the MoFe protein to be oxidized by only six electrons. This was established by polarographic titration with $S_2O_4^{2-}$. Controlled potential reduction confirmed this result and further established that the total of six electrons was distributed between two redox regions, each accepting three electrons with an n value of·1 and $E^{o\prime}$ values of −550 mV and −710 mV (both relative to the SCE) and clearly separated by a plateau (Figure 3).

The two oxidation states of the MoFe protein that accept a total of 12 and six electrons, respectively, are quite easily and reproducibly prepared by the methods described. Three other oxidation states of the MoFe protein have been prepared, but at the present time the conditions for routine preparation are not fully characterized and consequently success rates are variable.

Figure 4 shows two separately prepared oxidation states of the MoFe protein in which three electrons have been removed from two different redox regions. A comparison of the two curves of Figure 4 with Figure 3 suggests that the curves of Figure 4 are simply the two separately prepared redox regions comprising Figure 3. Indeed, a redox state comparable to the one on the right of Figure 4 can be prepared by selectively reducing the oxidized protein shown in Figure 3 at the plateau near −650 mV.

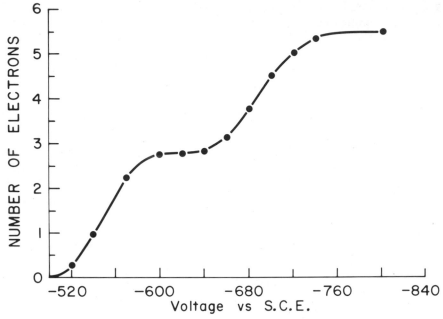

Figure 3.   Controlled potential electrolysis of dye-oxidized MoFe protein. Redox potentials of −550 and −700 mV versus the saturated calomel electrode were determined for redox regions I and II, respectively. N = 1 values were found for each region.

One other redox state has very recently been prepared but only partially characterized. This form of the oxidized MoFe protein can accept a total of nine electrons, as evidenced by both the polarographic and controlled potential reduction techniques. The controlled potential reduction experiment indicates just two redox regions, the more positive region accepting six electrons and the more negative accepting three.

The nitrogenase complex also undergoes oxidation by the oxidants used for the MoFe protein alone. Long exposure times (0.5 to 1 hr) to excess oxidant produce the complex in an oxidized state. Figure 5 indicates the reduction of the complex as a function of applied potential. Two redox regions are observed; the first, at −530 mV, accepts about four electrons with an $n = 1$ value, and the more negative region, at −690 mV, accepts approximately 12 electrons with $n = 2$. Polarographic titrations confirm that 16 electrons are required for total reduction of the oxidized nitrogenase complex by $S_2O_4^{2-}$.

Reduction Reactions in the Presence of Substrates

The MoFe protein oxidized by six electrons shown in Figure 3 was studied by controlled potential reduction and polarographic reduction under various gaseous atmospheres in an attempt to determine if the presence of any of these gases influenced the number of electrons required for reduction or the redox potential. The curves resulting from reduction of the oxidized MoFe protein under N_2, 10% carbon monoxide in argon, acetylene, or hydrogen atmospheres were identical to that shown in Figure 3 obtained under argon. Identical curves were also obtained when the reductions were carried out under argon at pH values 7.0 and 8.0. The MoFe protein oxidized by 12 electrons also showed no variation in redox properties when reduced under these same atmospheres and pH conditions.

We conclude from these measurements that these gaseous substrates are not interacting significantly with the redox centers undergoing reduction. Also, there are no proton-linked redox reactions occurring in the pH range 7 to 8 in these oxidized forms of the MoFe protein.

Spectral Properties of Oxidized MoFe Proteins

The primary emphasis so far has been on characterizing the redox properties of the various redox states of the MoFe protein. Their spectral characterization is still in progress, so only a partial set of data can be presented.

Table 1 lists extinction coefficients at 400 nm and some EPR results for the MoFe protein in several oxidized states. The MoFe protein in the

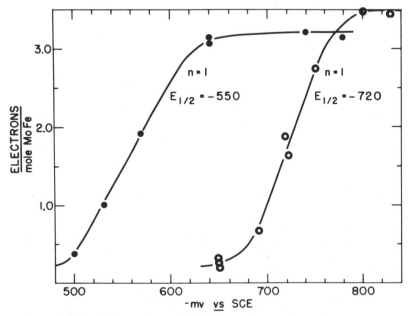

Figure 4. Two separately prepared oxidation states of the MoFe protein. Reducing equivalents were measured by controlled potential reduction measurements.

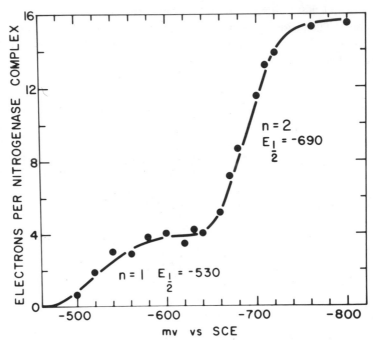

Figure 5. Controlled potential reduction of dye-oxidized nitrogenase complex. A molecular weight of 320,000 was used for the complex, which consisted of approximately 1 MoFe protein, 1 Fe protein, and 1–2 Shethna proteins.

oxidation state shown in Figure 3 and labeled as Ox(3), Ox(3) in Table 1 has no EPR signal. However, on controlled potential reduction in the voltage region −500 to −650 (region I of Figure 3), the typical EPR signal observed for the $S_2O_4^{2-}$-reduced MoFe protein grows in.

Table 2 contains extinction coefficients at 400 nm and the specific activities of the MoFe protein when higher oxidation states occur. These

Table 1.   Spectral properties of the oxidized MoFe protein[a]

I	II	$E_{400} \times 10^4$	EPR
Ox(3)	Ox(3)	7.2	—
RED	Ox(3)	6.7	4.3, 3.6, 2.01
Ox(3)	RED	4.3	N.D.
RED	RED	4.5	4.3, 3.6, 2.01

[a] Headings I and II refer to the redox regions indicated by these symbols in Figure 3. Ox and RED refer to oxidized and reduced states of the protein and the number in parentheses indicates the extent of oxidation in electrons. E_{400} is the molar extinction coefficient of the MoFe protein based on 230,000 daltons. N.D., not determined.

Table 2. Properties of highly oxidized MoFe
protein[a]

I	II	$E_{400} \times 10^4$	Specific activity
Ox(3)	Ox(3)	7.2	1860
Ox(6)	Ox(3)	5.3	(1000)
Ox(6)	Ox(6)	7.6	1450
Ox(6)	Ox(6)	~7	Inactive (air)

[a] The symbols I and II refer to the same redox regions given in Figure 3 except that in the last three entries more extensive oxidation has occurred than that shown in Figure 3, Specific activity is in nmoles hydrogen·min·mg of MoFe protein maximized by excess Fe protein (the specific activity for the number in parentheses was not maximized by excess Fe protein).

higher oxidation states still possess fairly high specific activities, but some loss in activity is usually observed.

DISCUSSION

Orme-Johnson et al. (1977) have recently summarized a number of previously reported redox studies of dye-oxidized MoFe protein and have shown that a wide range of redox potentials has been reported by different workers using different techniques. In part, this range can be understood if the various studies dealt with different redox states of the MoFe protein. In view of our results, which indicate that the conditions of oxidation can influence both the redox region undergoing oxidation and the extent of oxidation, this explanation seems quite likely. However, we have not observed any oxidation states in the MoFe protein that undergo reduction at potentials as positive as the −30 to −70 mV range reported by Albrecht and Evans (1973), Zumft, Palmer, and Mortenson (1973), and Walker and Mortensen (1974). Perhaps in our studies we have not as yet attained the proper conditions for expression of this redox center.

Mortenson, Walker, and Walker (1976) and Walker and Mortenson (1974) have reported that four electrons can be removed from MoFe protein by thionine oxidation, giving an EPR/silent state. Orme-Johnson et al. (1977) have found that a total of six electrons are removed from the MoFe protein by thionine oxidation and, by simultaneously following EPR and Mössbauer spectral changes, have shown that certain spectral features are fully expressed when three electrons are removed from the MoFe protein [see Orme-Johnson et al. (1977) for a detailed discussion]. The results in Figure 3 clearly show that a total of six electrons are removed from the MoFe protein, three each from two separate redox regions. Figure 4 indicates that under certain conditions the MoFe protein can be selectively oxidized by only three electrons, resulting in two distinguishable redox states being formed. In addition, controlled potential reduction of the oxidized MoFe protein at -650 mV yields a MoFe protein oxidized by three electrons. Thus, it appears that stable redox states of the MoFe protein are interrelated by electron transfer in groups of three electrons.

EPR studies of these oxidized states of the MoFe protein are still ongoing, but present data indicate that the usual EPR signal arises when the oxidized MoFe protein shown in Figure 3 is reduced in the voltage range -500 to -650 mV. Thus, the EPR signal is associated with redox region I of Figure 3 with an $E^{\circ\prime}$ value of -540 mV and an n value of 1. Oxidation beyond this EPR-silent state, which will accept six electrons, to states that will accept nine or 12 electrons is accompanied by the appearance of a new low-temperature EPR signal in the $g = 2$ and $g = 4.3$ regions. At the present time, neither the precise g values and their response to reduction at various potentials nor the temperature dependence of this signal have been carefully determined.

The optical spectral data for the various redox states of the MoFe protein show small but significant variation with the degree of oxidation. The molar extinction coefficients at 400 nm found in Table 1 demonstrate this effect. The decrease in absorbance when dye-oxidized MoFe protein undergoes reduction arises from contributions from each redox region in Figure 3. However, Table 1 shows that the most negative redox region (region II) gives rise to the majority of this decrease because the redox state formed with region I reduced and region II oxidized differs only slightly from the fully oxidized protein. Starting with the fully reduced MoFe protein, the optical data in Tables 1 and 2 show an alternating increase, then a decrease, in extinction as sets of three electrons are removed from the MoFe protein.

The reduction experiments of oxidized MoFe protein conducted at pH 7.0 or 8.0 and under various gaseous atmospheres gave no indication that the redox centers undergoing reduction were influenced by the conditions studied. The gaseous substrates studied are either not bound by the isolated MoFe protein or are bound by centers not participating in the redox reactions reported here. Orme-Johnson et al. (1977) have observed EPR signals associated with carbon monoxide binding to nitrogenase components under fixing conditions but not to the separately isolated Fe or MoFe proteins. Two types of CO binding sites were observed, one at low and one at higher CO concentrations. Quite significantly, the appearance of the EPR-detectable CO binding was extremely sensitive to the temperature at which CO and the component proteins were incubated. This may indicate that certain conditions are necessary to develop proper binding of the component proteins and thus form the substrate-inhibitor binding site.

The results shown in Figure 5 demonstrate that the MoFe protein bound in the nitrogenase complex has quite different redox behavior from that measured in its isolated states. The nitrogenase complex used in this study has an approximate Fe protein:MoFe protein:Shethna II protein ratio of 1:1:1–2. We tentatively suggest that the smaller and more positive redox region represents reduction of the Shethna II and Fe proteins, whereas the larger and more negative region corresponds to reduction of MoFe protein. Some redox components of the MoFe protein might also undergo redox reactions in the more positive region of Figure 5. What the exact contribution of each of the component proteins actually is to the redox regions of Figure 5 remains to be elucidated. Further study of both the nitrogenase complex as isolated and the complex resulting from a mixture of the separately isolated and purified Fe and MoFe proteins is clearly of interest and is presently being investigated.

ACKNOWLEDGMENTS

We thank Dr. James Fee of the University of Michigan for initial low-temperature EPR measurements. Drs. W. E. Newton, E. I. Stiefel, J. W. McDonald, and the late W. A. Bulen have provided many helpful and stimulating discussions. This paper represents contribution No. 675 from the Charles F. Kettering Research Laboratory.

REFERENCES

Albrecht, S. L., and M. C. W. Evans. 1973. Measurement of the oxidation reduction potential of the EPR detectable active center of the molybdenum iron protein of chromatium nitrogenase. Biochem. Biophys. Res. Commun. 55:1009–1014.
Bulen, W. A., and J. R. LeComte. 1972. Nitrogenase complex and its components, In: A. San Pietro (ed.), Methods in Enzymology, Vol. 24, Part B, pp. 456–470. Academic Press, Inc., New York.
Mortenson, L. E., M. N. Walker, and G. A. Walker. 1976. Effect of magnesium di- and triphosphates on the structure and electron transport function of the components of clostridial nitrogenase. In: W. E. Newton and C. J. Nyman (eds.), Proceedings of the First International Symposium on Nitrogen Fixation, Vol. 1, pp. 117–149. Washington State University Press, Pullman.
Orme-Johnson, W. H., W. D. Hamilton, T. Ljones, M.-Y. W. Tso, R. H. Burris, V. K. Shah, and W. J. Brill. 1972. Electron paramagnetic resonance of nitrogenase and nitrogenase components from Clostridium pasteurianum W5 and Azotobacter vinelandii OP. Proc. Natl. Acad Sci. U.S.A. 69:3142–3145.
Orme-Johnson, W. H., L. C. Davis, M. T. Henzl, B. A. Averill, N. R. Orme-Johnson, E. Munck, and R. Zimmerman. 1977. Pathways in biological N_2 reduction. In: W. E. Newton, J. R. Postgate, and C. Rodriguez-Barrueco (eds.), Recent Developments in Nitrogen Fixation, pp. 131–178. Academic Press, Inc., New York.
Palmer, G., J. S. Multani, W. C. Cretney, W. G. Zumft, and L. E. Mortenson.

1972. Electron paramagnetic resonance studies on nitrogenase. I. The properties of molybdoferredoxin and azoferredoxin. Arch. Biochem. Biophys. 153:325–332.

Shah, V. K., and W. J. Brill. 1973. Nitrogenase IV. Simple method of purification to homogeneity of nitrogenase components from *Azotobacter vinelandii*. Biochim. Biophys. Acta 305:445–454.

Smith, B. E., D. J. Lowe, and R. C. Bray. 1972. Nitrogenase of *Klebsiella pneumoniae:* Electron-paramagnetic resonance studies on the catalytic mechanism. Biochem. J. 130:641–643.

Walker, M., and L. E. Mortenson. 1974. Evidence for the existence of a fully reduced state of molybdoferredoxin during the functioning of nitrogenase, and the order of electron transfer from reduced ferredoxin. J. Biol. Chem. 249:6356–6358.

Watt, G. D., and W. A. Bulen. 1976. Calorimetric and electrochemical studies on nitrogenase. In: W. E. Newton and C. J. Nyman (eds.), Proceedings of the First International Symposium on Nitrogen Fixation, Vol. 1, pp. 248–256. Washington State University Press, Pullman.

Watt, G. D., and A. Burns. 1977. Kinetics of dithionite ion utilization and ATP hydrolysis for reactions catalyzed by the nitrogenase complex from *Azotobacter vinelandii*. Biochemistry 16:264–270.

Zumft, W. G., and L. E. Mortenson. 1975. The nitrogen-fixing complex of bacteria. Biochim. Biophys. Acta 416:1–52.

Zumft, W. G., G. Palmer, and L. E. Mortenson. 1973. Electron paramagnetic resonance studies on nitrogenase. II. Interaction of adenosine 5'-triphosphate with azoferredoxin. Biochim. Biophys. Acta 292:413–421.

Nitrogen Fixation, Volume I
Edited by W. E. Newton and W. H. Orme-Johnson
Copyright 1980   University Park Press   Baltimore

The Mechanism of Biological Nitrogen Fixation: Transient Complexes in Catalytic Cycles

R. N. F. Thorneley, J. Chatt,
R. R. Eady, D. J. Lowe, M. J. O'Donnell,
J. R. Postgate, R. L. Richards, and B. E. Smith

The theme of this paper is intermediates that have been detected during the catalytic cycles of nitrogenase. Those intermediates that have been identified by EPR spectroscopy are transient forms of nitrogenase with Fe_4S_4 clusters at the -1 and -3 oxidation levels. The responses of the intensities of these EPR signals to changes in substrate, product, and inhibitor concentrations, together with kinetic, redox titration, and Mössbauer data, have allowed us to formulate two schemes that relate these intermediates to each other in catalytic cycles. One purpose of this paper is to discuss these working hypotheses in more general terms than was possible in the originally published format (Lowe, Eady, and Thorneley, 1978) in the hope that others will be stimulated to criticize and develop these ideas.

The second type of intermediate is a nitrogenase-bound dinitrogen hydride that has been detected by quenching the enzyme while it is reducing dinitrogen to ammonia (Thorneley, Eady, and Lowe, 1978). Alternative structures for this intermediate, with implications for the mechanism of biological nitrogen fixation, are discussed with particular reference to the chemistry of well-characterized dinitrogen and dinitrogen-hydride complexes of molybdenum and tungsten.

EPR SIGNALS ASSOCIATED WITH INTERMEDIATES

Early EPR studies on the MoFe and Fe proteins of nitrogenases from various sources identified the two signals associated with these proteins when isolated in the presence of dithionite (signals I_a, I_b, and II in Table 1). Signals I_a and I_b are the high and low pH forms of the $\frac{3}{2}$ spin center of Kp1 (Smith, Lowe, and Bray, 1973) and probably arise from a Fe_4S_4 cluster at the −1 oxidation level (Smith and Lang, 1974), located in the FeMo cofactor (Rawlings et al., 1978). The pK_a for the interconversion of the two forms is about 8.7 for Kp1 (Smith, Lowe, and Bray, 1973) and less than 5.5 for Bp1. Whole cells of *Bacillus polymyxa* give signal I in a low pH form, which suggests either that the intracellular pH is less than 5 or that the conformation of Bp1 giving rise to the low pH form of signal I can be stabilized by interaction other than with protons. Signal I_c is a new signal that is only observed below 30°K after freeze-quenching enzyme that is reducing acetylene. It is not a simple Mo center because its temperature variation is the same as signals I_a and I_b, which have been shown to be associated with Fe in the MoFe protein. In addition, Mo(V) and Mo(III) EPR spectra should be observable above 80°K (Jarrett, 1957; Lowe, Lynden-Bell, and Bray, 1972).

Signals II_a, II_b, and II_c are the native, MgATP, and MgADP complexed forms of Kp2, respectively. Signal II_a probably arises from a Fe_4S_4 cluster at the −3 oxidation level. Its anisotropic linewidths and low integrated intensity (Figure 1) can be explained by assuming that it is located close to the rapidly relaxing paramagnetic center, which is not observable by EPR (Lowe, 1978). The model predicts that the effect of MgATP is not to change the symmetry at the Fe_4S_4 center, as suggested previously (Orme-Johnson and Davis, 1977; Eady and Smith, 1978), but to induce a gross conformation change in the protein that rotates the Fe_4S_4 center through 50°–60° relative to the line joining it to the second center. The effect of MgADP (Signal II_c) is produced by a smaller rotation. Evidence for such a MgATP- or MgADP-indueed conformation change is provided by the increased reactivities of sulfhydryl groups with 5,5′-dithiobis(2-nitrobenzoate) (Thorneley and Eady, 1973) and of the iron with α,α'-bipyridyl (Walker and Mortenson, 1974). The intensities of signals II_a, II_b, and II_c are, according to this model, a function of the distance (~ 10 Å) between the two centers. Hence, the observation that signal II is essentially bleached in the steady state does not necessarily mean that the Fe_4S_4 center is oxidized. It could mean that the conformation of the reduced Fe protein during turnover (or alone in 0.5 M urea; see Zumft, Palmer, and Mortenson, 1973) positions the two centers closer together than in the isolated protein. Alternatively, Thorneley, Yates, and Lowe (1976) have suggested that, since Ac2 is reduced more quickly by $SO_2^{\cdot-}$ than it is oxidized by Ac1, the high steady-state concentration of $Ac2_{ox}$ (assumed from the low intensity of signal II) is due to MgADP inhibition of reduction. This latter explanation suggests that MgADP is formed by hydrolysis of MgATP at site(s) on the Fe protein after or concurrent with electron transfer to the MoFe protein. Thorneley, Yates, and Lowe (1976) also showed, using stopped-flow spec-

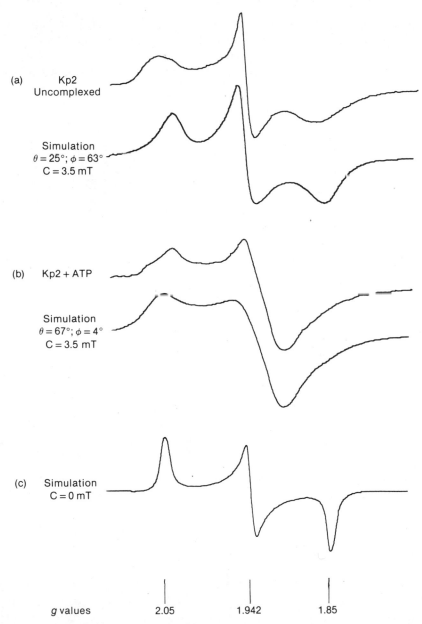

Figure 1. Experimental and simulated EPR spectra of Kp2 protein. (a) Kp2 (uncomplexed), 11 mg/ml in 25 mM Tris/HCl, 50 mM MgCl₂, 1 mM Na₂S₂O₄ (pH 8.1). (b) Same conditions and concentrations as in (a) with addition of 1.9 mM ATP. Spectra were recorded at 18°K at a microwave power and frequency of 20 mW and 9.1 GHz, respectively. The simulations use the interaction parameters C as shown, with θ and ϕ being the azimuthal and equatorial angles specifying the direction from the observed center to the second rapidly relaxing center. The simulation (c) with C = OmT gives the EPR spectrum as it would appear in the absence of interaction (gain 0.5x of that of the other simulations). See Lowe (1978) for further details.

Table 1. EPR signals associated with the nitrogenase proteins of *K. pneumoniae* as isolated in the presence of sodium dithionite and in mixtures of component proteins in the steady state during turnover

Designation	g Values	g_{av}	Protein (^{57}Fe substitution)	4Fe-4S oxidation level	Comments	References
I_a	4.32, 3.63, 2.009	—	Kp1	−1	Protein as isolated, low-pH form; decreased intensity during turnover	Smith, Lowe, and Bray (1973)
I_b	4.27, 3.73, 2.018	—	Kp1	−1	Protein as isolated, high-pH form; decreased intensity during turnover	Smith, Lowe, and Bray (1973)
I_c^a	4.67, 3.37, ~2.0		—	−1	Only observed during turnover under acetylene; intensity enhanced by ethylene	Lowe, Eady, and Thorneley (1978)
II_a	2.053, 1.942, 1.865	1.953	Kp2	−3	Protein as isolated	Smith, Lowe, and Bray (1973)
II_b	2.036, 1.929 (g_m)	1.965	Kp2	−3	MgATP-bound form	Smith, Lowe, and Bray (1973)
II_c	g Values poorly defined, intermediate between II_a and II_b		Kp2	−3	MgADP-bound form	Lowe, Smith, and Yates, unpublished work

					Comments	Reference
III	2.073, 1.969, 1.927	1.990	Kp1	-3?	CO bound tightly; only observed during turnover	Lowe, Eady, and Thorneley (1978)
IV	2.17, 2.06, 2.06	2.10	Kp1	-1	CO bound weakly; only observed during turnover	Lowe, Eady, and Thorneley (1978)
V	2.139, 2.001, 1.977	2.039	Kp1	-1?	Only observed during turnover; may be associated with H_2 evolution	Yates and Lowe (1976); Lowe, Eady, and Thorneley (1978)
VI[a]	2.125, 2.000, 2.000	2.042	Kp1	-1	Ethylene bound form only observed during turnover; decreased by acetylene	Lowe, Eady, and Thorneley (1978)
VII[a]	5.7, 5.4		Kp1	High spin Fe (III)?	Only observed during turnover	Lowe, Eady, and Thorneley (1978)
VIII[a]	2.092, 1.974, 1.933	2.000	Kp1	-3?	Only observed during turnover	Lowe, Eady, and Thorneley (1978)

[a] Not yet reported from other organisms.

trophotometry, that the reduction of $Ac2_{ox}$ protein by $SO_2^{\cdot-}$ proceeded in four phases. The first phase occurred with $k > 10^8$ $M^{-1} \cdot sec^{-1}$ and corresponded to a one-electron equivalent reduction with concomitant development of EPR signal II. If a second center does exist in the Fe protein, then the first phase would require two electrons for full development of signal II, which would not integrate to one or two electrons/mol Ac2. A convenient explanation of these data would be that the $Ac2_{ox}$ protein used by Thorneley, Yates, and Lowe (1976), with a specific activity of about 1600 nmol C_2H_2 reduced/min·mg, was only half active. This would also allow the three slow phases to be associated with damaged or inactive protein.

Signals III and IV arise from the MoFe protein and are observed under turnover conditions in the presence of 10^{-3} and 10^{-1} atm of CO, respectively (Yates and Lowe, 1976; Burris and Orme-Johnson, 1976). Orme-Johnson and Davis (1977) assigned them to Fe_4S_4 cluster(s) at the -3 oxidation level (low CO, signal III) and -1 oxidation level (high CO, signal IV). Neither of these clusters is thought to be part of the iron-molybdenum cofactor (FeMoco), nor to be in rapid (within turnover time) redox equilibrium with the $\frac{3}{2}$ spin cluster, which gives rise to signal I in FeMoco (Orme-Johnson et al., 1978). It was suggested that these two classes of Fe-S cluster occur on parallel electron transfer pathways. The data on which this hypothesis was based are discussed below in the context of alternative models.

Signal V, reported by Yates and Lowe (1976) for Ac and Kp nitrogenase and by Orme-Johnson and Davis (1977) for Av nitrogenase, is only observed during turnover under an atmosphere of argon or N_2. During acetylene reduction, signal V is absent (Yates and Lowe, 1976; Lowe, Eady, and Thorneley, 1978). This signal may be associated with an intermediate in an H_2 evolution cycle (Davis et al., 1978; Lowe, Eady, and Thorneley, 1978). The g_{av} of signal V of 2.039 for Kp, with one g value below 2.000, makes the assignment of an oxidation level to the Fe_4S_4 cluster from which this signal arises somewhat difficult. To overcome this, Lowe, Eady, and Thorneley (1978) provided alternative schemes for an H_2 evolution cycle in which signal V arose from a Fe_4S_4 cluster at either the -3 or -1 oxidation levels. We only discuss the second of these possibilities because we now feel that the -1 oxidation level is the more likely assignment and because it more easily accommodates coupling of the H_2 evolution cycle with the acetylene reduction cycle.

Signal VI is a new axial signal with all of its g values ≥ 2.000. Since it is broadened when [57]Fe-substituted Kp1 is used, it arises from a Fe_4S_4 cluster at the -1 oxidation level in the MoFe protein. It is only seen during turnover in the presence of added ethylene, and therefore arises from an enzyme intermediate with the product ethylene already bound. An apparent binding constant for ethylene of 1.3 mM is calculated from the signal intensity as a function of ethylene concentration. The decrease in signal VI intensity caused by increasing the acetylene concentration at a fixed ethylene concentration gives an apparent binding constant for acetylene of $< 8 \ \mu M$. A second, weak acetylene binding site (k = 13 mM) is detected using signal VIII (see below). [^{13}C]ethylene and [^{13}C]acetylene do not broaden either signal VI or VIII, respectively, thus providing no evidence for direct binding of this product or substrate to a Fe_4S_4 cluster. Similarly, Orme-Johnson and Davis (1977) have shown that signals III and IV are not broadened by ^{13}CO.

Signal VII is a new EPR signal that is observed during turnover. ^{57}Fe substitution shows it to originate in the MoFe protein. Between 12°K and 20°K there is less than a 10% change in intensity after correction for the Boltzmann distribution temperature term. Hence, it does not arise from the excited state of the $\frac{3}{2}$ spin system. Since we do not understand the origin of this signal, we have not attempted to accommodate it in the schemes discussed below. However, it is of interest to note that its g values are typical of high-spin ferric iron.

Signal VIII is very similar to, but not identical with, signal III, the most significant difference being that signal III but not signal VIII is observable at 30°K. ^{57}Fe substitution shows signal VIII to originate from a Fe_4S_4 cluster in the MoFe protein and, by analogy with signal III, we take its oxidation level to be -3 (Lowe, Eady, and Thorneley, 1978).

Schemes 1 and 2 were constructed following the assignment of the EPR signals to the MoFe or Fe proteins. The schemes are consistent with the responses of signal intensities to changes in the ethylene and acetylene concentrations (see Lowe, Eady, and Thorneley, 1978, for details). They utilize two Fe_4S_4 clusters that operate between the -4 and -1 oxidation levels. Before discussing the schemes in detail, it is relevant to consider the redox properties of the isolated MoFe protein. The center, which gives rise to signals I_a and I_b (center X in the schemes) and is located in the FeMoco (Rawlings et al., 1978), becomes further reduced during turnover. On the basis of Mössbauer studies on Kp1, Smith and Lang (1974) concluded that this reduction corresponded to $[Fe_4S_4]^{1-}$ being reduced to $[Fe_4S_4]^{2-}$. This center can also be oxidized, presumably to $[Fe_4S_4]^0$, without loss of activity. However, this redox potential is species dependent (O'Donnell and Smith, 1978), varying from -180 mV for Kp1 protein to 0 mV for Cp1 protein at pH 8, and is therefore very unlikely to be involved in the enzymic reaction and is not included in the schemes.

The Mössbauer studies of Smith and Lang (1974) indicated that all of the remaining iron in Kp1 was in the ferrous state. For this reason, center Y in the dithionite-reduced protein (Scheme I, species a) has been assigned a -4 oxidation level. Further support for this assignment comes from the oxidation of species (a) with ferricyanide, which produces a transient EPR-active species with g values of 2.054, 1.953, and 1.813 (O'Donnell, Smith, and Lowe, unpublished work). These are consistent with center Y being oxidized by ferricyanide to the -3 oxidation level. Fe_4S_4 clusters at the -4 oxidation level have not been detected elsewhere in biological systems and have only been observed transiently in experiments with model cluster complexes (DePamphilis et al., 1974; Cambray et al., 1978).

Equilibrium redox measurements on the MoFe protein do not give rise to EPR signals from $[Fe_4S_4]^{3-}$ centers. However, oxidation of the protein by the dye Lauth's violet changed all of the iron in the molecule (Smith and Lang, 1974). Part of this change is associated with the oxidation of center X (see above), but recent Mössbauer studies (Smith, O'Donnell, Lang, and Spartalian, unpublished work) have detected an additional redox process at -350 mV. In this process, all of the iron except that associated with center X is oxidized to a complex mixture of magnetic species in the Mössbauer spectra. The midpoint potential of this oxidation (-350 mV) is relatively close to the potential at which nitrogenase functions (-450 mV) (Evans and Albrecht, 1974; Watt and Bulen, 1976; Postgate and Elson, unpublished work). It is therefore not surprising that transient oxidation of these iron atoms occurs, under some conditions, during enzyme turnover.

SCHEME 1: HYDROGEN EVOLUTION CYCLE—INTERMEDIATE FORMS OF THE MoFe PROTEIN

This scheme accommodates all of the EPR signals (except signal VII) that are detected in the MoFe protein when nitrogenase is reducing protons to evolve H_2 under an atmosphere of argon. Species (a) is the MoFe protein as isolated in the presence of dithionite ion with center X at the -1 and center Y at the -4 oxidation levels. Species (b), which has centers X and Y at the -2 and -4 oxidation levels, is EPR silent and is concluded to account for up to 90% of the MoFe protein in the steady state because the integrated EPR signal intensities total less than 0.05 electron spin equivalents per mole of Kp1.

The important features of this cycle are summarized below. A conformation change occurs in species (a) that causes a spin state change in center X such that signals I_a and I_b convert to signal V in species (p). This step does not involve electron transfer from the Fe protein and hence we predict that it may be possible to perturb the conformation of center X in the MoFe protein to give signal V without enzymic turnover (i.e., use of mixed solvents, urea, and so on). Center X is reduced by two successive electron transfers from the Fe protein to give species (l) (whether one or two mole equivalents of Fe protein are required is not important in this context). The reduced species (l) is oxidized by protons via the alternative pathways (l) $\rightarrow$ (n) $\rightarrow$ (g) $\rightarrow$ (p) or (l) $\rightarrow$ (m) $\rightarrow$ (p) to evolve H_2. The single binding site

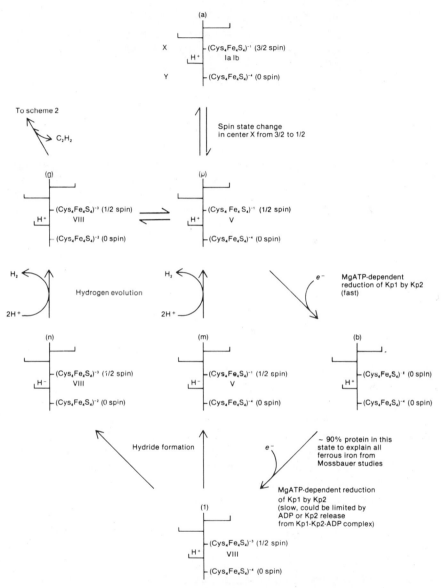

Scheme 1. H_2 evolution.

for the reducible proton has been used in species (a) to account for the protolytic equilibrium between the forms of (a) that give rise to signals I_a and I_b. This cycle is initiated on the addition of Fe protein plus MgATP and evolves H_2 immediately without a significant lag phase. This is not the case for acetylene reduction (scheme 2), which under certain conditions, i.e., low temperature, limiting Fe protein (see Thorneley, Eady, and Lowe, 1978, for details), or with the heterologous nitrogenase Kp1:Cp2 (Smith et al., 1976), only occurs after a lag phase that is considerably longer than the turnover time. The intermediate (g) is present in both schemes 1 and 2, and the slow transition from the H_2 evolution to the acetylene reduction cycle is regarded as a consequence of its low concentration in scheme 1. The initial burst of H_2 evolution is due to the l → m → p → b cycle predominating. The lag for acetylene reduction is accounted for if the l → n → g → p → b cycle is slower. This explanation of the "burst" of hydrogen evolution and "lag" in acetylene reduction (Thorneley, Eady, and Lowe, 1978) is an attractive alternative to the once-only activation model discussed by Smith et al. (1976). It differs from the model of Davis, Shah, and Brill (1975) in that scheme 2 (acetylene reduction) is a more oxidized cycle than scheme 1 (proton reduction). Their model implies that acetylene reduction occurs at a more reduced level than proton reduction.

SCHEME 2: ACETYLENE REDUCTION, ETHYLENE BINDING, AND CO INHIBITION CYCLE

The acetylene reduction cycle is entered via intermediate (g) from cycle 1. Intermediate (g), after binding acetylene, reduces it to give intermediate (h) with ethylene bound. The equilibria between intermediates (h) → (d) → (e) → (f) account for the competitive binding of acetylene and its reduction product ethylene as indicated by the increase of signal IV at the expense of signal VI intensity (Lowe, Eady, and Thorneley, 1978). Because signal VI is not observed in the steady state when ethylene is being produced from acetylene, the release of bound ethylene must be rapid relative to the rate-limiting step in acetylene reduction. Cycle 2 is completed by the reduction of intermediate (f) by electron transfer from the Fe protein. Inhibition by CO is explained by the accumulation of intermediates (j) and (i) at high and low CO concentrations, respectively. The assignment of signal IV to a Fe_4S_4 cluster at the −1 oxidation level at high CO [species (j)] and of signal III to this cluster at the −3 oxidation level [species (i)] is in agreement with the original proposal of Orme-Johnson and Davis (1977). Under conditions of limiting Fe protein or low temperature, which result in lags in acetylene reduction (Thorneley, Eady, and Lowe, 1978), scheme 2 predicts corresponding lags in the appearance of signals III and IV, which are associated with the CO-bound species (i) and (j). The conditions under which Davis et al. (1978) have reported signals III and IV to appear within the turnover time were not those that give significant lags for acetylene reduction in the Kp system. It is interesting to note that below 15°C these signals were reported not to appear. Since low temperature has been shown to induce lags in acetylene reduction, the question arises as to whether or

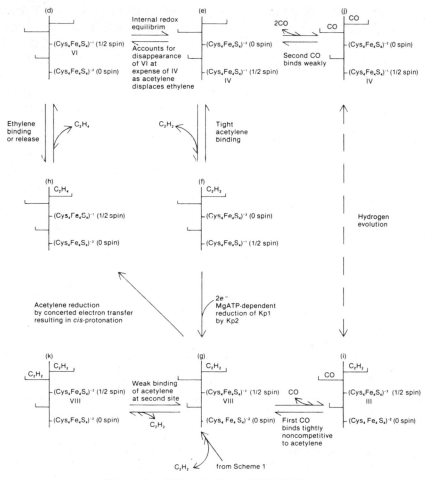

Scheme 2. C_2H_2 reduction and CO inhibition.

not Davis et al. (1978) waited long enough for these signals to develop at 15°C. Because at least 50% of the MoFe protein accumulates in (j) under high CO partial pressures, scheme 2 predicts that this occurs at the expense of the EPR-silent species (b) in cycle 1. This is a consequence of the sum total of EPR-active intermediates being only about 10% of the MoFe protein. This prediction could best be tested by Mössbauer studies under conditions that result in lags in acetylene reduction in the presence of high CO concentrations. A decrease in the intensity of the all-ferrous component of the Mössbauer spectra should be observed.

Since H_2 evolution is not inhibited by CO, a cycle must be present in scheme 2 that reduces protons to H_2. This has not been included for the sake of clarity, but could easily occur between species (i) and (j). This would also explain inhibition of H_2 evolution by acetylene only occurring after the lag phase (Thorneley and Eady, 1977).

We have assumed in schemes 1 and 2 that centers X and Y in all of the intermediates are in redox equilibrium. This contrasts with the suggestion of Orme-Johnson et al. (1978) that the center(s) giving rise to the $\frac{3}{2}$ spin EPR signal (I_a and I_b) are not in redox equilibrium with the centers that give rise to signals III and IV in the presence of CO. They based their model on the failure of dithionite to completely reduce Av1 that had been oxidized with the dye thionine (Mössbauer spectroscopy showed that only the center giving rise to signals I_a and I_b was reduced). However, when a similar experiment was done with Kp1 protein (Smith and Lang, 1974), dithionite slowly reduced both types of iron centers. The concentrations of dithionite, bisulfite impurity, and pH could account for these differences between Kp1 and Av1 because these factors cause the effective redox potential of dithionite to vary by about 100 mV (Mayhew, 1978). In this context, it is interesting to note that oxidized Av1 was fully reduced by methyl viologen.

The final description of the electron transfer reactions between intermediates involving 32 iron atoms, two molybdenum atoms, and various substrates, products, and inhibitors is likely to be extremely complicated. Fully active, tetrameric MoFe protein probably contains two Fe_4S_4 clusters that give rise to signals I_a and I_b (center X). Hence, the above schemes provide a role for 16 of the approximately 32 iron atoms of the MoFe protein. The usefulness of the two schemes lies not only in their ability to correlate a considerable amount of EPR, Mössbauer, and kinetic data, but also in the experimentally testable predictions they contain.

A DINITROGEN HYDRIDE INTERMEDIATE BOUND TO NITROGENASE AND THE RELATED CHEMISTRY OF THE REDUCTION OF DINITROGEN AND DINITROGEN HYDRIDES LIGATING MOLYBDENUM AND TUNGSTEN

The procedure used to detect the dinitrogen hydride intermediate produced by nitrogenase is shown in Figure 2. The technique relies on the high sensitivity and specificity of p-dimethylaminobenzaldehyde (PDMAB) for the detection of hydrazine (N_2H_4) (Watt and Crisp, 1952). Nitrogenase, while it is reducing N_2 to NH_3, is quenched with ethanol containing HCl (1 M) and PDMAB (70 mM). After centrifugation to remove precipitated protein and other assay components, the absorbance of the supernatant at 458 nm is used to calculate the concentration of N_2H_4. The intermediate is detectable in the nitrogenase assay system via the N_2H_4 that it yields at concentrations >0.1 μM. The visible absorption spectrum of the PDMAB derivative of N_2H_4 (Figure 3) is essentially identical to that of the derivative formed with the quenched enzyme solution.

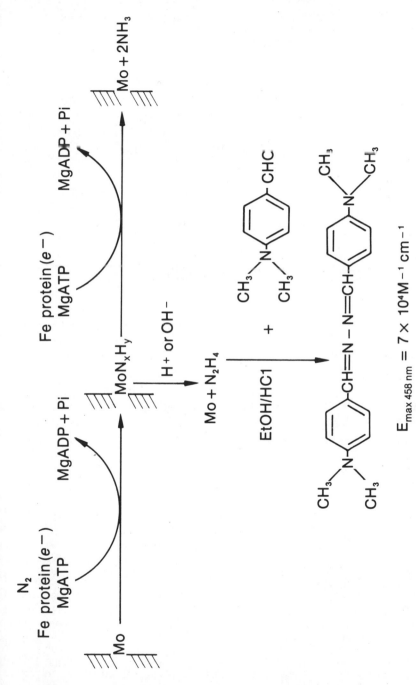

Figure 2.   Schematic representation of the method used to trap the dinitrogen-hydride intermediate of *Klebsiella pneumoniae* nitrogenase.

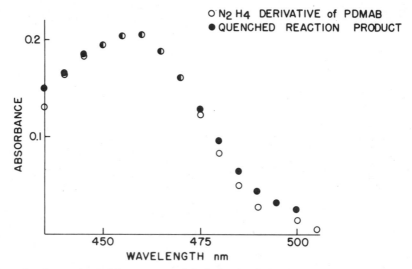

Figure 3.   Comparison of the spectrum of the hydrazine derivative of p-dimethylaminobenzal-dehyde (0.75 μM) with an equivalent concentration of the quenched reaction product.

The correlation between the concentration of intermediate and the rate of NH_3 production is shown in Figure 4. The maximum concentration of intermediate is reached with a time constant of about 3 sec, which is close to the turnover time for the enzyme under these conditions. The concentration of intermediate decreases at long times because of inhibition of nitrogenase activity by MgADP. The effects of some alternative substrates and inhibitors of N_2 reduction on the maximum concentration of N_2H_4 detected are given in Table 2. One-hundred percent corresponds to a concentration of

Table 2.   The effect of alternative reducible substrates and inhibitors on the dinitrogen hydride intermediate

Substrate	Alternative substrate or inhibitor	Quench	Percentage of N_2H_4 detected
N_2 (1.0 atm)	—	Acid	100
N_2 (1.0 atm)	—	Alkali (IN)	100
N_2 (0.08 atm)	—	Acid	69
N_2 (0.86 atm)	C_2H_2 (0.14 atm)	Acid	8
N_2 (0.71 atm)	C_2H_4 (0.29 atm)	Acid	86
N_2 (0.99 atm)	CO (0.01 atm)	Acid	16
N_2 (0.71 atm)	CO (0.29 atm)	Acid	0
N_2 (1.0 atm)	fumarate (24 mM)	Acid	100
N_2 (1.0 atm)	succinate (24 mM)	Acid	100
N_2 (1.0 atm)	allyl alcohol (230 mM)	Acid	100
(under argon)	CN^- (2 mM)	Acid	0
(under argon)	H^+ (pH 7.4)	Acid	0

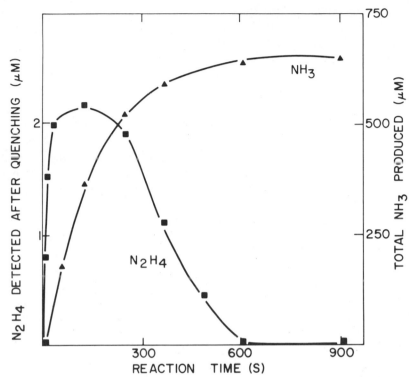

Figure 4.   Time course for concentrations of N_2H_4 (■) and NH_3 (▲) produced during catalytic reduction of N_2 by *K. pneumoniae* nitrogenase (pH 7.4) at 30°C. Assays contained 20 μM Kp1 and 18 μM Kp2. N_2H_4 was determined in assays (0.55-ml volume) quenched with 2 ml of ethanol containing 1 M HCl and 70 mM PDMAB (see Figure 2). NH_3 was determined by the indophenol method of Chaykin (1969).

intermediate equivalent to 10% of the total Kp1 concentration present in the assay. The data, together with Figure 4, show that the concentration of intermediate depends on 1) the degree of saturation of the enzyme with N_2; and 2) the relative concentrations of MgATP and MgADP. The absence of N_2H_4-producing intermediate when CN^- or H^+ is the reducible substrate (Table 2) shows that the intermediate is intimately involved with N_2 reduction. The failure of the intermediate to accumulate in the presence of CO or MgADP shows that the rate of formation of the intermediate must be inhibited at least as strongly as is its rate of further reduction. A detailed mechanistic interpretation of the kinetics of formation of the intermediate is not appropriate at this stage because of uncertainties in the concentrations of active enzyme and the consequent difficulty in the calculation of the turnover time. However, since the maximum concentration of intermediate is only 10% of the total enzyme present, the rate-limiting step in N_2 reduction to NH_3 is unlikely to be the further reduction of this intermediate.

Schrauzer et al. (1976) suggested free diazene (diimide) as the first intermediate in NH_3 formation from N_2 using $[Mo(O)(H_2O)(CN)_4]^{2-}$ with $NaBH_4$ as reductant. This conclusion was based on the observed 50% decrease in the yield of N_2H_4 and NH_3 (measured as N_2 after OBr^- oxidation) when reduction of N_2 was carried out in the presence of fumarate (24 mM). In contrast to this model system, we have shown that neither fumarate nor allyl alcohol decreases the maximum concentration of the intermediate (Table 2) or inhibits NH_3 formation by nitrogenase. This suggests that disproportionation of free diazene is not involved in the enzyme mechanism.

In order to discuss the possible structures of the intermediate and the nature of the active site of N_2 reduction, we now compare the reactions of the intermediate with the chemistry of dinitrogen and dinitrogen hydride complexes.

The reduction of N_2 to NH_3 and N_2H_4 has been shown to occur at early transition metal (titanium, zirconium, molybdenum, and tungsten) sites. In all cases, the coordination of N_2 to the metal as a bridging ligand (titanium and zirconium) and/or as a terminal ligand (zirconium, molybdenum, and tungsten) causes it to become susceptible to electrophilic attack by protons. Electrons are fed from the metal into the dinitrogen as this attack occurs, with the consequent formation of N_2H_4 and/or NH_3. In the case of the binuclear Ti(II) and Zr(II) complexes (Bercaw, 1977), the number of electrons available from the two metal atoms is only four, so that N_2H_4 is the major product. In the case of the mononuclear Mo(0) and W(0) complexes, however, each metal atom can supply the six electrons necessary for the formation of two molecules of NH_3 from an N_2 ligand. Ammonia is usually the major product, but N_2H_4 can also be obtained depending on the reaction conditions (Chatt, Pearman, and Richards, 1977a). Complexes of molybdenum and tungsten that contain N_2 in intermediate stages of reduction have been isolated. We concentrate on their chemistry as a guide to intermediate species that may occur during the enzymic reduction of N_2 to NH_3. Our discussion is based on the reasonable assumption that molybdenum binds dinitrogen at the nitrogenase site (see Smith, 1977).

REDUCTION OF DINITROGEN IN COMPLEXES $[M(N_2)_2(PR_3)_4]$ (A) WHERE M = MOLYBDENUM OR TUNGSTEN, AND PR_3 = PMe_2Ph OR $PMePh_2$

When complexes (A) are treated with an excess of sulfuric acid in methanol, one of the N_2 ligands is liberated and the other is converted to NH_3 with yields of about 1.9 mol NH_3/tungsten atom, or about 0.7 mol/molybdenum atom (Chatt, Pearman, and Richards, 1977a). A variety of other acids may replace sulfuric acid in this reaction, including phosphoric, hydrobromic, and organic acids, but yields of ammonia are generally lowered. For (A: M = tungsten, PR_3 = PMe_2Ph), methanol alone will produce good yields of

NH_3 ($\sim$1.6 mol/tungsten atom) provided that the solution is irradiated or heated under reflux (Chatt, Pearman, and Richards, 1976a). The presence of sodium methoxide decreases the yield of NH_3 slightly from this reaction, but *no* NH_3 is obtained by treatment of complexes (A) with an excess of aqueous KOH or NaOH (Chatt, Pearman, and Richards, 1977a). This latter observation is relevant to speculation concerning the nature of the intermediate in the enzyme system, because nitrogenase gives a similar quantity of N_2H_4 whether it is quenched with acid or alkali (Table 2), implying that bound dinitrogen is not the intermediate that is the direct source of hydrazine from the enzyme system.

COMPLEXES OF DINITROGEN HYDRIDE LIGANDS OBTAINED FROM DINITROGEN COMPLEXES

The Diazenido Ligand (—N=NH)⁻

This ligand may be considered to represent the first stage of dinitrogen reduction at a single metal site. It has only been isolated in the complexes *trans*-$[MX(N_2H)(dppe)_2]$ (M = molybdenum or tungsten, X = F⁻, Cl⁻, Br⁻, or I⁻, dppe = $Ph_2PCH_2CH_2PPh_2$) (Chatt, Pearman, and Richards, 1976b). It undergoes a facile, reversible protonation giving the hydrazido(2⁻) complexes that represent a stable stage of reduction (equation 1).

$$trans\text{-}[MX(N_2H)(dppe)_2] + HX = [MX(NNH_2)(dppe)_2]X \qquad (1)$$

Hydrazido(2⁻)(⁝⁝N⁝⁝NH₂⁻²) Complexes

These are prepared as above for the diphosphine complexes or as in equation 2 when monophosphine complexes are involved.

$$cis\text{-}[Mo(N_2)_2(PMe_2Ph)_4] \xrightarrow[\text{HX (excess)}]{\text{MeOH}}$$
$$[MoX_2(NNH_2)(PMe_2Ph)_4] + [PMe_2PhH]X + N_2 \qquad (2)$$

Thus complexes of the general types $[MX(NNH_2)(dppe)_2]Y$ (B) (Y = X⁻, BF_4^-, ClO_4^-, and so on) and $[MX_2(NNH_2)(PMe_2Ph)_3]$ (C) have been prepared (Chatt, Pearman, and Richards, 1975). Whereas the N_2H_2 in complexes (B) undergoes no further reduction at room temperature by treatment with acid, compounds (C) readily react with sulfuric acid in methanol to give essentially the same yields of NH_3 as are obtained from the parent dinitrogen complexes. They also undergo ligand replacement reactions, e.g., equations 3 and 4.

$$[WBr_2(NNH_2)(PMe_2Ph)_3] + CH_3C_5H_4N \rightarrow$$
$$[WBr(NNH_2)(CH_3C_5H_4N)(PMe_2Ph)_3]Br \qquad (3)$$

$$[WI_2(NNH_2)(PMe_2Ph)_3] + 2\ 8\text{-hqH} \rightarrow$$
$$[W(8\text{-hq})(NNH_2)(PMe_2Ph)_3]I\ (D) + 8\text{-hqH}_2I \qquad (4)$$
$$\text{where } 8\text{-hqH} = 8\text{-hydroxyquinoline}$$

The structures of a number of these complexes have been determined (I. A. Hanson, personal communication). We show the relevant parameters for Compound D as an example in Figure 5. This complex shows the rather short nitrogen-nitrogen and tungsten-nitrogen distances [1.37 (1) and 1.795 (11) Å], consistent with the $N—NH_2^{2-}$ ligand acting as essentially a four-electron donor (Chatt, Heath, and Richards, 1974). On treatment with sulfuric acid in methanol or with concentrated alkali, complex D gives N_2H_4 (0.4 mol/tungsten atom) and only trace quantities of NH_3 in contrast to compound (C) (Chatt, Fakley, and Richards, unpublished data). Obviously, the type of nitrogen hydride produced from the $=N—NH_2^{2-}$ ligand depends greatly on its coligands.

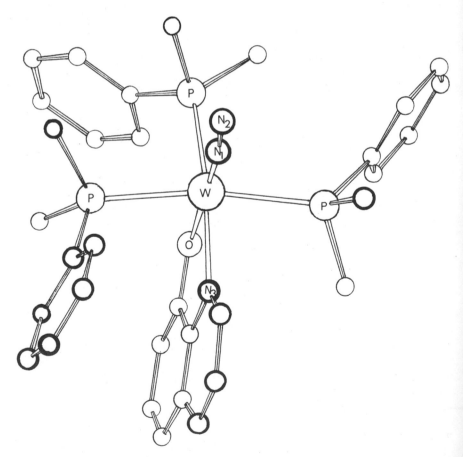

Figure 5. X-ray structure of [W(NNH$_2$)(8-hq)(PMe$_2$Ph)$_3$]I. Tungsten-nitrogen distance, 1.795(11) Å; N$_1$-N$_2$ distance, 1.37(1) Å. Hydrogen atoms not located.

Hydrazido (1–) complexes

These may be prepared from a bis-dinitrogen complex as in equation 5 (Chatt, Pearman, and Richards, 1977b)

$$trans\text{-}[W(N_2)_2(PMePh_2)_4] \xrightarrow[\text{CH}_2\text{Cl}_2]{\text{HCl}}$$
$$[WCl_3(NHNH_2)(PMePh_2)_2](E) + N_2 + 2[PMePh_2H]Cl \quad (5)$$

or by further protonation of complexes (C) and (D) with stoichiometric amounts of HX. Complex (E) gives substantial amount of N_2H_4 and NH_3 in about equal molar quantities on treatment with acid. However, it provides NH_3 (0.76 mol/tungsten atom) and only a trace of N_2H_4 on treatment with base (Chatt, Pearman, and Richards, 1977b).

CONCLUSIONS CONCERNING THE STRUCTURE OF THE DINITROGEN HYDRIDE INTERMEDIATE IN NITROGENASE

The above complexes that contain dinitrogen ligands in intermediate stages of reduction, although not all prepared with the same phosphine or anionic ligands, may be used as a reasonable basis for scheme 3, which shows intermediates in the stepwise reduction of N_2 to NH_3. We have shown that enzyme-bound dinitrogen is unlikely to be the intermediate that yields hydrazine directly after quenching with acid or alkali. The ligand —N≡N—H$^-$ protonates so readily to give the =N—NH$_2{}^{2-}$ ligand that it is unlikely to be present in significant concentration. The =N—NH$_2{}^{2-}$ ligand represents a stable stage of reduction in the artificial metal complex systems, and if it occurs in the natural system it is likely to be a sufficiently long-lived intermediate to be caught by the acid or alkali quench. Its state in the above tungsten complexes, where it gives about equal yields of N_2H_4 whether it is treated with acid or alkali, may well represent its state in nitrogenase.

The next stage of reduction would lead to a hydrazido (1–) (NHNH$_2{}^-$), or a hydrazinimido (1–) ([=N̈—N̈H$_3$]$^-$) ligand. Complex (D), which contains the former, gives no hydrazine on treatment with alkali. The [=N̈—N̈H$_3$]$^-$ ligand would be expected to degrade rapidly according to equation 6.

$$M\overset{\frown}{=}N\overset{\frown}{\cdots}NH_3{}^+ \rightarrow M\equiv N + NH_3 \quad (6)$$

Therefore, these ligands appear to be unlikely candidates for the hydrazine-producing intermediate, if they occur in the enzyme. Nevertheless, reaction 6 does provide a likely mechanism for nitrogen-nitrogen bond cleavage on the route from dinitrogen to ammonia. Thus, of the possible intermediates up to the N_2H_3 stage, chemical evidence favors =N—NH$_2{}^{2-}$ as the most likely to be responsible for hydrazine formation after acid or alkali quench.

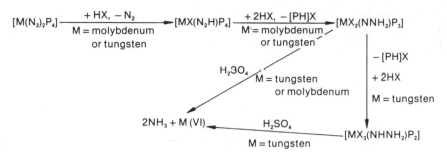

Scheme 3.   Reduction of dinitrogen in molybdenum and tungsten complexes.

However, it is possible that the pertinent intermediate could be enzyme-bound N_2H_4, but the high apparent K_m for N_2H_4 (about 15 mM) (Bulen, 1976) and the failure of N_2H_4 to accumulate in solution at low concentrations suggest that such an N_2H_4 complex is not the intermediate (Thorneley, Eady, and Lowe, 1978).

At the present stage of development, it seems highly probable that the dinitrogen reduction occurs on molybdenum and tungsten as in scheme 3. It is, however, highly improbable that in the enzyme the oxidation state of molybdenum changes by six units, as does that of tungsten in these complexes that give ammonia so efficiently. It is more likely that the molybdenum is in some intermediate oxidation state, capable both of binding N_2 and of transferring electrons into it from the Fe-S centers of the enzyme, leading to the degradation of N_2 in the manner that is outlined in scheme 4.

Other schemes involving the reduction of dinitrogen bridging two metal atoms, or sideways-bonded N_2 on one metal atom, leading to diazene intermediates, have been proposed (Nikonova and Shilov, 1977; Schrauzer, 1977). Obviously, such intermediates could lead to hydrazine as the result of acid or base quenching, and none of our experiments excludes them. On the other hand, the mechanism of reduction of dinitrogen that we propose is the first to offer a detailed sequence of reduction steps through to the unstable

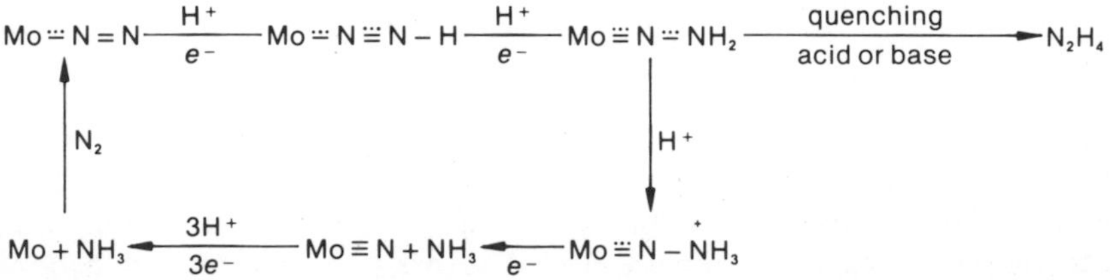

Scheme 4.   Possible reduction cycle of dinitrogen on molybdenum in nitrogenase.

$[-\overset{2-}{N}-\overset{+}{NH_3}]^-$, which should rapidly degrade to give ammonia. This is the mechanism that we prefer at the present time and it accounts neatly for the production of hydrazine by either acid or base quenching of the active enzyme system.

REFERENCES

Bercaw, J. E. 1977. Reduction of molecular nitrogen to hydrazine at titanium and zirconium. In: W. Newton, J. R. Postgate, and C. Rodriguez-Barrueco (eds.), Recent Developments in Nitrogen Fixation, pp. 25–40. Academic Press, Inc., New York.

Bulen, W. A. 1976. Nitrogenase from *Azotobacter vinelandii* and reactions affecting mechanistic interpretations. In: W. E. Newton and C. J. Nyman (eds.), Proceedings of the First International Symposium on Nitrogen Fixation, Vol. 1, pp. 177–186. Washington State University Press, Pullman.

Burris, R. H., and W. H. Orme-Johnson. 1976. Mechanism of biological N_2 fixation. In: W. E. Newton and C. J. Nyman (eds.), Proceedings of the First International Symposium on Nitrogen Fixation, Vol. 1, pp. 208–233. Washington State University Press, Pullman.

Cambray, J., R. W. Lane, A. G. Wedd, R. W. Johnson, and R. H. Holm. 1978. Chemical and electrochemical inter-relationships of the 1-Fe, 2-Fe, and 4-Fe analogues of the active sites of iron-sulfur proteins. Inorg. Chem. 16:2565–2671.

Chatt, J., G. A. Heath, and R. L. Richards. 1974. Diazene-N and hydrazido(2−) complexes. J. Chem. Soc. (Dalton), pp. 2074–2082.

Chatt, J., A. J. Pearman, and R. L. Richards. 1975. Diazenido, hydrazido(2−) and hydrazido(1−) ligands as intermediates in the reduction of ligating dinitrogen to ammonia. J. Organometallic Chem. 101:C45–47.

Chatt, J., A. J. Pearman, and R. L. Richards. 1976a. Relevance of oxygen ligands to reduction of ligating dinitrogen. Nature 259:204.

Chatt, J., A. J. Pearman, and R. L. Richards. 1976b. Diazenido-complexes of molybdenum and tungsten. J. Chem. Soc. (Dalton), pp. 1520–1524.

Chatt, J., A. J. Pearman, and R. L. Richards. 1977a. Conversion of dinitrogen in its molybdenum and tungsten complexes into ammonia and possible relevance to the nitrogenase reaction. J. Chem. Soc. (Dalton), pp. 1852–1860.

Chatt, J., A. J. Pearman, and R. L. Richards. 1977b. Preparation, oxidation and protonation reactions of *trans*-bis(dinitrogen)tetrakis(methyldiphenylphosphine)-tungsten. J. Chem. Soc. (Dalton), pp. 2139–2142.

Chaykin, S. 1969. Assay of nicotinamide deaminase, determination of ammonia by the indophenol reaction. Anal. Biochem. 31:375–382.

Davis, L. C., M. T. Henzl, R. H. Burris, and W. H. Orme-Johnson. 1978. Iron sulfur clusters in the molybdenum-iron protein component of nitrogenase. EPR of the CO-inhibited state. Biochemistry. In press.

Davis, L. C., V. K. Shah, and W. J. Brill. 1975. Nitrogenase VII. Effect of component ratio, ATP and H_2 on the distribution of electrons to alternative substrates. Biochim. Biophys. Acta 403:67–78.

DePamphilis, B. V., B. A. Averill, T. Herskovitz, L. Que, and R. H. Holm. 1974. Synthetic analogs of the active sites of iron sulfur proteins. VI. Spectral and redox characteristics of the tetranuclear clusters $[Fe_4S_4(SR)_4]^{2-}$. J. Amer. Chem. Soc. 96:4159–4169.

Eady, R. R., and B. E. Smith. 1978. Physico-chemical properties of nitrogenase and its components. In: R. W. F. Hardy (ed.), Dinitrogen Fixation, Section 2, pp. 401–490. Wiley Interscience, New York.

Evans, M. C. W., and S. L. Albrecht. 1974. Determination of the applied oxidation-reduction potential required for substrate reduction by chromatium nitrogenase. Biochem. Biophys. Res. Commun. 61:1187–1192.

Jarrett. 1957. J. Chem. Phys. 27:1298–1304.

Lowe, D. J. 1978. Simulation of the electron-paramagnetic-resonance spectrum of the Fe-protein of nitrogenase: A prediction of the existence of a second paramagnetic centre. Biochem. J. 175:955–957.

Lowe, D. J., R. R. Eady, and R. N. F. Thorneley. 1978. Electron-paramagnetic-resonance studies on nitrogenase of *Klebsiella pneumoniae*. Evidence for acetylene- and ethylene-nitrogenase transient complexes. Biochem. J. 173: 277–290.

Lowe, D. J., R. Lynden-Bell, and R. C. Bray. 1972. Biochem. J. 130:239–249.

Mayhew, S. G. 1978. The redox potential of dithionite and SO_2^- from equilibrium reactions with flavodoxins, methyl viologen and hydrogen plus hydrogenase. Eur. J. Biochem. 85:535–547.

Nikonova, L. A., and A. E. Shilov. 1977. Dinitrogen fixation in homogenous protic media. In: W. Newton, J. R. Postgate, and C. Rodriguez-Barrueco (eds.), Recent Developments in Nitrogen Fixation, pp. 41–52, Academic Press, Inc., New York.

O'Donnell, M. J., and B. E. Smith. 1978. The molybdenum-iron protein of nitrogenase. Electron paramagnetic resonance studies on the redox properties between +50 and −450 mV. Biochem. J. 173:831–839.

Orme-Johnson, W. H., and L. C. Davis. 1977. Current topics and problems in the enzymology of nitrogenase. In: W. Lovenberg (ed.), Iron Sulfur Proteins III, pp. 16–58. Academic Press, Inc., New York.

Orme-Johnson, W. H., E. Münck, R. Zimmermann, W. J. Brill, V. K. Shah, J. Rawlings, M. T. Henzl, B. A. Averill, and N. R. Orme-Johnson. 1978. On the metal centres in nitrogenase. In: T. P. Singer and R. N. Ondarza (eds.), Mechanism of Oxidizing Enzymes. Elsevier, New York.

Rawlings, J., V. K. Shah, J. R. Chisnell, W. J. Brill, R. Zimmermann, E. Münck, and W. H. Orme-Johnson. 1978. Novel metal cluster in the iron molybdenum cofactor of nitrogenase. J. Biol. Chem. 253:1001–1004.

Schrauzer, G. N. 1977. Nitrogenase model systems and the mechanism of biological nitrogen reduction: Advances since 1974. In: W. Newton, J. R. Postgate, and C. Rodriguez-Barrueco (eds.), Recent Developments in Nitrogen Fixation, pp. 109–118. Academic Press, Inc., New York.

Schrauzer, G. N., P. R. Robinson, E. L. Moorehead, and T. M. Vickey. 1976. The chemical evolution of a nitrogenase model. XI. Reduction of molecular nitrogen in molybdocyanide systems. J. Amer. Chem. Soc. 98:2815–2820.

Smith, B. E. 1977. The structure and function of nitrogenase: A review of the evidence for the role of molybdenum. J. Less-Common Metals 54:465–475.

Smith, B. E., and G. Lang. 1974. Mössbauer spectroscopy of the nitrogenase proteins from *Klebsiella pneumoniae*. Structural assignments and mechanistic conclusions. Biochem. J. 137:169–180.

Smith, B. E., D. J. Lowe, and R. C. Bray. 1973. Studies by electron paramagnetic resonance on the catalytic mechanism of nitrogenase of *Klebsiella pneumoniae*. Biochem. J. 135:331–341.

Smith, B. E., R. N. F. Thorneley, R. R. Eady, and L. E. Mortenson. 1976. Nitrogenase from *Klebsiella pneumoniae* and *Clostridium pasteurianum*. Kinetic investigations of cross-reactions as a probe of the enzyme mechanism. Biochem. J. 157:439–447.

Thorneley, R. N. F., and R. R. Eady. 1973. Nitrogenase of *Klebsiella pneumoniae:* Evidence for an adenosine triphosphate-induced association of the iron-sulphur protein. Biochem. J. 133:405–409.

Thorneley, R. N. F., and R. R. Eady. 1977. Nitrogenase of *Klebsiella pneumoniae.* Distinction between proton-reducing and acetylene-reducing forms of the enzyme: Effect of temperature and component protein ratio on substrate-reduction kinetics. Biochem. J. 167:457–461.

Thorneley, R. N. F., R. R. Eady, and D. J. Lowe. 1978. Biological nitrogen fixation by way of an enzyme-bound dinitrogen-hydride intermediate. Nature 272: 557–558.

Thorneley, R. N. F., M. G. Yates, and D. J. Lowe. 1976. Nitrogenase of *Azotobacter chroococcum.* Kinetics of reduction of oxidised iron-protein by sodium dithionite. Biochem. J. 155:137–144.

Walker, G. A., and L. E. Mortenson. 1974. Effect of magnesium adenosine 5'-triphosphate on the accessibility of the iron of clostridial azoferredoxin, a component of nitrogenase. Biochemistry 13:2382–2387.

Watt, G. D., and W. A. Bulen. 1976. Calorimetric and electrochemical studies on nitrogenase. In: W. E. Newton and C. J. Nyman (eds.), Proceedings of the First International Symposium on Nitrogen Fixation, Vol. 1, pp. 248–273. Washington State University Press, Pullman.

Watt, G. W., and J. D. Crisp. 1952. A spectrophotometric method for the determination of hydrazine. Anal. Chem. 24:2006–2008.

Yates, M. G., and D. J. Lowe. 1976. Nitrogenase of *Azotobacter chroococcum:* A new electron-paramagnetic-resonance signal associated with a transient species of the Mo-Fe protein during catalysis. FEBS Lett. 72:121–126.

Zumft, W. G., G. Palmer, and L. E. Mortenson. 1973. Biochim. Biophys. Acta 292:413–421.

Nitrogen Fixation, Volume I
Edited by W. E. Newton and W. H. Orme-Johnson
Copyright 1980   University Park Press   Baltimore

Structure and Mechanism of Action of Nitrogenase Active Center

G. I. Likhtenstein

Although considerable experimental data were obtained on nitrogenase following the pioneering work in the 1960s, the structure of the enzyme remains one of the most challenging problems to biochemists and biophysicists. This review is intended to provide an outline of the main features of the nitrogenase active center based on the facts and theoretical considerations resulting from the research at the Laboratory for the Kinetics of Enzyme Action over the past 10 years. The study involved complex physicochemical approaches, including theoretical, kinetic, and thermodynamic analyses and traditional biochemical and chemical techniques, as well as the most recent methods of physical labeling. The basic idea underlying these methods is the chemical modification of selected functional groups of the enzyme by special compounds (labels) whose properties allow monitoring of the state of the biological matrices by physicochemical methods (Likhtenstein, 1976, 1979; Kulikov and Likhtenstein, 1977). Several variations of spin and electron-scattering labeling were first introduced in attacking the nitrogenase activity center problem.

PROPERTIES OF THE NITROGENASE PREPARATIONS

Nitrogenase was isolated from cell-free preparations of *Azotobacter vinelandii*. The purification procedure included: heating (10 min, 55°C); fractionation with protamine sulfate; chromatography on Sephadex G-150; and fractionation with protamine sulfate or chromatography on Biogel A 1.5. The specific activity of the final preparation, designated as nitrogenase complex I, was 400–450 nmol of C_2H_2/min/mg of protein. The activity may be increased by a factor of 1.5–2.0 through the addition of Fe protein. Thus, the nitrogenase complex I preparation involves an admixture containing MoFe protein. The procedure of obtaining nitrogenase with maximum activity (nitrogenase complex II) consists of: fractionation with protamine sulfate; chromatography on Whatman cellulose 32-DE with elution by a linear concentration gradient of $MgCl_2$ (the nitrogenase is eluted by 0.09 M $MgCl_2$); and chromatography on Sephadex G-150. The resulting nitrogenase preparation has a specific activity of about 1000 nmol of C_2H_2/min/mg of protein, a sedimentation coefficient 14.5 ± 5 S and shows homogeneity in polyacrylamide gel electrophoresis. The method of separating Fe protein and MoFe protein relies on chromatography on Whatman cellulose DE-52. The optimum specific activities of Fe protein and MoFe protein were 2000 and 1400 nmol C_2H_2/min/mg of protein, respectively.

SOME KINETIC PECULIARITIES

The method for determining the electron transport sequence and the functional role of each of the nitrogenase components, Fe protein and MoFe protein, was based on the treatment of the individual components by specific inhibitors of a definite function of the enzyme followed by recombination to give the nitrogenase complex I (Syrtsova et al., 1971; Levchenko et al., 1977).

The experiments show that nitrogenase activity was suppressed if MoFe protein (but not Fe protein) was incubated with CO, an inhibitor of the dinitrogen binding site. The ATPase activity of the reconstituted complex disappeared after incubation of Fe protein (but not MoFe protein) with reagents specific for free SH— groups, e.g., *para*-chloromercury benzoate (PCMB), iodoacetamide or a polymercury compound, the so-called Hoffman reagent.

On the basis of such studies, it was assumed that electrons transfer from the Fe component to the MoFe component through ATP hydrolysis and the reduction of N_2 is accomplished on MoFe component (Syrtsova et al., 1971). Conclusive evidence for the transfer was forthcoming from a series of remarkable ESR experiments (see, for example, the review by Orme-Johnson et al., 1977).

For elucidating the question of whether the binding of substrates to nitrogenase requires an oxidative or reductive state of the enzyme, a photoaffinity labeling approach has been introduced (Syrtsova et al., 1977). Azidodithiocarbonate (ATC), which serves as a noncompetitive inhibitor with $K_i = 5 \cdot 10^{-3}$ M (see Figure 1) and undergoes the following chemical

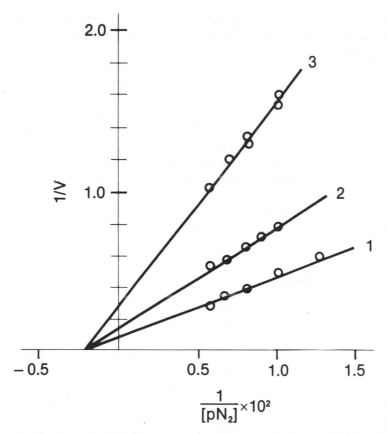

Figure 1. The effect of azidodithiocarbonate (ATC) concentration on the nitrogen fixation activity of nitrogenase complex II. A plot of the reciprocal of the N_2 concentration (pN_2, mm partial pressure) against the reciprocal of the specific nitrogen fixation activity (V, relative units) at different levels of ATC: (1) none; (2) $5 \cdot 10^{-3}$ M; and (3) 10^{-2} M.

transformation on illumination,

$$N_3 - C \overset{S}{\underset{S^-}{\diagdown}} \quad \xrightarrow{h\nu} \quad N_2 + :N - C \overset{S}{\underset{S^-}{\diagdown}}$$

(I)

was used as the affinity label for the nitrogenase active center. Because of its extremely high reactivity, the nitrene, I, binds to the surrounding matrices in close proximity on illumination; to the active center, in this case. The nitrogenase complex II preparation, without ATP, Mg^{2+}, and reductant, was illuminated with light of λ_{max} = 313 nm in the presence of $5 \cdot 10^{-4}$ M ATC. The activity of this illuminated preparation was suppressed to a greater extent than control preparations either in the dark or with illumination in the absence of ATC. In this way, it was found that ATC can interact with the native oxidized form of nitrogenase without preliminary reduction and that this interaction shows no requirement for ATP and Mg^{2+}.

If the kinetic behavior of nitrogenase-catalyzed reactions is analyzed on a basis that allows the interaction of some substrates and inhibitors with the oxidized form of nitrogenase but limits the N_2 interaction to the reduced form, then the existence in nitrogenase of a single active site, binding N_2, inhibitors, and electron interceptors may be assumed (Syrtsova et al., 1977; Likhtenstein et al., 1978). Using this assumption, the majority of the apparently inconsistent data on inhibition may be correlated without the conception of a plurality of binding centers in nitrogenase.

STRUCTURE OF THE IRON CLUSTER SYSTEM

The first evidence for the existence of nonheme iron-sulfur centers in nitrogenase (up to 30 atoms of iron in each macromolecule) came to light in early biochemical work. Several methods were proposed for solving the problem of the relative arrangement of the iron atoms. Do the atoms make up a single multinuclear complex (cluster) or are they distributed over the entire macromolecule (Syrtsova et al., 1971)?

One of the methods includes chemical modification of the cysteine residues that complex these iron atoms in such proteins, and makes use of a specific reagent (a mercurial) that contains a nitroxide radical. The spin label dislodges the iron from the complex and replaces it. If the iron atoms form a single multinuclear complex, the spin labels, now covalently bound to the cysteine residues, must be close to one another. In fact, the introduction of the spin labels into the iron centers of nitrogenase complex I and its components (as well as those of pea ferredoxin and of a model nonheme protein), under optical and ESR control, produces a singlet that is indicative of an intense exchange interaction between closely arranged spins. The interaction disappears on unfolding the proteins with urea. A similar approach employing a spin label based on an isocyanide has been used to study the cluster arrangement of the surface iron atoms in nitrogenase (Kulikov et al., 1975).

The fact that the nonheme iron in nitrogenase is bound to the protein by means of cysteine residues makes it possible to use the electron microscope (Hitachi Hu-125) to determine the location of the iron, since the specific reagent for cysteine groups, PCMB, contains the heavy mercury atom (Levchenko et al., 1973; Levchenko et al., 1977). After determining

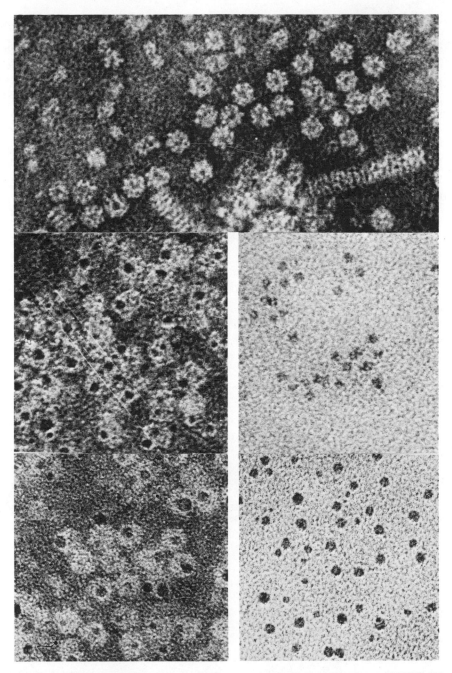

Figure 2.   Electron photomicrograph of nitrogenase complex I (magnification × 700,000). (A) N₂ase stained by 2% phosphoric-tungstic acid (PTA). (B) N₂ase modified by 100 equivalents of PCMB: (i) negative staining by 1% Na₂MoO₄; (ii) without staining. (C) N₂ase modified successively by two equivalents of HR and then by 100 equivalents of PCMP: (i) negative staining by 0.2% PTA; (ii) without staining.

the location of the cysteine groups on the nitrogenase macromolecule from the position of the mercury replacing the iron, the original location of the nonheme iron can be revealed. The resultant distinct electron photomicrographs of nitrogenase complex I and MoFe protein display the regular grouping of the mercury granules into clusters having the shape of tetrahedrons (Figure 2). The size of the granules is 6 ± 1 Å and the distance between them is about 8 Å. According to the dimensions of the granules, we may assume that each corner of the clusters is formed by a minimum of four mercury atoms. Thus, the Fe_4S_4 clusters seem to be gathered together in assemblies.

"ATPase" PORTION OF THE ACTIVE CENTER

It is known that Mg^{2+} can be replaced in the nitrogenase reactions by the paramagnetic ion, Mn^{2+}. So, Mn^{2+} may serve as a spin label in physical experiments. The interaction of nitrogenase complexes I and II and the components with ATP, ADP, and Mn^{2+} has been investigated by means of NMR measurements of the rate of longitudinal relaxation of the water protons $(1/T_1)$ (Syrtsova et al., 1972; 1977). The formation of ternary complexes of nitrogenase and of Fe protein with ATP-Mn^{2+} and ADP-Mn^{2+} was detected by the increase of $1/T_1$. Blocking of the free sulfhydryl groups of the enzyme eliminated the accelerating relaxation effect. All ternary complexes produce different effects on $1/T_1$ and therefore have different structures. Probing the surfaces of oxidized and reduced nitrogenase preparations by measuring $1/T_1$ has not revealed any paramagnetic centers that are accessible to solvent molecules.

A labeling approach was applied to evaluate the distance between the "ATPase" center and the nonheme iron clusters. The approach is based on the assignment of the free SH— groups to the "ATPase" part of the enzyme. In each case, the attachment of the label suppressed the "ATPase" activity, but the inhibited enzyme could be reactivated by 1,4-dithiothreitol.

Addition of the radical (SL I)

to the free SH— groups of nitrogenase produces some alteration in the line shape of the ESR spectrum of the label because of a spin-exchange interaction with Fe_nS_n. This alteration was most evident in the saturation parameters of the spectra (Kulikov et al., 1975; Kulikov and Likhtenstein, 1977). Because the exchange interaction is important only when the distance between spins is less than 6–8 Å, we conclude that the label (as well as ATP) is disposed in the vicinity of the paramagnetic part of the oxidized enzyme. The evidence in favor of a nonheme nature for this part of the enzyme was provided by further experiments. The incorporation of fluorescent derivatives of mercurials or of iodoacetamide into the ATP center was accompanied by intense fluorescence quenching, which, according to theory of inductive resonance energy migration (by Förster), corresponds to a close disposition of iron-sulfur clusters.

Similar results were obtained by affinity labeling of the "ATPase" center using the luminescent etheno-derivatives of ATP (E-ATP) (Gvosdev et al., 1975a). The luminescent residue was covalently attached to the active center of nitrogenase complex I. A greater than fiftyfold quenching in the luminescence of the E-ATP fragment of the MoFe component (obtained after the dissociation of the labeled nitrogenase complex I) was observed (Figure 3). According to the Förster theory, this quenching indicates that the E-ATP fragment and the iron cluster are spaced less than 7 Å apart in MoFe protein and about 30 Å apart in Fe protein.

For a study of the topography of free SH— groups and cysteine residues in the iron cluster, the method of electron scattering labels has been employed (Likhtenstein et al., 1973; 1978; Levchenko et al., 1977). One molecule of the label (Hoffman reagent, HR) contains a group of closely arranged heavy mercury atoms and appears on the electron photomicrograph in the form of an individual point. Treatment of the nitrogenase complex I for modification of free SH— groups by HR and then with PCMP for modification of iron-sulfur cluster results in the formation of closely arranged ensembles of mercury atoms (Figure 2). This result is naturally treated as a consequence of the close arrangement of the free SH— groups and the iron-containing clusters.

For establishing the functional relationships between the "ATPase" center and the electron-carrying chain in nitrogenase, the redox affinity label (SL II)

was used (Gvosdev et al., 1975b). SL II serves as a competitive inhibitor of ATP hydrolysis and therefore may be addressed to the ATPase center. As

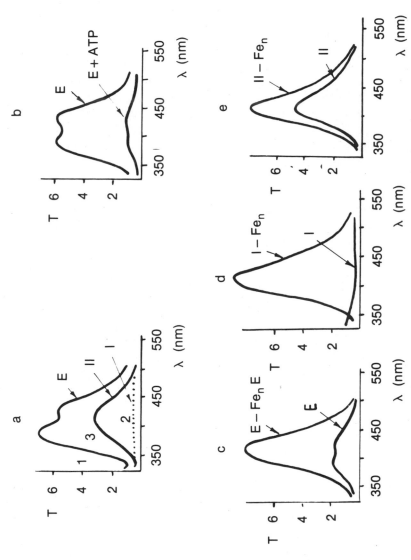

Figure 3.   Fluorescence spectra of nitrogenase complex I and the protein components labeled by $1,N^6$ etheno-ATP. For graph a: 1) labeled nitrogenase complex (E); 2) labeled MoFe protein (I); and 3) labeled Fe protein (II). Graph b shows the effect of added ATP at $4 \cdot 10^{-3}$ M. Graph c shows spectra of E before and after removing iron atoms. Graph d shows effect of removing iron atoms (I-Fe) on spectrum of MoFe protein (I). Graph e shows effect of removing iron atoms (II-Fe) on Fe protein (II) spectrum.

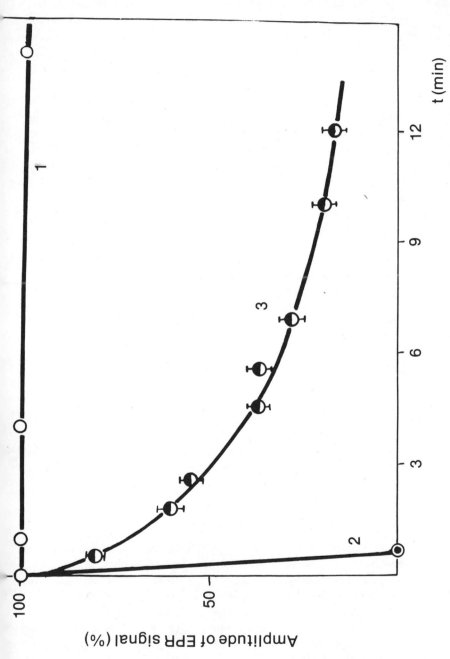

Figure 4.   The kinetics of redox affinity probe reduction by nitrogenase, pH 7, $t = 25°C$: 1) nitroxide radical derivative (free label) (10^{-3} M); 2) radical SL II (10^{-3} M); 3) radical SL II in the presence of ATP (10^{-2} M).

shown in Figure 4, the redox affinity label SL II is reduced by nitrogenase complex I at a rate more than 100-fold faster than the free (i.e., lacking the triphosphate chain) label. The addition of ATP slows down the reduction of SL II. This observation implies that favorable conditions exist in the ATPase center for reduction of the nitroxide group in the label, which replaces the ATP molecule directly from the iron-containing cluster.

Proton magnetic resonance has been used to determine the relative positions of the ATPase and dinitrogen binding centers and to establish the magnetic state of the iron or molybdenum atoms in nitrogenase (Syrtsova et al., 1974; Likhtenstein, 1976). The specific substrate analog was acrylonitrile, which inhibits the reduction of N_2 and is simultaneously reduced at its nitrile group, apparently in the dinitrogen binding center ($K_m \sim 10^{-2}$ M). The experiment involved measurement of the dipole-dipole paramagnetic contribution to the broadening ($\Delta H_{1/2}$) of the PMR lines of acrylonitrile as it enters the active center of the enzyme. Application of the equation

$$r = a(f_m/\Delta H_{1/2})^{1/6}$$

(where a is a coefficient depending on the properties of the paramagnetic sites, and f_m is the fraction of bound substrate) makes it possible to estimate the distance (r) between substrate protons and various electron spin sites on nitrogenase. We find that: 1) from the intrinsic paramagnetic center in the oxidized and reduced state of the enzyme, $r > 6$ Å; 2) from a spin-label (SL I) on the free SH— groups of the ATPase center, $r > 11$ Å; 3) from a cluster of isonitrile spin labels that block the ATPase activity by binding to the iron atoms, $r > 7$ Å; and 4) from the ternary complexes E-Mn^{2+}-ATP and E-MN^{2+}-ADP, $r > 10$ Å.

According to the foregoing data, all of the paramagnetic centers in native and spin-labeled preparations of nitrogenase, including the multinuclear iron-containing complex, are separated from the acrylonitrile protons by distances greater than 6 Å. These data suggest that the ATPase center and the nitrogen binding segment are not in direct contact with each other.

MECHANISM OF THE DINITROGEN REDUCTION

As suggested by Likhtenstein and Shilov (1970), the reduction of dinitrogen, when effected by a reagent with a redox potential close to that for dihydrogen evolution, may be performed in the mild conditions only by multielectron (four or six) concerted mechanisms. One- or two-electron mechanisms, with formation of N_2H or N_2H_2, require overcoming very high thermodynamic barriers (73 Kcal/mol and 48 Kcal/mol, respectively) and appear not to be realizable under these conditions. According to a recent quantum-mechanical calculation by A. F. Schestakov (unpublished data), the stabilization energy of diimide in the coordinate sphere of transition

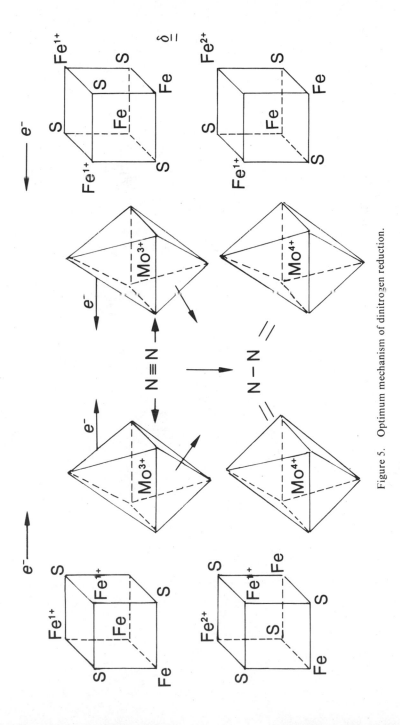

Figure 5.   Optimum mechanism of dinitrogen reduction.

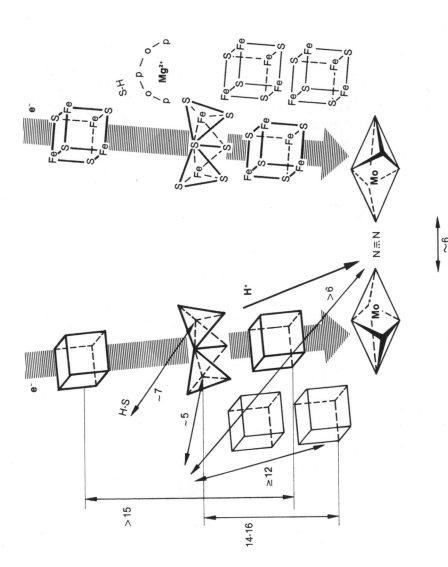

Figure 6.   Proposed model of nitrogenase active center. Distances are given in Å.

metals (requiring 40–50 Kcal/mole to compensate for this barrier) can be attained only at the expense of one or two additional electron transfers from the metal orbitals to the antibonding diimide orbitals.

It could be imagined that simultaneous changes of valence of the metals forming a binuclear dinitrogen complex in nitrogenase (more probably molybdenum) take place, for example, with the formation of nitride derivatives of the metals. If so, the essential rearrangement of ligand positions may be expected. In the present case, at least 14 nuclei must be reoriented. As was demonstrated (Likhtenstein, 1976), such reorientation in the course of the single elementary act has the extremely low value of synchronization probability of 10⁻ᶜ. From the viewpoint of the developing concept of optimum motion (Likhtenstein, 1976; 1979; Likhtenstein and Mullocandov, 1977), the four-electron mechanism of N_2 reduction is much more preferable. The electronic compensation of transfer from adjacent iron clusters can provide the variability of the ligand positions of the binuclear complex (Figure 5). The binuclear nature of the catalytic complex, the requirement of adjacent clusters, and hydrazine formation as the intermediate product have been demonstrated for chemical model systems (see, for example, the review by Nikonova and Shilov, 1977).

CONCLUSION

As a result of the analysis of the data presented above and of the existing literature, the following model of nitrogenase action has been suggested (Figure 6). The electrons from one-electron donors (ferredoxin, flavodoxin) are accepted by the Fe component and then are transferred, by virtue of the energy-yielding ATP hydrolysis, to the MoFe component. In so doing, the higher reduction potential of the MoFe component is generated. The coupled ATPase-electron-transferring site is situated on the boundary between the two components near to assemblies of iron clusters. Electrons enter the electron-carrying chain of the MoFe protein and proceed stepwise through iron cluster carriers to the binuclear molybdenum center. The molybdenum atoms are reduced to the +3 valence state. The terminal step of the process, reduction of dinitrogen, seems to occur as a four-electron concerted attack with two electrons transferring from the molybdenum atoms and two electrons from the iron clusters (Figure 5). In this manner, the multinuclear transition metal system in nitrogenase serves as a "switching over device" converting a one-electron process to a four-electron one.

ACKNOWLEDGMENTS

I express my sincerest gratitude to my colleagues from the Laboratory for the Kinetics of Enzyme Action, Doctors R. I. Gvosdev, L. A. Syrtsova, L. A.

Levchenko, A. P. Sadkov, T. N. Pysarskaya, A. V. Kulikov, and A. I. Kotelnikov, for their collaboration, and to Professor A. E. Shilov for his generous help.

REFERENCES

Gvosdev, R. I., A. I. Kotelnikov, A. P. Pivovarov, A. P. Sadkov, and A. A. Kost. 1975a. Use of the affinity luminescent labeling for the study of nitrogenase structure. Bioorganichescaya Khim. 1:1207–1214.

Gvosdev, R. I., A. I. Kotelnikov, A. P. Sadkov, and G. I. Likhtenstein. 1975b. The study of nitrogenase from *Azotobacter vinelandii* by means of a redox affinity probe. Dokl. Acad. Nauk SSSR 220:1453–1457.

Kulikov, A. V., L. A. Syrtsova, G. I. Likhtenstein, and T. N. Pysarskaya. 1975. Study of nitrogenase by means of spin-labeling. Molec. Biol. (Russian) 9:203–212.

Kulikov, A. V., and G. I. Likhtenstein. 1977. The use of spin relaxation phenomena in the investigation of the structure of model and biological systems by the method of spin labels. Adv. Molec. Relaxation Interaction Processes 10:47–79.

Levchenko, L. A., A. V. Raevsky, A. P. Sadkov, and G. I. Likhtenstein. 1973. Study of iron-clusters in nitrogenase by electron density labels. Dokl. Akad. Nauk SSSR 211:238–242.

Levchenko, L. A., A. V. Raevsky, G. I. Likhtenstein, A. P. Sadkov, and T. S. Pivovarova. 1977. Study of the topography of the nitrogenase active center by the electron microscopy method with the use of the electron density labels. Biokhimiya 42:1755–1764.

Likhtenstein, G. I. 1976. Spin Labeling Methods in Molecular Biology. Wiley-Interscience, New York.

Likhtenstein, G. I. 1979. Multi-Nuclei Redox Metalloenzymes. Nauka, Moscow.

Likhtenstein, G. I., and A. E. Shilov. 1970. Thermodynamic and kinetic peculiarities of dinitrogen reductive fixation. Zh. Fizichescoy Khim. 44:849–856.

Likhtenstein, G. I., L. A. Levchenko, A. V. Raevsky, A. P. Sadkov, T. S. Pivovarova, and R. I. Gvosdev. 1973. The application of electron density labels to the study of the nitrogenase active center topography. Dokl. Acad. Nauk SSSR 213:1442–1446.

Likhtenstein, G. I., and E. A. Mullocandov. 1977. About the principle of "optimal motion" in elementary steps of chemical and biochemical reactions. II. Mechanism of dinitrogen reduction. Kinetica i Kataliz 18:883–888.

Likhtenstein, G. I., R. I. Gvosdev, L. A. Levchenko, and L. A. Syrtsova. 1978. Structure and mechanism of action of the nitrogenase active center. Isv. Akad. Nauk SSSR (Ser. Biol.), pp. 1–28.

Nikonova, L. A., and A. E. Shilov. 1977. Dinitrogen fixation in homogeneous protic media. In: W. Newton, J. R. Postgate, and C. Rodriguez-Barrueco (eds.), Recent Developments in Nitrogen Fixation, pp. 41–52. Academic Press, Inc., New York.

Orme-Johnson, W. H., L. C. Davis, M. T. Henzl, B. A. Averill, N. R. Orme-Johnson, E. Münck, and R. Zimmerman. 1977. Components and pathways in biological nitrogen fixation. In: W. Newton, J. R. Postgate, and C. Rodriguez-Barrueco (eds.), Recent Developments in Nitrogen Fixation, pp. 131–178. Academic Press, Inc., New York.

Syrtsova, L. A., L. A. Levchenko, E. N. Frolov, G. I. Likhtenstein, T. N. Pysarskaya, L. V. Vorob'ev, and V. A. Gromoglasova. 1971. Structure and function of the nitrogenase components from *Azotobacter vinelandii*. Molec. Biol. (Russian) 5:726–734.

Syrtsova, L. A., I. I. Nazarova, T. N. Pysarskaya, and W. B. Nazarov. 1972. NMR investigation of ATP binding by nitrogenase from *Azotobacter vinelandii*. Dokl. Acad. Nauk SSSR 206:367–369.

Syrtsova, L. A., G. I. Likhtenstein, T. N. Pysarskaya, W. L. Berdinsky, W. P. Lezina, and A. U. Stepanyantz. 1974. NMR study of nitrogenase topography. Molec. Biol. (Russian) 8:824–831.

Syrtsova, L. A., T. N. Pysarskaya, I. I. Nazarova, A. M. Uzenscaya, and G. I. Likhtenstein. 1977. Coupling of ATPase and substrate binding sites in *Azotobacter vinelandii*. Bioorganichescaya Khim. 3:1251–1260.

Nitrogen Fixation, Volume I
Edited by W. E. Newton and W. H. Orme-Johnson
Copyright 1980   University Park Press   Baltimore

Azotobacter vinelandii Biochemistry: H_2 (D_2)/N_2 Relationships of Nitrogenase and Some Aspects of Iron Metabolism

E. I. Stiefel, B. K. Burgess, S. Wherland, W. E. Newton, J. L. Corbin, and G. D. Watt

Azotobacter vinelandii has been one of the most studied of the nitrogen-fixing species. This paper discusses three aspects of azotobacter biochemistry. First, we address the relationships of $H_2(D_2)$ and N_2 for the pure nitrogenase components. Second, we consider the evidence that *A. vinelandii* excretes siderophores under conditions of low-iron stress. Finally, we discuss the nonheme iron cytochrome $b_{557.5}$ and present evidence that it is in fact a ferritin-cytochrome molecule, and as such is the first bacterial occurrence of this class of molecule. Its remarkable redox behavior suggests intriguing possibilities for its biological role.

$H_2(D_2)/N_2$ RELATIONSHIPS IN NITROGENASE

The dihydrogen reactions of nitrogenase have been extensively studied in vitro (Jackson, Parshall, and Hardy, 1968; Bulen, 1976; Newton et al., 1977; Stiefel et al., 1977) and in vivo (e.g., Schubert and Evans, 1976). Here, we focus on the H_2 inhibition reaction where H_2 (and of course D_2) have been found to be specific inhibitors of N_2 reduction catalyzed by nitrogenase. In the presence of D_2 and N_2, this inhibition manifests itself not only in lowered levels of NH_3 formation, but also in the production of HD in the gas phase. Although the observation of HD was originally attributed to an exchange reaction, recent work from our laboratory (Bulen, 1976; Newton et al., 1977; Stiefel et al., 1977) has indicated that HD production is an electron-requiring process. However, the detailed studies of HD production have all used either crude preparations or the *A. vinelandii* (Av) complex. In view of the importance of the HD production patterns with respect to mechanistic interpretations of nitrogenase action, we present here some of our recent results on the formation of HD under N_2 and D_2 by pure Av nitrogenase component proteins.

The component proteins were prepared by modifications of published procedures developed in our laboratory (Burgess, Jacobs, and Stiefel, 1979) that yield homogeneous proteins with specific activities of 2,500 nmole of e^- pairs/min/mg of Av1 and 2,000 nmole of e^- pairs/min/mg of Av2. These maximum specific activities are determined by varying component ratios over a wide range in the experiments described below.

The experimental design involves two features that distinguish it from earlier work. First, although the ratio of component proteins is the critical variable, the total protein is kept constant at 1 mg/ml, a level well above that where the "dilution effect" becomes significant. Second, all products are measured on the same reaction vessel. Thus, HD and H_2 are measured mass spectrometrically, whereas NH_3 is measured spectrophotometrically by the Chaykin (1969) method after microdiffusion. The data points (representing triplicate determinations) shown in Figure 1 allow one to follow all of the reaction products of a single run as a function of molar component ratio.

The experimental results displayed in Figure 1 reveal that pure component proteins produce significant level of HD under conditions where H_2 evolution and NH_3 production are also occurring. The fact that all curves reach maxima at values of component ratios between 1 and 5 ([Av2]:[Av1]) is consistent with other experiments performed under 100% N_2 or 100% D_2 atmospheres (Burgess et al., 1979). Minimum rates of production for each product occur at the extremes of component ratios and simply reveal the limiting Av2 or Av1 levels present at the two ends of the abscissa. Most significantly, these results establish HD production as a property of the component proteins of nitrogenase under turnover conditions and show clearly that earlier results (Bulen, 1976; Newton et al., 1977) were not artifacts that resulted from working with impure proteins or the nitrogenase complex of *A. vinelandii*.

The results in Figure 1 can be analyzed to yield information complementary to that achieved in the earlier work (Bulen, 1976; Newton et al., 1977). Figure 2 shows the calculations of total electron flow rates of nitrogenase as a function of component protein ratio. The open circles represent a separate experiment using 100% N_2 in the gas phase wherein all electrons go to either production of H_2 ($2e^-/H_2$) or NH_3 ($3e^-/NH_3$), with both being measured in the same reaction vessel. In this experiment, 60% of the total flow goes to NH_3 and 40% goes to H_2 at [Av2]:[Av1] ratios above 1. It

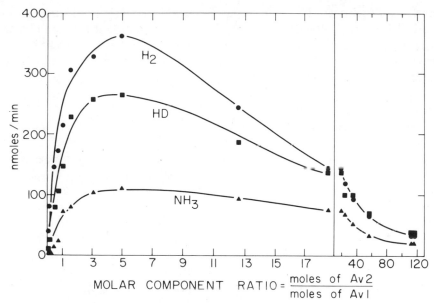

Figure 1.   Formation of NH₃ (▲), HD (■), and H₂ (●) by nitrogenase components under a gas phase of 50% D₂, 40% N₂, and 10% argon. All assays represent triplicate runs and NH₃, HD, and H₂ are each analyzed for each reaction vessel. Product formed is per mg of total protein, which is held constant at 1 mg/ml as the ratio is varied. Molecular weights of 230,000 and 64,000 were assumed for Av1 and Av2, respectively (Burgess et al., 1979).

must be noted that even at 6 atm N₂ the electron flow to H₂ cannot be suppressed (Hadfield and Bulen, 1969), and all experiments to date seem to require the limiting stoichiometry of one mole of H₂ evolved for each mole of N₂ reduced (Rivera-Ortiz and Burris, 1975). For the present purposes, this experiment is important insofar as it sets the level of electron flow through nitrogenase, which other experiments (Watt and Burns, 1977) show to be largely independent of the particular substrate being reduced. Therefore, for the experiments of Figure 1 calculated total electron flow rates should yield a curve similar to that found under 100% N₂. The solid triangles in Figure 2 reveal that, when electrons are only attributed to H₂ and NH₃, the total electron flow appears to be reduced under the D₂-inhibited conditions. However, when one electron is added for each HD produced, then the solid circle curve, wherein total electron flow has remained largely unaltered, is obtained. It is clear that the results with component proteins are fully consistent with the work on Av complex, which showed unequivocally a requirement of one electron for the formation of each HD molecule.

Additional insight into nitrogenase turnover occurs on inspection of the relative electron flows at low [Av2]:[Av1] ratios. At very low values only H₂ evolution occurs, and as the [Av2:Av1] ratio increases first HD and

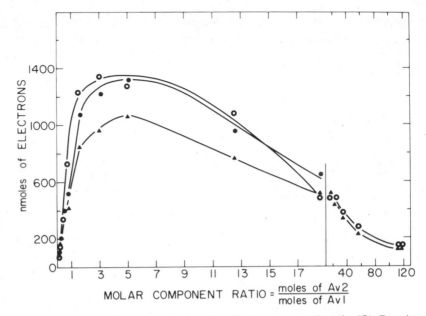

Figure 2.   Total electron flow rates as a function of component protein ratio. (O), Experiment with 100% N_2 gas phase, calculated by attributing two electrons per H_2 and three electrons per NH_3 formed; (▲), experiments with 50% D_2, 40% N_2, 10% argon gas phase and calculated by attributing two electrons per H_2, three electrons per NH_3, but no electrons for HD; (●) experiment with less than 50% D_2, 40% H_2, 10% argon gas phase and calculated by attributing two electrons per H_2, three electrons per NH_3, and one electron per HD.

finally NH_3 production is observed. These results are consistent with earlier studies that indicated that H_2 evolution thrived at low [Av2]:[Av1] levels (Eady et al., 1972) or low ATP levels (Silverstein and Bulen, 1970) compared to N_2 reduction. These observations have been meaningfully interpreted (see, for example, Orme-Johnson et al., 1977) in terms of lower electron flow rates through Av1 being required for the two-electron process of H_2 evolution compared to the six-electron process of N_2 reduction. Clearly, the N_2-dependent HD formation reaction has an electron flow requirement intermediate between that for N_2 reduction and that for H_2 evolution.

The above electron balance studies specify the following stoichiometry for the D_2 inhibition–HD production reaction of nitrogenase:

$$\underbrace{E(N_2 + 2e^-) + 2H^+} + D_2 \to 2HD + E + N_2$$

where E is an active form of nitrogenase. This reaction explains the 1 e^-/HD result and takes into account the need for N_2 and active enzyme in HD production. It is tempting to combine the first two terms in the equation such that an $E(N_2H_2)$ intermediate is specifically responsible for the

HD formation reaction. Then, the reaction of E(N₂H₂) with D₂(or H₂) to form N₂ and 2HD (or 2H₂) becomes a valid stoichiometric interpretation of dihydrogen inhibition of the N₂ fixation reaction of nitrogenase. Our results with component proteins thus confirm and extend our earlier conclusions that HD formation and H₂ inhibition of N₂ fixation are manifestations of the same molecular process, which involves the reaction of an enzyme-bound reduced dinitrogen intermediate. Although the simplest species to which this reactivity may be attributed is a bound diimide-level moiety, there are other possibilities. However, a bound N₂H₄ species seems to be eliminated based on the lack of inhibition of N₂H₄ reduction by D₂ and the lack of HD formation during N₂H₄ reduction (Bulen, 1976; Newton et al., 1977). The relationship between the reduced dinitrogen intermediate postulated in this work and that postulated by Thorneley, Eady, and Lowe (1978), based on identification of N₂H₄ after quenching of an N₂-fixing nitrogenase reaction mixture, remains to be established.

SIDEROPHORES

The high levels of iron in nitrogenase and in other proteins supportive of N₂ fixation suggest that N₂-fixing organisms may have evolved efficient mechanisms for the accumulation, storage, and mobilization of iron. This section summarizes evidence that *A. vinelandii* excretes siderophore-like molecules in response to low-iron stress, and the next section discusses a ferritin-like molecule that is isolated from the bacterium under iron-sufficient N₂ fixing conditions.

Iron-deficient cultures of *A. vinelandii* produce a yellow-green fluorescent peptide (YGFP) (Bulen and LeComte, 1962), which was previously isolated and shown by hydrolysis to consist of a chromophoric unit in addition to nine amino acids, some of which are unusual (Corbin and Fry, unpublished data). The chromophore was found (Corbin, Karle, and Karle, 1970) by x-ray crystallography (of its trimethyl derivative) to be a novel heterocyclic system (I). The nonapeptide is connected through an L-aspartate amino group to the carboxylate function of the chromophore.

(I)

The remaining eight amino acids consist of D-citrulline, D-β-hydroxyaspartate, glycine, D-serine, L-serine, and three L-homoserine residues. Although Japanese workers have reported a sequence (Fukasawa et al., 1972), their assignment is not consistent with data obtained in our laboratory (Corbin and Fry, unpublished data) and we consider the amino acid sequence to be an open question.

A second novel catechol compound, also isolated (Corbin and Bulen, 1969) from iron-deficient cultures of *A. vinelandii*, is N,N'-bis(2,3-dihydroxybenzoyl)-L-lysine (DHBL) (system II), whose structure was confirmed by synthesis.

(II)

The accumulation of I and II in the medium is greatly enhanced by iron deficiency and under such conditions there is a correlation of growth with the detection of the catecholate. The chemical nature of these small molecules and their physiological response to iron makes them strong candidates for siderophores (Neilands, 1977).

Recent results in our laboratory (Corbin, unpublished data) show that both I and II strongly bind Fe(III), based on changes in visible spectra and the observation of typical ferric-siderophore EPR signals. Crude measurements of the stability of the complexes show them to be at least as stable as the Fe(III)-EDTA complex ($K_f = 10^{25}$). Interestingly, both compounds are also found by spectrophotometric titration to bind Mo(VI), added as MoO_4^{2-} (for similar binding see Ketchum and Owens, 1975). It seems likely that these compounds, elaborated under iron-deficient conditions, are involved in the transport of iron. The fact that they strongly bind molybdenum as well, although probably coincidental, nevertheless raises the possibility that catecholate siderophores are also involved in the transport of metals other than iron.

THE FERRITIN-CYTOCHROME MOLECULE

In 1973, Bulen, LeComte, and Lough reported the purification and crystallization from N_2-fixing *A. vinelandii* cells of a molecule given the interim designation "non-heme iron cytochrome $b_{557.5}$ (*Azotobacter vinelandii*)." The molecule was reported to contain from 2.8–3.6 μmole of iron/mg of protein, and the heme was identified as protoheme. The protein is crystallized by addition of Mg^{2+} and purifies in parallel with the nitrogenase complex of *A. vinelandii*. The material appears homogeneous on polyacrylamide and SDS gel electrophoresis. The very large iron content (e.g., Avl contains ~0.1 μmole of iron/mg protein) and the preliminary indications of redox activity for both the heme and nonheme iron have led us to investigate this molecule further. These investigations lead us to conclude that the molecule is a "ferritin-cytochrome" molecule and as such is the first bacterial example of a ferritin-like molecule.

The most striking proof for the ferritin nature of the molecule comes from inspection of electron micrographs of the crystalline preparation fixed with glutaraldehyde, dried with ethanol, and embedded in plastic. The high magnification of the unstained preparation (Figure 3) clearly reveals the presence of ~55 Å–diameter cores within nearly spherical protein shells of diameter ~105 Å (estimated from center to center distance of the cores). These dimensions are virtually identical to those found for mammalian ferritin (Harrison et al., 1974). Similarly, the apparent variable occupancy of the inner cores parallels that found in ferritin. Comparison of the electron

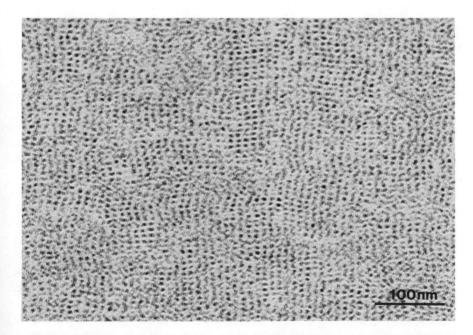

Figure 3.   Electron micrography of crystals of the ferritin-cytochrome. The crystals were fixed with glutaraldehyde, dehydrated with ethanol, and embedded in plastic. No staining was used.

micrographs of horse spleen ferritin (Harrison et al., 1974) and that of the *A. vinelandii* molecule reveals a degree of similarity verging on indistinguishability.

Samples of the Av molecule were studied by Mössbauer spectroscopy of ^{57}Fe at natural abundance by Lang and Spartalian (unpublished data) and reveal a quadrapole-split doublet characteristic of Fe^{3+} at 77°K. As the temperature is lowered to 1.5°K, a six-line magnetically split multiplet appears. The limiting low and high temperature spectra of the Av molecule are virtually identical to those found previously (Boas and Window, 1966) for horse ferritin. Although the details of the temperature dependence are somewhat different, the similar limiting spectra and temperature dependence are nonetheless characteristic of an iron oxide–hydroxide core of particle size compatible with that found in the electron micrographs.

The mode of preparation, the high iron content, the electron micrographs, the Mössbauer spectra, and other evidence (Stiefel and Watt, 1979) lead us to conclude that this molecule represents the first example of a bacterial ferritin and we give it the new tentative name "bacterioferritin-cytochrome $b_{557.5}$ (*Azotobacter vinelandii*)."

The ferritin nature of the molecule being established, we turn now to its cytochrome nature, insofar as it was the heme absorption spectrum that allowed Bulen, LeComte, and Lough (1973) to originally purify the protein. The oxidized spectrum contains a Soret band at 417 nm as the only truly distinct peak. On reduction with $S_2O_4^{2-}$/methyl viologen, a reduced cytochrome spectrum appears with α, β, and Soret bands at 557.5, 525, and 425 nm, respectively (Bulen, LeComte, and Lough, 1973).

To investigate the redox properties of the ferritin cytochrome, coulometric titrations using the technique described by Watt and Bulen (1976) and Watt (1979) were performed. Typically, clean reductions occur (Figure 4) that reveal a reversible Nernst plot (for $n = 1$) with a potential of −416 mV at pH 7, using methyl viologen as the mediator. Coulometry permits the counting of electrons transferred to the redox active centers and, most remarkably, the titration reveals that one electron is taken up for *each* iron atom present in the molecule. Thus, both the heme and the large excess of nonheme iron (roughly 100 iron atoms/heme) are reduced in this process. Significantly, potentiometric titration of the protein, performed by monitoring the appearance of the α band of the reduced heme, reveals that the heme has a redox behavior similar to that of bulk iron.

The ferritin-cytochrome protein is clearly capable of taking up exceedingly large numbers of electrons and the question is raised as to whether the fully reduced form of the molecule remains intact. Our preliminary experiments show that, at least under certain conditions, the reduced form can be passed through an anaerobic Biogel P-2 column such that >90% of the (ferrous) iron is maintained in the reduced ferritin-cytochrome, which elutes

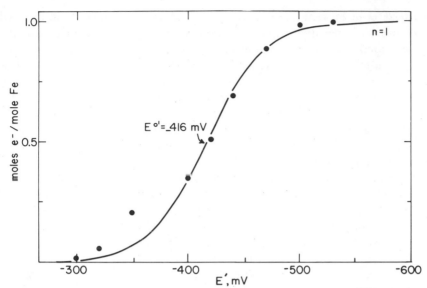

Figure 4.   Coulometric titration of ferritin-cytochrome at pH 7 in Tes buffer by the technique of Watt (1979).

with the solvent front. Separate experiments demonstrate that the column would easily separate reduced ferritin-cytochrome from added ferrous ions. The fully reduced ferritin-cytochrome can be reoxidized by air to a species that is spectroscopically and analytically indistinguishable from the original oxidized material. The Av ferritin-cytochrome thus has the remarkable capability of storing reducing equivalents at -416 mV at pH 7 (at least for short periods of time).

The above properties inspire us to speculate on possible roles for this protein. Obviously, the high iron content makes this molecule a prime candidate for an iron storage protein of $A.$ $vinelandii$, much as ferritins perform this function in mammalian, plant, and fungal systems (Crichton, 1973). If iron storage is indeed its physiological raison d'être then, as has been postulated in mammalian systems, ferritin iron mobilization may be affected by the reduction of Fe^{3+} to the more labile Fe^{2+} state. Perhaps the presence of the cytochrome in the same molecule facilitates the reduction of the internal ferritin iron, although our preliminary experiments with $S_2O_4^{2-}$ seem to indicate that, at least with this reductant, bulk iron is reduced much more slowly than heme iron. If Fe^{2+} formation is indeed required, then perhaps some specific Fe^{2+} chelator or other effector is required to facilitate iron removal from the core. Why is such a low potential required to mobilize the iron? An intriguing hypothesis is that the iron in the ferritin-cytochrome is a specific iron-storage depot for nitrogenase and its low redox potential ensures that only when the local redox potential in the cell approaches the negative values required for nitrogenase turnover will the iron be mobilized from the ferritin-cytochrome.

An alternative or possibly additional function of the ferritin-cytochrome could involve its functioning as an electron storage protein. Thus, the ability of a single molecule to take up hundreds of electrons at a low redox potential, in a reversible manner, suggests the possibility that that molecule may function to supply low-potential redox equivalents to be used in respiration, biosynthesis, or nitrogen fixation. These intriguing possibilities are under active investigation.

We further speculate as to the possible ubiquity of bacterial ferritin-cytochromes of the type discussed above. Spectroscopically, the heme absorptions found in the *A. vinelandii* molecule are extremely similar to those found in b_1-type cytochromes of unknown physiological function from various bacterial sources (Hagihara, Sato, and Yamanaka, 1975). Although none of the other b_1-type cytochromes has been reported as containing nonheme iron, these do share with the *A. vinelandii* ferritin-cytochrome a very low redox potential and a marked tendency to be present in high molecular weight aggregates. Since cytochrome b_1 preparations often involve use of detergents, which cause loss of the Fe^{3+} core and leave apoferritin, the possibility remains that these are the bacterioferritin-cytochrome analogs of apoferritin. If this is so, then the ferritin-cytochrome may be a ubiquitous protein in aerobic bacteria and we would have yet another example of a fascinating class of molecules being discovered as a side product of research in nitrogen fixation.

ACKNOWLEDGMENTS

We thank Dr. Harry Calvert for his advice and for execution of the electron microscopy experiments, and Deloria Jacobs and Sam Lough for expert technical assistance. We are grateful to George Lang and Kevos Spartalian of Pennsylvania State University for performing the Mössbauer experiments. This paper represents contribution No. 676 from the Charles F. Kettering Research Laboratory.

REFERENCES

Boas, J. F., and B. Window. 1966. Mössbauer effect in ferritin. Aust. J. Phys. 79:573–576.

Bulen, W. A. 1976. Nitrogenase from *Azotobacter vinelandii* and reactions affecting mechanistic interpretations. In: W. E. Newton and C. J. Nyman (eds.), Proceedings of the First International Symposium on Nitrogen Fixation, Vol. 1, pp. 177–186. Washington State University Press, Pullman.

Bulen, W. A., and J. R. LeComte. 1962. Isolation and properties of a yellow green fluorescent peptide from Azotobacter medium. Biochem. Biophys. Res. Commun. 9:523–528.

Bulen, W. A., J. R. LeComte, and S. Lough. 1973. A hemoprotein from *Azotobacter* containing non-heme iron: Isolation and characterization. Biochem. Biophys. Res. Commun. 54:1274–1281.

Burgess, B. K., D. B. Jacobs, and E. I. Stiefel. 1979. Large-scale purification of high-activity *Azotobacter vinelandii* nitrogenase. Submitted for publication.

Burgess, B. K., S. Wherland, W. E. Newton, and E. I. Stiefel. 1979. H_2-inhibition and HD-formation reactions of nitrogenase from *Azotobacter vinelandii*: Mechanistic Interpretations. In preparation.

Chaykin, S. 1969. Assay of nicotinamide deamidase. Determination of ammonia by the indo-phenol reaction. Anal Biochem. 31:375–382.

Corbin, J. L., and W. A. Bulen. 1969. The isolation and identification of 2,3-dihydroxybenzoic acid and 2-*N*,6-*N*-Di-(2,3-dihydroxybenzoyl)-L-lysine formed by iron-deficient *Azotobacter vinelandii*. Biochemistry 8:757–762.

Corbin, J. L., I. L. Karle, and J. Karle. 1970. Crystal structure of the chromophore from the fluorescent peptide produced by iron deficient *Azotobacter vinelandii*. Chem. Commun.:186–187.

Crichton, R. R. 1973. Ferritin. Structure and Bonding 17:67–134.

Eady, R. R., B. E. Smith, K. A. Cook, and J. R. Postgate. 1972. Nitrogenase of *Klebsiella pneumoniae*: Purification and properties of the component proteins. Biochem. J. 128:655–675.

Fukasawa, K., M. Goto, K. Sasaki, Y. Hirata, and S. Sato. 1972. Structure of the yellow-green fluorescent peptide produced by iron-deficient *Azotobacter vinelandii* strain O. Tetrahedron 28:5359–5365.

Hadfield, K. L., and W. A. Bulen. 1969. Adenosine triphosphate requirement of nitrogenase from *Azotobacter vinelandii*. Biochemistry 8:5103–5108.

Hagihara, B., N. Sato, and T. Yamanaka. 1975. Type *b* cytochromes. In: P. D. Boyer (ed.), The Enzymes, Vol. XI, Part A, pp. 550–593. Academic Press, Inc., New York.

Harrison, P., R. J. Hoare, T. G. Hoy, and I. G. Macara. 1974. Ferritin and haemosiderin: Structure and function. In: A. Jacobs and M. Worwood (eds.), Iron in Biochemistry and Medicine, pp. 73–114. Academic Press, Inc., New York.

Jackson, E. K., G. W. Parshall, and R. W. F. Hardy. 1968. Hydrogen reactions of nitrogenase. Formation of the molecule HD by nitrogenase and by an inorganic model. J. Biol. Chem. 243:4952–4958.

Ketchum, P. A., and M. S. Owens. 1975. Production of a molybdenum-coordinating compound by *Bacillus Thuringiensis*. J. Bacteriol. 122:412–417.

Neilands, J. B. 1977. Siderophores: Biochemical ecology and mechanisms of iron transport in enterobacteria. In: K. N. Raymond (ed.), Advances in Chemistry, No. 162, Bioinorganic Chemistry II, pp. 3–32. American Chemical Society, Washington, D.C.

Newton, W. E., W. A. Bulen, K. L. Hadfield, E. I. Stiefel, and G. D. Watt. 1977. HD formation as a probe for intermediates in N_2 reduction. In: W. Newton, J. R. Postgate, and C. Rodriguez-Barrueco (eds.), Recent Developments in Nitrogen Fixation, pp. 119–130. Academic Press, Inc. New York.

Orme-Johnson, W. H., L. C. Davis, M. T. Henzl, B. A. Averill, N. R. Orme-Johnson, E. Münck, and R. Zimmerman. 1977. Components and pathways in biological nitrogen fixation. In: W. Newton, J. R. Postgate, and C. Rodriguez-Barrueco (eds.), Recent Developments in Nitrogen Fixation, pp. 131–178. Academic Press, Inc., New York.

Rivera-Ortiz, J. M., and R. H. Burris. 1975. Interactions among substrates and inhibitors of nitrogenase. J. Bacteriol. 123:537–545.

Schubert, K. R., and H. J. Evans. 1976. Hydrogen evolution: A major factor affect-ing the efficiency of nitrogen fixation in nodulated symbionts. Proc. Natl. Acad. Sci. U.S.A. 73:1207–1211.

Silverstein, R., and W. A. Bulen. 1970. Kinetic studies of the nitrogenase-catalyzed hydrogen evolution and nitrogen reduction reactions. Biochemistry 9:3809–3815.

Stiefel, E. I. 1977. The mechanisms of nitrogen fixation. In: W. Newton, J. R. Postgate, and C. Rodriguez-Barrueco (eds.), Recent Developments in Nitrogen Fixation, pp. 67–108. Academic Press, Inc., New York.

Stiefel, E. I., W. E. Newton, G. D. Watt, K. L. Hadfield, and W. A. Bulen. 1977. Molybdoenzymes: The role of electrons, protons, and dihydrogen. In: K. N. Raymond (ed.), Advances in Chemistry, No. 162, Bioinorganic Chemistry II, pp. 353–388. American Chemical Society, Washington, D.C.

Stiefel, E. I., and G. D. Watt. 1979. *Azotobacter* cytochrome $b_{557.5}$ is a bacteriofer-ritin. Nature 279:81–83.

Thorneley, R. N. F., R. R. Eady, and D. J. Lowe. 1978. Biological nitrogen fixation by way of an enzyme-bound dinitrogen-hydride intermediate. Nature 272: 557–558.

Watt, G. D. 1979. An electrochemical method for measuring redox potentials of low potential proteins by microcoulometry at controlled potentials. Anal. Biochem. In press.

Watt, G. D., and W. A. Bulen. 1976. Calorimetric and Electrochemical Studies on Nitrogenase. In: W. E. Newton and C. J. Nyman (eds.), Proceedings of the First International Conference on Nitrogen Fixation, pp. 248–256. Washington State University Press, Pullman.

Watt, G. D., and A. Burns. 1977. Kinetics of dithionite ion utilization and ATP hydrolysis for reactions catalyzed by the nitrogenase complex from *Azotobacter vinelandii*. Biochemistry 16:264–270.

Nitrogen Fixation, Volume I
Edited by W. E. Newton and W. H. Orme-Johnson
Copyright 1980   University Park Press   Baltimore

Cyclopropenes: New Chemical Probes of Nitrogenase Active Site Interactions

C. E. McKenna, C. W. Huang, J. B. Jones, M.-C. McKenna, T. Nakajima, and H. T. Nguyen

The structural complexity of the nitrogenase system—which stands in striking contrast to its functional raison d'être, a diatomic molecule—is continuing to emerge, but a detailed picture of the active site for substrate reduction does not yet exist. While processes of into-, inter-, and intra-componental electron transfer and concomitant ATP hydrolysis leading to generation of active states of the enzyme constitute essential parts of the overall mechanism, the binding, activation, and reduction of N_2 and other unsaturated substrates comprise a particularly crucial mechanistic sequence from the catalytic standpoint. Clarification of these steps requires further insight into the dynamics of substrate-site interactions, as well as more information about the composition, structure, and states of the reducible substrate binding moiety.

Chemical probes are often valuable tools for investigation of enzyme active sites. Information may be derived, for example, from a physically detectable functional group introduced into the active site, from chemical modification of the active site, or from some transformation effected by the active site on the probe. Usually the chemical probe is a substrate analog capable of selective binding to the active site. Such approaches are hindered with nitrogenase because N_2 has no covalent derivatives. Indirect evidence

This research was supported by grant # HFF-77 from the Herman Frasch Foundation and grant # 10185-AC3 from the Petroleum Research Fund. J. B. J. and M.-C. McK. acknowledge support from NSF grants # PCM 77-17637 and # BMS 13608, respectively.

for diazene as a bound $2e^-$ reduction intermediate in nitrogenase-catalyzed N_2 reduction has been inferred from N_2-dependent HD formation from D_2 and H_2O (Bulen, 1976; Newton, Corbin, and McDonald, 1976; Newton et al., 1977; Stiefel et al., this volume). However, the instability of free N_2H_2 complicates direct demonstration that it can interact with nitrogenase. Hydrazine, a possible $4e^-$ intermediate, is reduced by the enzyme to NH_3 (Bulen, 1976) and can be trapped from nitrogenase mixtures actively fixing N_2 (Thorneley, Eady, and Lowe, 1978). However, N_2H_4 is a relatively poor substrate at pH 7, and some time ago Schöllhorn and Burris (1967a) reported that several N,N'-disubstituted hydrazines had little or no effect on nitrogenase activity, thus discouraging use of N_2H_4 derivatives as active site probes.

The well known ability of nitrogenase to catalyze reduction of certain unsaturated compounds other than N_2 in principle offers a means to probe nitrogenase interactions chemically. These compounds include the diverse functional groups represented by C_2H_2, N_3^-, N_2O, and HCN. Work by Burris, Hardy, Postgate, and others in this area has created a substantial body of mechanistically significant facts and interpretations that has been extensively reviewed elsewhere (Burris and Orme-Johnson, 1976; Burns and Hardy, 1975). A plot of the date of discovery for a given substrate versus the log of its $1/K_m$ (as an approximate measure of substrate affinity) show that virtually all of these adventitious substrates were first described about one decade ago (Figure 1). A possible reason for the dearth of subsequent substrate analog development is suggested by the plot. It shows that replacement of a terminal proton in C_2H_2 or HCN by a methyl group drastically increases the K_m; in 2-butyne, with a second methyl group, the K_m becomes so large that reduction is virtually undetectable. Burns and Hardy (1975) have attributed these phenomena to steric effects. As a result of this decreased affinity of nitrogenase for even comparatively small substrate homologs, the design of new active site probes based on such structures has evidently been inhibited.

If the steric hypothesis is correct, one can readily perform a trivial gedanken experiment in which the binding affinity of methylacetylene to the enzyme is increased merely by inducing a Maxwell's demon to bend the obtrusive methyl carbon away from the $C≡C$ bond axis (Figure 2a). This in turn suggested that the sterically offending carbon could be tied back chemically, resulting (with some rearrangement of hydrogen atoms) in the strained ring compound cyclopropene (Figure 2a). The unusual geometry and orbital properties of cyclopropenes make them especially interesting potential chemical probes of nitrogenase (see the comparison with C_2H_2 and C_2H_4 in Figure 2b). Whereas substitution on the $C≡C$ bond must be coaxial, cyclopropene geometry allows substitutional steric incrementation either in the plane of the ring at one or both ends of the formal $C=C$ bond (H-1, H-2) or in a plane perpendicular to the ring (H-3 and H-3') (Figure

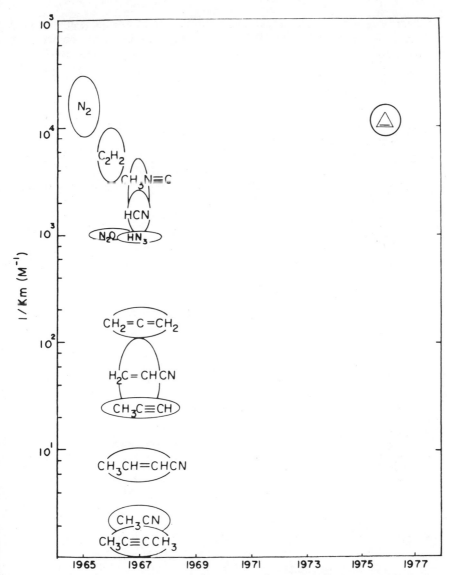

Figure 1.   Log ($1/K_m$) values versus discovery chronology of unsaturated nitrogenase substrates (*Azotobacter* nitrogenase under standard assay conditions; for diagram clarity, a few substrates from the end-of-sixties "burst" period have been omitted).

2b). In addition, C-1—C-2 and C-3 are hybridized differently, thus conferring possibly different functional properties on their substituents. Ring strain in cyclopropene endows it with chemical reactivity intermediate between that of the nitrogenase substrate, C_2H_2, and the nonsubstrate, C_2H_4. Alternative steric and electronic arguments advanced by Schöllhorn and Burris (1967b) to explain this C_2H_2/C_2H_4 selectivity have not yet been

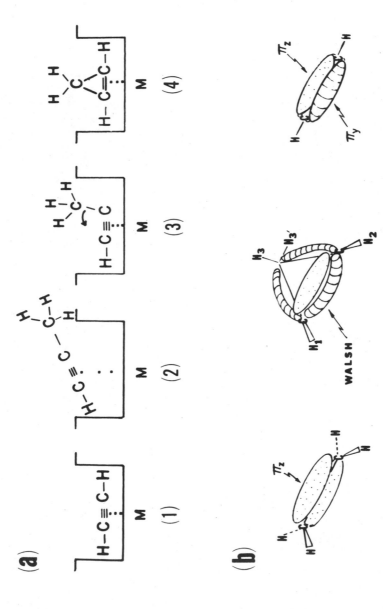

Figure 2. (a) Heuristic application of the nitrogenase active site steric hindrance hypothesis: 1) compact C_2H_2 molecule binds; 2) longer $CH_3C\equiv CH$ molecule binds poorly; 3) gedanken experiment—binding restored by bending back $CH_3C\equiv CH$ methyl group with aid of Maxwell's demon (not shown); 4) cyclopropene predicted to fit active site. M represents active site metal(s) involved in binding; cyclopropene orientation is depicted for schematic convenience only and is not meant to imply an edge versus face-on interaction with M. (b) Comparison of structural geometry and higher occupied orbital symmetry of C_2H_4, c-C_3H_4 and C_2H_2.

resolved. In the sense that its Walsh orbitals are π-like, cyclopropene is more like C_2H_2, although as a planar molecule it more resembles C_2H_4 sterically (Figure 2b).

Cyclopropene also resembles C_2H_2 in that both have possible reduction products readily analyzed by FID gas chromatography, which facilitates kinetic studies. However, only one $2e^-$ product is possible with C_2H_2, whereas two $2e^-$ products of quite different structure are possible with cyclopropene. In addition, the number of stereochemical possibilities of $2e^-$ reductions in D_2O is substantially larger with cyclopropene. We are intrigued by the possibilities of cyclopropene and related molecules as chemical probes for nitrogenase, and are therefore undertaking a comprehensive investigation of their interactions with this enzyme. We present here an account of our current results with this approach.

EXPERIMENTAL METHODS

[Details of the preparation and analysis of highly purified cyclopropene will be given elsewhere.] *Azotobacter vinelandii* OP was continuously cultured and harvested in log phase. For purification of the enzyme, no existing procedure for two-component separation seems to satisfy simultaneously the desiderata of high yield, high final activity, adaptability to scale-up, and ease of implementation. We have found that aerobic protamine sulfate fractionation and linear gradient DEAE-cellulose chromatography adapted to purification by the method of Shah and Brill (1973) afford 2–3 g/batch of crystalline Av1 having a specific activity $\approx$ 2100–2200. Av2 of specific activity $\approx$ 2000 can be prepared on a gram scale by chromatography of the appropriate linear gradient DEAE-cellulose fraction on Sephadex G-100, followed by ultrafiltration. This technique avoids the scale-limiting preparative electrophoresis step in Shah and Brill's Av2 preparation (1973). Some experiments in our laboratory have utilized Av "complex," and Av1/Av2 prepared by modifications of earlier published procedures (McKenna, McKenna, and Higa, 1976).

RESULTS

Nitrogenase-Catalyzed Cyclopropene Reduction

We have demonstrated (McKenna, McKenna, and Higa, 1976) that incubation of 0.05 atm or less of cyclopropene in helium or argon with *A. vinelandii* nitrogenase in a standard assay mixture results in the initially linear formation of both propene and cyclopropane (Figure 3). The products have been identified by gas chromatography retention times, mass spectrometry, ^{1}H NMR (Figure 4a), and infrared spectroscopy (Figure 4b).

Formation of propane (by $4e^-$ reduction of cyclopropene) accounts for less than 1% of total products. Conversion of cyclopropene to propene and cyclopropane is dependent on ATP, dithionite (Figure 3), and both active enzyme components. H_2 at low pressure cannot replace dithionite as electron donor, thus ruling out the possibility that nitrogenase-generated free H_2 could account for the observed reductions (Figure 3a). Propene does not detectably interact with nitrogenase, and cyclopropane has a very weak inhibitory activity (McKenna, McKenna, and Huang, unpublished data). Cyclopropene undergoes a very slow decomposition (with respect to the assay time scale) over assay mixture minus enzyme, but decomposition products as a source of either propene or cyclopropane can easily be discounted on the basis of stoichiometry alone. Furthermore, with aged samples the reduction still requires unreacted cyclopropene and its initial rate remains a function of the effective cyclopropene gas phase concentration.

About twice as much propene as cyclopropane is made in these reductions, seemingly independently of component purity, which component is saturating to maximal activity, and whether the nitrogenase used consists of copurified or isolated recombined components (McKenna, McKenna, and Higa, 1976). This experimental product ratio corresponds to the partitioning predicted by random reduction of the cyclopropene ring with respect to its edges: reduction of the C-1—C-2 edge would yield cyclopropane, but reduction of the C-2—C-3 and C-1—C-3 edges would yield twice as much propene on a statistical basis. Although the coincidence between the observed product partitioning ratio and that predicted by "edge analysis" could be fortuitous, the relative invariability of this ratio under many assay conditions suggests that it reflects an intrinsic catalytic property of the nitrogenase active site. Since in vivo *Azotobacter* nitrogenase differs from the purified in vitro component mixtures, notably with regard to O_2 lability, yet represents an ultimate reference point as "native" enzyme, we attempted to establish whether or not the isolation process or the in vitro assay conditions affected product partitioning. Such a comparison can be made because cyclopropene, unlike HCN, for example, has little or no toxicity for the bacteria. We find (McKenna and Huang, 1979) that: 1) NH_3-grown *A. vinelandii*, having no C_2H_2 reduction activity, has no detectable cyclopropene reduction activity; 2) N_2-grown *A. vinelandii* incubated with cyclopropene-argon mixtures possesses little or no cyclopropene reduction activity; and 3) N_2-grown *A. vinelandii* incubated with cyclopropene-argon mixtures containing an optimal concentration of O_2 (as determined by parallel experiments with C_2H_2 as substrate) forms propene and cyclopropane in a ratio of 2.0 ± 0.25, i.e., the same product ratio is obtained for both in vivo and in vitro nitrogenase. Kinetic parameters [K_m and $V_m/V_m(C_2H_2)$] determined for the bacterial reduction are similar to those determined for purified in vitro enzyme (*vide infra*). The ability to

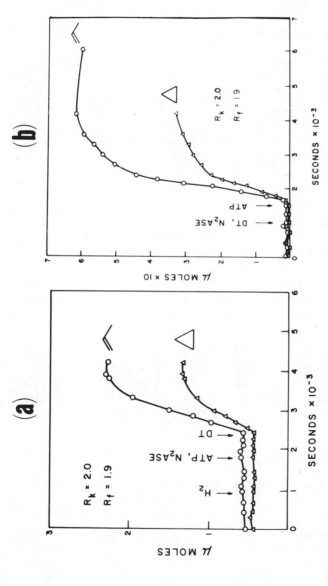

Figure 3. Time course studies of propene and cyclopropane evolution from cyclopropene-nitrogenase assay mixtures, R_k = initial rate ratio of propene:cyclopropane; R_f = final ratio of propene:cyclopropane. (a) Inability of H_2 to replace dithionite as electron donor. (b) ATP requirement.

reduce cyclopropene to propene and cyclopropane in a ratio of about 2:1 appears to be an inherent property of active *A. vinelandii* nitrogenase irrespective of whether it is functioning under normal physiological or artificial fixing conditions.

Kinetic Characteristics of Cyclopropene Reductions

K_m and V_m data can be used to establish whether or not cyclopropene is a "good" substrate of nitrogenase. One can arbitrarily define a good nitrogenase substrate as having a $K_m \leq 1$ mM and $V_m \geq 0.25 \cdot V_m$ of C_2H_2. We have measured K_m and V_m values for cyclopropene reductions by *Azotobacter* nitrogenase "complex," highly active recombined Av1 and Av2 components, and in vivo enzyme (Table 1) (McKenna and Huang, 1979). The V_m of cyclopropene is approximately half of the V_m of C_2H_2, or about twice that for N_2, uncorrected for electron transfer stoichiometry. The $V_m : V_m$ ratio for propene and cyclopropane as products is close to 2, and values for K_m with respect to either product are 0.01–0.015 atm. These data mean that the product partitioning ratio remains about 2 over cyclopropene concentrations extending from 0.004 atm to near the extrapolated pressure limit at V_m. In order to compare the cyclopropene K_m with the K_m of N_2 or C_2H_2, a suitable solubility correction must be made. Our estimates of the Henry's Law constant in H_2O at 30°C for cyclopropene correspond to a concentration of 0.009 M (in equilibrium with cyclopropene at 1 atm). Solubility correction gives a K_m for cyclopropene of $\sim 1 \times 10^{-4}$ M, which is comparable to that of N_2 itself. Cyclopropene thus rivals C_2H_2 (K_m 1–2 $\times$ 10^{-4} M) as a "good" substrate and has an order of magnitude smaller K_m than the "fairly good" substrates HCN, N_3^-, and N_2O. Its K_m is several orders of magnitude smaller than that of allene, which also has $V_m \sim 0.5 \cdot$ V_m of C_2H_2 (Burns, Hardy, and Phillips, 1975). This relationship of cyclopropene to other nitrogenase substrates is shown graphically in Figure 1.

Table 1.   Cyclopropene K_m, V_m, and R values for *Azotobacter* nitrogenase

| Enzyme state | Propene formation | | Cyclopropane formation | | |
	K_m (atm)	V_m (c–C$_3$H$_4$) V_m (C$_2$H$_2$)	K_m (atm)	V_m (c–C$_3$H$_4$) V_m (C$_2$H$_2$)	R[a]
In Vivo	0.014	0.38	0.015	0.20	1.9
In Vitro	0.012	0.29	0.011	0.14	2.1
[Fe + MoFe (2:1)]					

[a] Ratio of V_m of propene formation : V_m of cyclopropane formation.

Stereochemical and Mechanistic Pathways
of Cyclopropane and Propene Formation

Our studies of mechanistic aspects of cyclopropene-nitrogenase interactions are at an early stage, but certain preliminary findings can be discussed.

Cyclopropene Reduction Stereochemistry   As discussed by Burns and Hardy (1975), the predominant product of nitrogenase-catalyzed C_2H_2 reduction in D_2O is *cis*-1,2-d_2-ethylene, i.e., the stereochemistry of proton (deuteron) addition is *syn*. Because cyclopropene and cyclopropane are planar, *syn* and *anti* proton additions across the cyclopropene C=C bond are easily differentiated by infrared spectroscopy of the products *cis* and *trans*-d_2-cyclopropane (Schlag, 1958). We have isolated and obtained the infrared spectrum of cyclopropane produced from D_2O nitrogenase fixing systems. The *cis*-d_2 isomer is 10 times or more as abundant and mass spectral analysis confirms formation of a d_2 product (McKenna, McKenna, and Huang, 1979).

Possible Intermediates in Propene Formation   A conceivable pathway to the propene product involves ring opening to form a bound vinyl carbene species that isomerizes to allene or methyl acetylene prior to reduction. Both acyclic C_3H_4 isomers are reduced by the enzyme to propene (Burns and Hardy, 1975; Burns, Hardy, and Phillips, 1975). Inasmuch as the K_m values for allene and methyl acetylene are fairly large, it might be expected that some fraction of such intermediates would be released and therefore detected, but these values might also be misleading in this respect since they include the process of free substrate entry into the active site. Allene is reduced by nitrogenase in D_2O to 2,3-d_2-propene (McKenna, McKenna, and Huang, 1979; cf. Burns, Hardy, and Phillips, 1975) and methyl acetylene is reduced to 1,2-d_2-propene (McKenna, McKenna, and Huang, 1979). These deuterated compounds are unambiguously distinguishable from each other, and from other d_2-propene isomers, by [^1]H NMR. Specifically, they lack a C-2 methine proton that has a characteristic resonance at τ = 4.27 (cf. Figure 4a). Integration data for 100-MHz [^1]H NMR spectra for cyclopropene-derived propene, generated with D_2O-fixing systems, are consistent with a mixture of d_2-propenes that cannot be readily resolved because [^2]H broadening obscures multiplet fine structure. The spectra do clearly establish that formation of a substantial portion of the propene cannot involve catalyzed isomerization of cyclopropene to either of the linear C_3H_4 isomers, since they show a prominent signal at the C-2 methine proton position. By a combination of [^2]H-decoupled 220-MHz [^1]H NMR, [^1]H-decoupled 30.7-MHz [^2]H NMR, and infrared spectroscopy, however, the principal component of this mixture has since been found to be 1,3-d_2-propene, formed with limited stereoselectivity (McKenna, McKenna, and Huang, 1979).

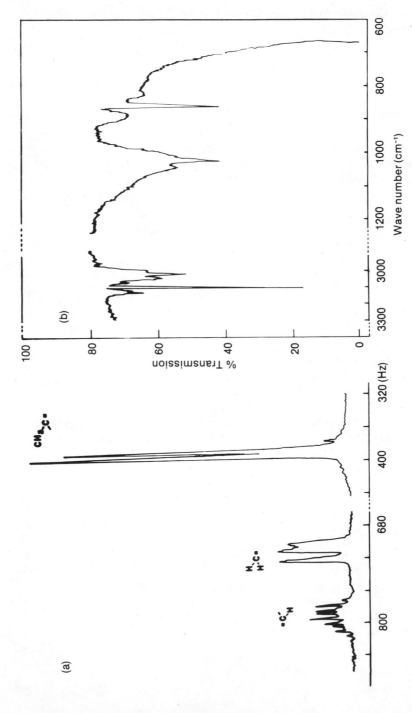

Figure 4.   (a) 100-MHz FT ^{1}H NMR spectrum of propene derived from nitrogenase (H_2O) reduction of cyclopropene. Offset from TMS is +220 Hz.
(b) Infrared spectrum of cyclopropane derived from nitrogenase (H_2O) reduction of cyclopropene.

Table 2.   Comparison of C_3H_4 isomers and 1-methylcyclopropene as inhibitors for nitrogenase

Inhibitor	Partial pressure (atm)	Substrate	Partial pressure (atm)	Percentage of inhibition
Cyclopropene	0.050	C_2H_2	0.010	67
	0.020	N_2	0.100	52
Allene	0.050	C_2H_2	0.010	~ 0
Methyl acetylene	0.050	C_2H_2	0.010	~ 0
1-Methylcyclopropene	0.13	C_2H_2	0.010	< 10

Cyclopropene as an Inhibitor of C_2H_2 or N_2 Reduction by Nitrogenase

Cyclopropene at 0.05 atm partial pressure substantially inhibits C_2H_2 reduction ($P_{C_2H_2}$ = 0.01 atm) or N_2 reduction (P_{N_2} = 0.10 atm). Under the same conditions, the two linear C_3H_4 isomers (allene and methyl acetylene) have no effect on C_2H_2 reduction and 1-methylcyclopropene shows little inhibitory activity at 0.13 atm (Table 2) (McKenna, McKenna, and Huang, 1979). When relative H_2O solubility of these gases is taken into account, the difference in inhibitory effectiveness between cyclopropene and its cumulated diene and alkyne isomers, or 1-methyl derivative, is unchanged or increased. Under the conditions thus far explored, cyclopropene is a competitive inhibitor of N_2, but displays a mixed type of inhibition for C_2H_2 (McKenna and McKenna, unpublished data).

DISCUSSION AND SUMMARY

Consideration of the steric limitations associated with previous chemical probes of the active site has led us to the concept of cyclopropene as a new nitrogenase substrate and inhibitor. We have demonstrated that cyclopropene is in fact a substrate of nitrogenase, undergoing ATP-dependent $2e^-$, but not $4e^-$, reduction in the obligatory presence of active Fe and MoFe proteins and with dithionite as electron donor. The following conclusions have emerged.

The reaction is unique among known nitrogenase catalytic reductions in that two isomeric products are formed, one involving alkene-alkane conversion (to cyclopropane) and one formally involving a reductive ring cleavage (to propene). Initial rate studies under standard assay conditions suggest that propene is formed both in vitro and in vivo about twice as fast as cyclopropane. This finding supports the important assumption that in vitro nitrogenase assayed under artificial conditions is catalytically unchanged relative to the in vivo enzyme functioning with an endogenous electron donor under intracellular conditions. The insensitivity of the product ratio suggests a new criterion for evaluation of active site functional similarity in nonenzymatic systems intended to model nitrogenase catalysis. It does not exclude the possibility that, under special conditions (e.g., non-standard pH or other change in medium, presence of inhibitors, or suboptimal component ratios), some deviation in the product ratio may be observable and interpretatively useful.

The molar K_m of cyclopropene is comparable to those of C_2H_2 and N_2; the V_m of cyclopropene is about half of the V_m of C_2H_2. Cyclopropene is a good nitrogenase substrate in contrast to its linear isomers, methyl acetylene and allene, which have much higher K_m values. Cyclopropene can be regarded as a combination of an alkene double bond and strained cyclopropane ring. These structures individually show weak or undetectable activity with nitrogenase, demonstrating that they must be combined in a single structural entity in order to create a strong nitrogenase active site interaction. This suggests that active site C_2H_2/C_2H_4 selectivity is electronic rather than steric in origin. In conditions under which cyclopropene is an effective inhibitor of C_2H_2, allene and methyl acetylene have no detectable inhibitory effect. K_i values for the three C_3H_4 isomers reflect their corresponding K_m values. Replacement of a C-1 proton with a methyl group results in a less effective inhibitor. Thus, as with $HC{\equiv}CR$ and $N{\equiv}CR$, an increase in molecular bulk along the cyclopropene C-C multiple bond axis apparently generates steric hindrance, consistent with comparable orientation of these substrates. On the other hand, the bulky "hinge" of the cyclopropene ring seems easily accommodated by the active site; it requires at least one relatively unhindered region of the active site situated along the axis perpendicular to, and bisecting, the $C{=}C$ bond.

Continued development of the chemical probe approach, and introduction of new physical probes, such as near infrared-visible-near ultraviolet CD and MCD (Stephens et al., 1979), should be useful in further elucidation of nitrogenase enzymology.

ACKNOWLEDGMENTS

The authors wish to thank Mr. Michael Herson for preliminary cyclopropene-N_2 inhibition experiments, and Professor M. D. Kamen for his continuing interest and encouragement.

REFERENCES

Bulen, W. A. 1976. Nitrogenase from *Azotobacter vinelandii* and reactions affecting mechanistic interpretations. In: W. E. Newton and C. J. Nyman (eds.), Proceedings of the First International Symposium on Nitrogen Fixation, Vol. 1, pp. 177–186. Washington State University Press, Pullman.

Burns, R. C., and R. W. F. Hardy. 1975. Nitrogen fixation in bacteria and higher plants. In: A. Kleinzeller, G. F. Springer, and H. G. Wittman (eds.), Molecular Biology, Biochemistry, and Biophysics, Vol. 21. Springer-Verlag, New York.

Burns, R. C., R. W. F. Hardy, and W. D. Phillips. 1975. *Azotobacter* nitrogenase: Mechanism and kinetics of allene reduction. In: W. D. P. Stewart (ed.), N_2 Fixation by Free-Living Microorganisms, pp. 447–452. Cambridge University Press, London.

Burris, R. H., and W. H. Orme-Johnson. 1976. Mechanism of biological N_2 fixation. In: W. E. Newton and C. J. Nyman (eds.), Proceedings of the First International Symposium on Nitrogen Fixation, Vol. 1, pp. 208–233. Washington State University Press, Pullman.

McKenna, C. E., M.-C. McKenna, and M. Higa. 1976. Chemical probes of nitrogenase. 1. Cyclopropene. Nitrogenase-catalyzed reduction to propene and cyclopropane. J. Am. Chem. Soc. 98:4657–4659.

McKenna, C. E., M.-C. McKenna, and C. W. Huang. 1979. Low stereoselectivity in methylacetylene and cyclopropene reductions by nitrogenase. Proc. Natl. Acad. Sci., U.S.A. 76:4773–4777.

McKenna, C. E., and C. W. Huang. 1979. In vivo reduction of cyclopropene by *A. vinelandii* nitrogenase. Nature 280:609.

Newton, W. E., W. A. Bulen, K. L. Hadfield, E. I. Stiefel, and G. D. Watt. 1977. HD formation as a probe for intermediates in N_2 reduction. In: W. E. Newton, J. R. Postgate, and C. Rodriquez-Barrueco (eds.), Recent Developments in Nitrogen Fixation, pp. 119–130. Academic Press, Inc., New York.

Newton, W. E., J. L. Corbin, and J. W. McDonald. 1976. Nitrogenase: Mechanism and models. In: W. E. Newton and C. J. Nyman (eds.), Proceedings of the First International Symposium on Nitrogen Fixation, Vol. 1, pp. 53–74. Washington State University Press, Pullman.

Schlag, W. 1958. Unimolecular isomerization of *trans*-cyclopropane-d_2. Ph.D. thesis, University of Washington.

Schöllhorn, R., and R. H. Burris. 1967a. Reduction of azide by the N_2-fixing system. Proc. Natl. Acad. Sci. U.S.A. 57:1317–1323.

Schöllhorn, R., and R. H. Burris. 1967b. Acetylene as a competitive inhibitor of N_2 fixation. Proc. Natl. Acad. Sci. U.S.A. 58:213–216.

Shah, V. K., and W. J. Brill. 1973. Simple method of purification to homegeneity of nitrogenase components from *A. vinelandii*. Biochim. Biophys. Acta 305:445–454.

Stephens, P. J., C. E. McKenna, B. E. Smith, H. T. Nguyen, M.-C. McKenna, A. J. Thompson, F. Devlin, and B. J. Jones. 1979. Circular dichroism and magnetic circular dichroism of nitrogenase proteins. Proc. Nat. Acad. Sci. U.S.A. 76:2585–2589.

Thorneley, R. N. F., R. R. Eady, and D. J. Lowe, 1978. Biological nitrogen fixation by way of an enzyme-bound dinitrogen-hydride intermediate. Nature 272: 557–558.

Nitrogen Fixation, Volume I
Edited by W. E. Newton and W. H. Orme-Johnson
Copyright 1980   University Park Press   Baltimore

Iron-Molybdenum Cofactor of Nitrogenase

V. K. Shah

Activation of inactive nitrate reductase in extracts of *Neurospora crassa* mutant strain *nit-1* by acid-treated molybdoenzymes (Ketchum et al., 1970; Nason et al., 1971) offered a new concept of a cofactor common to different molybdoenzymes. When extracts of mutant strains of *Azotobacter vinelandii* and *Klebsiella pneumoniae*, which were defective in component I (MoFe protein) of nitrogenase, were screened for activation by acid-treated component I, strains analogous to *Nit-1* were found (Shah et al., 1973; Nagatani, Shah, and Brill, 1974; St. John et al., 1975). Activation of defective component I was observed in extracts of *A. vinelandii* mutant strain UW45, *K. pneumoniae* mutant strain UN106, and wild-type *A. vinelandii* that was derepressed for nitrogenase synthesis in tungsten-containing medium (Nagatani, Shah, and Brill, 1974; St. John et al., 1975). All efforts to isolate the activating factor by using conventional biochemical methods were unsuccessful; however, a unique method (Shah and Brill, 1977) yields stable preparations of an iron-molybdenum cofactor (FeMoco) suitable for electron paramagnetic resonance and Mössbauer spectroscopy studies.

This method can be used for the isolation of FeMoco from component I from a variety of nitrogen-fixing organisms. Molybdenum cofactors (Moco) from other molybdoenzymes also can be isolated by this basic technique (Pienkos, Shah, and Brill, 1977; Shah and Brill, 1977). Isolation of MoFeco, separation of MoFeco and Moco, and some interesting properties are described here.

This research was supported by the College of Agricultural and Life Sciences, University of Wisconsin, Madison, and by NSF grant PCM 76-24271 and NIH grant GM 22130.

RESULTS AND DISCUSSION

All operations were carried out under an H_2 atmosphere at 0–4°C. The reagents used were deoxygenated on a gassing manifold by evacuating and flushing with deoxygenated H_2 with constant mixing. These deoxygenated reagents contained 1.2 mM sodium dithionite, added as a 0.1 M aqueous solution (in 0.013 N NaOH) just before use. The syringe-needle assembly used for each step was flushed with H_2 before use.

Approximately 75 mg of nitrogenase component I in 4 ml of 0.25 M NaCl in 0.025 M Tris-HCl buffer (pH 7.4) was diluted with 8 ml of water in a glass centrifuge tube containing four glass beads (4-mm size). To this diluted solution of component I, citric acid solution was added to a final concentration of 15 mM, mixed thoroughly, and allowed to stand for 3 min. Disodium hydrogen phosphate solution was added to a final concentration of 25 mM; the contents were mixed thoroughly and allowed to stand for 25–30 min. The citrate/phosphate-treated component I was centrifuged at 8000 × g for 10 min and the supernatant solution was removed with a 4-in-long hypodermic needle on a syringe. This citrate/phosphate supernatant solution contains only traces of FeMoco activity but approximately 50% of the iron present in component I, and can be discarded. The citrate/phosphate-treated pellet of component I was washed twice by thoroughly suspending it in 8 ml of N,N-dimethylformamide, followed by centrifugation at 8000 × g for 10 min. This N,N-dimethylformamide supernatant solution contains less than 1% of the total FeMoco activity and can be discarded. The N,N-dimethylformamide–washed pellet was extracted three times by suspending it in 4 ml of N-methylformamide containing 5 mM Na_2HPO_4 (added as a 0.2 M aqueous solution) for 30 min, followed by centrifugation at 8000 × g for 10 min. The three N-methylformamide supernatants contain, respectively, 70%, 15%, and 5% of both the FeMoco activity and the molybdenum present in component I. Representative data for the distribution of FeMoco activity and of molybdenum and iron in different fractions are shown in Table 1 (Shah and Brill, 1977).

The FeMoco (about 90,000 units; see below) was applied to an anaerobic Sephadex G-100 column (77 × 1 cm) in N-methylformamide to remove any contaminating iron centers, Mo, Fe, or denatured component I protein that might be present. The column was equilibrated and eluted at a flow rate of 12 ml/hr with N-methylformamide containing 5 mM Na_2HPO_4 and 1.2 mM $Na_2S_2O_4$. Fractions were collected anaerobically and analyzed for FeMoco activity, Mo, and Fe. The FeMoco has a distinct brown color that was followed visually on the column. The FeMoco activity was eluted as a single peak, and the recovery of FeMoco activity, Mo, and Fe in this peak was 97%–98%. The coincident distribution of Fe and Mo with FeMoco activity in different fractions suggests that Fe is an integral part of

Table 1.   Isolation of FeMoco from nitrogenase component I

Fraction	Volume (ml)	Total activity (units)	Total Fe (nmoles)	Total Mo (nmoles)	Activity:Mo ratio	Fe:Mo ratio	Yield (%)
Azotobacter vinelandii							
Crystalline component I	5	121,100	7,273	441	275	16.5	100
Citrate-phosphate supernatant solution	17.2	113	3,614	1.3	—	—	—
DMF supernatant solution[a]	16.4	1,702	88	13.4	—	—	—
NMF supernatant solution[b] (FeMoco)	12.4	106,300	3,074	391	272	7.9	88
Residual pellet	5	4,125	188	31.2	—	—	—
Clostridium pasteurianum							
NMF supernatant solution (FeMoco)	2	63,800	1,790	225.4	283	7.9	—

[a] Values are for the pooled two-DMF supernatant solutions.
[b] Values are for the pooled three-NMF supernatant solutions.

the FeMoco. The Fe:Mo ratio in the FeMoco is 8:1. Dextran blue, FeMoco, and FeCl$_3$ are eluted at about 20, 35, and 50 ml, respectively.

The acetylene reduction assays for nitrogenase activity were performed under H$_2$ atmosphere in 9-ml serum bottles. All of the reagents were thoroughly evacuated and flushed with H$_2$ before use. Cell-free extracts (6–7 mg of protein) of *A. vinelandii* mutant strain UW45 or *K. pneumoniae* mutant strain UN106, derepressed for nitrogenase synthesis in a medium containing molybdate, or the wild-type *A. vinelandii*, derepressed in a medium containing tungstate instead of molybdate, were incubated with varying amounts of FeMoco for 30 min at room temperature. After this preincubation, 0.8 ml of reaction mixture (made anaerobically) containing 2.5 μmoles ATP, 30 μmoles creatine phosphate, 0.2 mg creatine phosphokinase, 5 μmoles MgCl$_2$, 15 μmoles Tris-Hcl buffer (pH 7.4), and 20 μmoles Na$_2$S$_2$O$_4$ was added. The reaction bottles were brought to 1 atmosphere. Acetylene (0.5 ml) was injected into each assay bottle, which was incubated at 30°C for 15 min in a water-bath shaker. The reaction was terminated by injecting 0.1 ml of 30% (w/v) trichloroacetic acid. Ethylene formed was measured after 30 min with a Packard 407 gas chromatograph with Porapak N column (Shah, Davis, and Brill, 1972; Nagatani, Shah, and Brill, 1974). Specific activity is expressed as the nmoles of ethylene formed/min/nmole of molybdenum. One unit of FeMoco is defined as the amount required to produce 1 nmole of ethylene/min under the reconstitution conditions used.

The FeMoco contains eight iron atoms and six acid-labile sulfides per molybdenum atom. The specific activity of FeMoco is 272 (nmoles of ethylene formed/min/nmol of Mo). There is better than 98% reconstitution between FeMoco and inactive component I in *A. vinelandii* mutant strain UW45. Activation of inactive component I in extracts of mutant strain UW45 was dependent on the FeMoco concentration (Figure 1) and followed saturation kinetics. Similar results were obtained when extracts of *K. pneumoniae* mutant strain UN106 or wild-type *A. vinelandii*, derepressed for nitrogenase synthesis in tungsten-containing medium, were titrated with FeMoco. Per unit of FeMoco, activation of inactive component I in extracts of wild-type *A. vinelandii*, derepressed for nitrogenase synthesis in tungsten-containing medium, was about 20% lower than that obtained with strain UW45. This lower activation might be due to the competition by a corresponding tungsten-containing moiety with the FeMoco (Shah and Brill, 1977).

The FeMoco isolated from nitrogenase component I of *Clostridium pasteurianum*, *Rhodospirillum rubrum*, *Bacillus polymyxa*, and *K. pneumoniae* also activates the defective component I in the extracts of *A. vinelandii* mutant strain UW45. These observations show that FeMoco from aerobic, anaerobic, facultative, and photosynthetic N$_2$-fixing or-

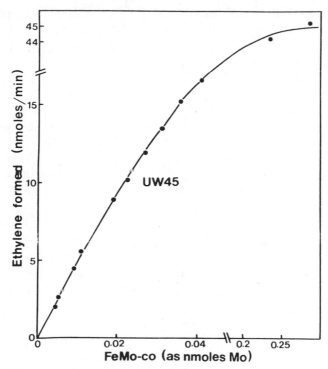

Figure 1.   FeMoco-dependent activation of inactive component I in extracts of *A. vinelandii* mutant strain UW45.

ganisms is very similar, even though component I from *C. pasteurianum* does not cross-complement with component II from *A. vinelandii* (Detroy et al., 1968; Emerich and Burris, 1976). The Fe:Mo ratio in FeMoco from *C. pasteurianum* component I was also 8:1 (Shah and Brill, 1977).

The FeMoco is extremely sensitive to O_2—brief exposure to air completely abolished its activating ability. The storage temperature had no effect on the activating ability of FeMoco (in N-methylformamide) as long as it was kept anaerobic. The FeMoco is unstable in aqueous medium even under strictly anaerobic conditions, and the storage temperature has a pronounced effect on its stability in aqueous media. Nitrogenase component I is stable, whereas FeMoco is unstable in an aqueous environment, which suggests that the protein of component I may be responsible for maintaining an aprotic environment at the FeMoco site in component I (Shah and Brill, 1977).

The visible spectrum of FeMoco has no distinctive features (Figure 2). There appears to be a steady decline in absorbance from about 400 to 700 nm. On progressive oxidation of the FeMoco, the absorbance decreases,

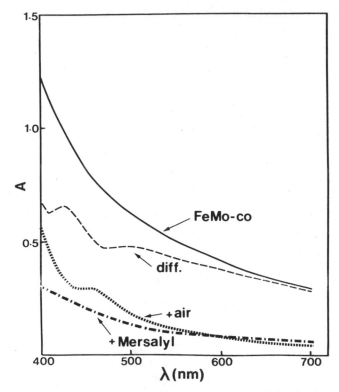

Figure 2.   Visible absorbance spectra of FeMoco before (——) and after (· · ·) exposure to air. Difference (−−−) spectra of the preceding two spectra. Effect of two equivalents of sodium mersalyl/Fe (−·−·−) on the visible absorbance spectrum of FeMoco. The FeMoco used for these studies had 26,500 units of activity per ml.

suggesting destruction of the chromophore. The destroyed chromophore is represented by the difference spectrum of FeMoco before and after exposure to air. Addition of sodium mersalyl to FeMoco resulted in a substantial reduction in visible absorbance with concomitant loss of activating ability (Shah and Brill, 1977).

The EPR spectrum of FeMoco (with two g values near 4 and one near 2) is of the same fundamental type as that observed for component I; the broad features and g shifts reflect differences in the local environment (Rawlings et al., 1978). The FeMoco from component I of *A. vinelandii, C. pasteurianum,* and *K. pneumoniae* give identical EPR signals. Combined Mössbauer and EPR studies (Münck et al., 1975) have shown four spectroscopic classes of iron centers in component I. One of these centers, labeled M_{EPR} (about 40% of the total Fe), is the EPR-active center, which undergoes reversible oxidation-reduction reactions during catalytic turnover of the enzyme system (Münck et al., 1975). M_{EPR} is a metal cluster containing iron (and perhaps molybdenum) atoms. This center is recovered in excellent yields in the FeMoco. This M_{EPR} center is not an Fe-S cluster of the type recognized so far in ferredoxins, because attempts to displace such clusters with thiophenol failed. Quantitative analyses show that both the native protein and the FeMoco contain one $\frac{3}{2}$ spin M_{EPR} center per molybdenum atom and that each center contains about six iron atoms in a novel spin-coupled structure. On brief exposure to air, FeMoco loses the EPR signal with concomitant loss of activating ability (Rawlings et al., 1978).

X-ray absorption spectroscopy of FeMoco and component I provided important information on primary ligation to Mo in nitrogenase. The absorption edge spectra strongly indicate the primary ligation of sulfur to Mo and the absence of doubly bound oxygen ($Mo=O$) from the Mo site (Cramer et al., 1978). The EXAFS data were best explained by postulating a Mo-Fe-S cluster.

Contrary to the earlier reports (Nason et al., 1971; McKenna et al., 1974; Zumft, 1974; Lvov et al., 1975), acid treatment of highly purified component I and FeMoco from component I of nitrogenase (Pienkos et al., 1977) failed to activate inactive nitrate reductase in extracts of *N. crassa* mutant strain *nit-1* (Table 2). However, partially purified preparations of component I contain *both* FeMoco and Moco and this explains the activation of nitrate reductase in strain *nit-1* by acid-treated component I of nitro-

Table 2.   In vitro activation of inactive nitrogenase and nitrate reductase

Source of addition[a]	Reconstituted nitrogenase activity C_2H_4 formed (nmoles/ 15 min)	Reconstituted nitrate reductase activity NO_2^- formed (nmoles/ 15 min)
Component I (80) (FeMoco)	331	0.0
Component I (80) aerobic	71	0.0
Xanthine oxidase (1.4) (Moco)	0	31.7
Xanthine oxidase (1.4) aerobic	0	8.1
Moco + FeMoco mixture from partially purified component I (28.5)	90	5.3

[a] The values in parentheses are pmoles of molybdenum in the source added to the activation assay.

genase. The Moco can be separated from FeMoco in these preparations by column chromatography on Sephadex G-100 in *N*-methylformamide (Figure 3). The FeMoco is synthesized only when nitrogenase is derepressed, whereas Moco is synthesized even when nitrogenase is repressed during growth on medium containing excess NH_4^+ (Pienkos et al., 1977). The FeMoco is specific for nitrogenase, whereas Moco is found in xanthine oxidase, nitrate reductase, sulfite oxidase, aldehyde oxidase, and in wild-type *A. vinelandii* grown in medium containing excess NH_4^+ (Nason et al., 1971; Johnson, Jones, and Rajagopalan, 1977; Pienkos et al., 1977).

Moco and FeMoco have similar properties: both are stable in *N*-methylformamide, sensitive to O_2, and unstable in aqueous environments. However, there are differences between these two cofactors—for example, their molecular weights as judged by Sephadex G-100 column chromatography—and a most intriguing question remains to be answered. Since Moco and FeMoco appear to coexist in nitrogen-fixing organisms, what is the relationship between these two cofactors? Is Moco a direct precursor of FeMoco?

The acetylene reduction assay for N_2 fixation has greatly simplified studies of N_2 fixation at the basic and applied levels. One of the interesting properties of FeMoco is that it catalyzes the reduction of acetylene to ethylene in absence of the enzyme protein and ATP (Shah, Chisnell, and Brill, 1978). The FeMoco catalyzes the reduction of acetylene to ethylene in the presence of sodium borohydride. The dependence of acetylene reduction

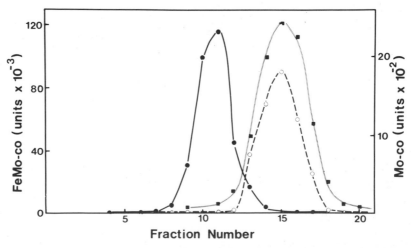

Figure 3.   Elution profile of FeMoco and Moco from a Sephadex G-100 column in NMF. The column was eluted with NMF at a flow rate of 12 ml/hr; approximately 3.5 ml fractions were collected anaerobically. ●——●, FeMoco; O---O, Moco from partially purified component I; ■···■, Moco from buttermilk xanthine oxidase.

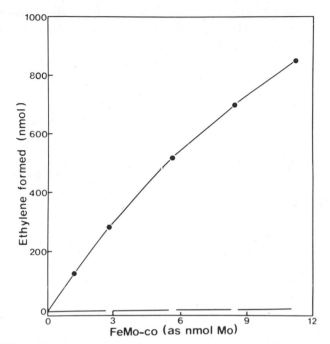

Figure 4.   Effect of FeMoco concentration on acetylene reduction. ●——●, 0.75 atm H₂ + 0.25 atm C₂H₂; O——O, 0.75 atm CO + 0.25 atm C₂H₂. The incubation time is 3 min.

on the concentration of FeMoco is shown in Figure 4. From the initial rates of the reaction, the specific activity of the iron-molybdenum cofactor is 34 nmoles of ethylene formed/min/nmole of Mo in the cofactor. On the basis of Mo content, the apparent turnover of FeMoco is 8% of that of nitrogenase in the acetylene-reduction assay (Shah and Brill, 1977; Shah, Chisnell, and Brill, 1978). Nitrogenase component I did not reduce acetylene under the conditions used for the FeMoco-catalyzed reaction. Like nitrogenase (Lockshin and Burris, 1965), FeMoco-dependent acetylene reduction (Figure 4) is strongly inhibited by carbon monoxide (Shah, Chisnell, and Brill, 1978).

Physiological and genetic studies of nitrogenase (Davis et al., 1972; Shah et al., 1973) showed that an electron paramagnetic resonance signal with $g = 3.65$ was caused by an active site on component I. FeMoco is responsible for this signal (Rawlings et al., 1978). The data on acetylene reduction further support the idea that FeMoco is an active site of nitrogenase, possibly even the site to which N₂ binds and is subsequently reduced. An important question that remains, however, is whether this acetylene-reduction activity is catalyzed by the same site on FeMoco as in nitrogenase. If the sites and mechanisms are the same, then insight into the mechanism of acetylene reduction by FeMoco should yield important clues into the mechanism of N_2 fixation by nitrogenase. This information should be helpful for designing new catalysts for industrial N_2 fixation.

REFERENCES

Cramer, S. P., W. O. Gillum, K. O. Hodgson, L. E. Mortenson, E. I. Stiefel, J. R. Chisnell, W. J. Brill, and V. K. Shah. 1978. The molybdenum site of nitrogenase. 2. A comparative study of Mo-Fe proteins and the iron-molybdenum cofactor by x-ray absorption spectroscopy. J. Am. Chem. Soc. In press.

Davis, L. C., V. K. Shah, W. J. Brill, and W. H. Orme-Johnson. 1972. Nitrogenase II. Changes in the EPR signal of component I (iron-molybdenum protein) of *Azotobacter vinelandii* nitrogenase during repression and derepression. Biochim. Biophys. Acta 256:512–523.

Detroy, R. W., D. F. Witz, R. A. Parejko, and P. W. Wilson. 1968. Reduction of N_2 by complementary functioning of two components from nitrogen-fixing bacteria. Proc. Natl. Acad. Sci. U.S.A. 61:537–541.

Emerich, D. W., and R. H. Burris. 1976. Interactions of heterologous nitrogenase components that generate catalytically inactive complexes. Proc. Natl. Acad. Sci. U.S.A. 73:4369–4373.

Johnson, J. L., H. P. Jones, and K. V. Rajagopalan. 1977. *In vitro* reconstitution of dimolybdosulfite oxidase by a molybdenum cofactor from rat liver and other sources. J. Biol. Chem. 252:4994–5003.

Ketchum, P. A., H. Y. Cambier, W. A. Frazier III, C. H. Madansky, and A. Nason. 1970. *In vitro* assembly of *Neurospora* assimilatory nitrate reductase from protein subunits of a *Neurospora* mutant and the xanthine oxidizing or aldehyde oxidase systems of higher animals. Proc. Natl. Acad. Sci. U.S.A. 66:1016–1023.

Lockshin, A., and R. H. Burris. 1965. Inhibition of nitrogen fixation in extracts from *Clostridium pasteurianum*. Biochim. Biophys. Acta 111:1–10.

Lvov, N. P., V. L. Ganelin, Z. Alikulov, and V. L. Kretovich. 1975. On the nature of a low molecular factor common to the molybdenum containing enzymes. J. Acad. Sci. USSR 3:371–376.

McKenna, C., N. P. Lvov, V. L. Ganelin, N. S. Sergeev, and V. L. Kretovich. 1974. Existence of a low molecular weight factor common to various molybdenum containing enzymes. Dokl. Akad. Nauk SSSR 217:228–231.

Münck, E., H. Rhodes, W. H. Orme-Johnson, L. C. Davis, W. J. Brill, and V. K. Shah. 1975. Nitrogenase. VIII. Mössbauer and EPR spectroscopy. The MoFe-protein component from *Azotobacter vinelandii* OP. Biochim. Biophys. Acta 400:32–53.

Nagatani, H. H., V. K. Shah, and W. J. Brill. 1974. Activation of inactive nitrogenase by acid-treated component I. J. Bacteriol. 120:697–701.

Nason, A., K. Y. Lee, S. S. Pan, P. A. Ketchum, A. Lamberti, and J. DeVries. 1971. *In vitro* formation of assimilatory reduced nicotinamide adenine dinucleotide phosphate: Nitrate reductase from a *Neurospora* mutant and a component of molybdenum-enzymes. Proc. Natl. Acad. Sci. U.S.A. 68: 3242–3246.

Pienkos, P. T., V. K. Shah, and W. J. Brill. 1977. Molybdenum cofactors from molybdoenzymes and *in vitro* reconstitution of nitrogenase and nitrate reductase. Proc. Natl. Acad. Sci. U.S.A. 74:5468–5471.

Rawlings, J., V. K. Shah, J. R. Chisnell, W. J. Brill, R. Zimmerman, E. Münck, and W. H. Orme-Johnson. 1978. Novel metal cluster in the iron-molybdenum cofactor of nitrogenase. Spectroscopic evidence. J. Biol. Chem. 253:1001–1004.

St. John, R. T., H. M. Johnston, C. Seidman, D. Garfinkel, J. K. Gordon, V. K. Shah, and W. J. Brill. 1975. Biochemistry and genetics of *Klebsiella pneumoniae* mutant strains unable to fix N_2. J. Bacteriol. 121:759–765.

Shah, V. K., L. C. Davis, and W. J. Brill. 1972. Nitrogenase. I. Repression and derepression of the iron-molybdenum and iron proteins of nitrogenase in *Azotobacter vinelandii*. Biochim. Biophys. Acta 256:498–511.

Shah, V. K., and W. J. Brill. 1977. Isolation of an iron-molybdenum cofactor from nitrogenase. Proc. Natl. Acad. Sci. U.S.A. 74:3249–3253.

Shah, V. K., J. R. Chisnell, and W. J. Brill. 1978. Acetylene reduction by the iron-molybdenum cofactor from nitrogenase. Biochem. Biophys. Res. Commun. 81:232–236.

Shah, V. K., L. C. Davis, J. K. Gordon, W. H. Orme-Johnson, and W. J. Brill. 1973. Nitrogenase III. Nitrogenaseless mutants of *Azotobacter vinelandii:* Activities, cross-reactions and EPR spectra. Biochim. Biophys. Acta 292:246–255.

Zumft, W. G. 1974. Separation of low molecular weight components from the molybdenum iron protein of clostridial nitrogenase. Ber. Dtsch. Bot. Ges. 87:135–143.

Nitrogen Fixation, Volume I
Edited by W. E. Newton and W. H. Orme-Johnson
Copyright 1980   University Park Press   Baltimore

Molybdenum Cofactor

J. L. Johnson, B. Hainline, H. P. Jones, and K. V. Rajagopalan

In 1964, Pateman et al. described a group of mutants of *Aspergillus nidulans* for which neither nitrate nor hypoxanthine could serve as the sole source of nitrogen, and that failed to induce nitrate reductase and xanthine dehydrogenase activities when exposed to nitrate and hypoxanthine, respectively. Genetic analysis of the mutants implicated at least six different loci, five of which were postulated to be involved in the synthesis of a molybdenum cofactor (*cnx*) required for the normal expression of the two enzymatic activities. The extensive studies on the analogous *nit-1* mutants of *Neurospora crassa* by Nason and his collaborators lent further credence to the existence of the common molybdenum cofactor. As in the case of the *cnx* mutants in *A. nidulans*, inactive nitrate reductase is induced in *Neurospora nit-1* on exposure to nitrate. Nason and coworkers (1970) found that mixing of *nit-1* extracts with extracts of uninduced nonallelic *nit* mutants or wild-type *N. crassa* led to in vitro complementation and expression of nitrate reductase activity. Subsequently, constitutive cofactor activity was found in *A. nidulans* (Ketchum and Downey, 1975) and in a wide variety of bacteria (Ketchum and Sevilla, 1973; Ketchum and Swarin, 1973). Convincing evidence for the uniqueness of the molybdenum cofactor was provided by the finding of Nason and coworkers (1971) that acid-denatured samples of all tested molybdoenzymes from diverse sources served as donors of cofactor for the activation of nitrate reductase in *nit-1* extracts. Significantly, neither inorganic molybdate nor a variety of organic complexes of the metal could generate nitrate reductase activity in the extracts.

Nason and coworkers (1971) had reported that preparations of the molybdenum-containing component of nitrogenase served as donor of the cofactor to *nit-1* nitrate reductase. However, more recently, Shah and Brill

(1977) isolated the molybdenum-containing prosthetic group of nitrogenase and showed it to contain iron, sulfide, and molybdenum in the ratio of 8:6:1. They also found that the purified nitrogenase cofactor did not complement *nit-1* nitrate reductase and that the cofactor derived from acid-denatured xanthine oxidase was incapable of activating cofactor-deficient nitrogenase (Pienkos, Shah, and Brill, 1977). Our laboratory has been interested in a group of molybdoenzymes of animal origin, especially sulfite oxidase and xanthine oxidase. Liver sulfite oxidase has been isolated from various sources and contains equimolar molybdenum and heme and no Fe-S; xanthine oxidase from all sources contains molybdenum, FAD, and Fe-S in the ratio of 1:1:4. These values are obviously incompatible with the existence in these enzymes of a molybdenum cofactor identical to that of nitrogenase.

Nason and coworkers always effected reconstitution in crude extracts of *N. crassa nit-1*. Although these studies showed that cofactor derived from sulfite oxidase or xanthine oxidase was capable of complementing *nit-1* nitrate reductase, the possibility remained that different types of cofactor structures provided by the donor enzymes were modified by the *nit-1* extract to the specific type required for nitrate reductase. Clearly, more definitive information on the molybdenum cofactor could only be derived through the use of purified apoenzymes. Such studies have been performed in our laboratory using purified demolybdo sulfite oxidase from rat liver.

A technique for production of molybdenum deficiency in rats through the use of tungsten as a competitive molybdenum antagonist has been developed (Johnson, Rajagopalan, and Cohen, 1974). The demolybdo forms of sulfite oxidase and xanthine oxidase have been purified from the livers of such animals (Johnson, Cohen, and Rajagopalan, 1974; Johnson et al., 1974) and shown to be immunologically and chemically identical to the respective native enzymes except in molybdenum content. Demolybdo sulfite oxidase preparations were found to contain up to 40% tungsten in the molybdenum site (Johnson, Cohen, and Rajagopalan, 1974). The availability of purified demolybdo sulfite oxidase has enabled us to carry out a systematic study of the reconstitution of the enzyme activity by inorganic molybdate, by acid-denatured xanthine oxidase, and by sources of preformed molybdenum cofactor. The results of these studies are outlined below.

IN VITRO RECONSTITUTION OF
SULFITE OXIDASE BY MOLYBDATE

The ability of inorganic molybdate to reconstitute demolybdo sulfite oxidase is shown in Figure 1. The activation process showed a pH optimum of 7.4 and a temperature optimum of 37°C, and required about 60 min for

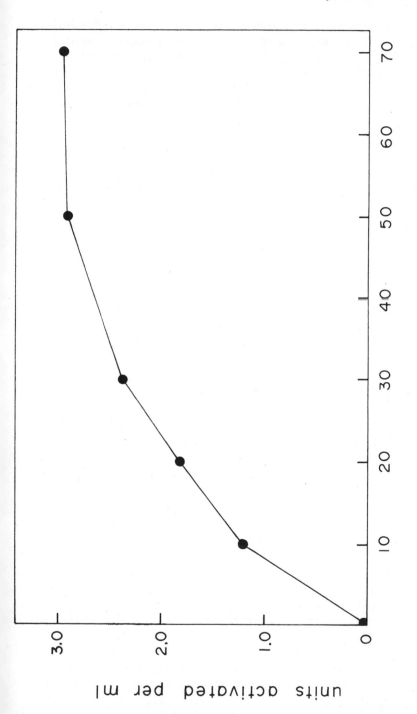

Figure 1.   Reconstitution of purified rat liver sulfite oxidase by molybdate in 0.01 M phosphate buffer (pH 7.4) at 37°C. Full reconstitution would have yielded 10 units/ml.

completion. The extent of activation by molybdate never exceeded the proportion of sulfite oxidase molecules containing tungsten. Significantly, prior incubation of the enzyme at 37°C for 60 min rendered it incompetent for subsequent activation by molybdate.

Anaerobic incubation of demolybdo sulfite oxidase with sulfite leads to the quantitative generation of a tungsten (V) EPR signal (Johnson and Rajagopalan, 1976). This signal was no longer observable in preparations incubated at 37°C for 60 min in the presence or absence of molybdate. When samples of demolybdo enzyme were rendered partially incompetent for reconstitution with molybdate, it was found that the extent of loss of the specific tungsten (V) EPR signal was greater than the extent of loss of reconstitutibility (Jones, Johnson, and Rajagopalan, 1977).

IN VITRO RECONSTITUTION OF
SULFITE OXIDASE BY MOLYBDENUM COFACTOR

The fact that the reconstitution of demolybdo sulfite oxidase by inorganic molybdate was only partial led to a search for a mechanism for reconstituting those molecules that contained neither molybdenum nor tungsten at the active site. Initial evidence for the existence of preformed molybdenum cofactor was obtained through in vitro complementation of homogenates of livers from control rats and livers from tungsten-treated rats to generate sulfite oxidase activity. The material in control rat liver was equally capable of reconstituting purified demolybdo sulfite oxidase. Studies on the intracellular localization of the reconstituting cofactor showed that it was localized exclusively in the mitochondria, and specifically on the outer membrane of the organelle (Johnson, Jones, and Rajagopalan, 1977).

Cofactor-dependent reconstitution showed linear dependence on sulfite oxidase and cofactor concentrations, and proceeded optimally at pH 7.4 and 37°C. In contrast to the molybdate activation process, enzyme previously incubated at 37°C for 60 min showed no loss of reconstitutibility by the cofactor-dependent mechanism. Maximum activation by the cofactor corresponded to 60% to 70% of the total population of demolybdo enzyme, suggesting that the rat liver molybdenum cofactor reconstitutes sulfite oxidase molecules with vacant molybdenum sites. The data presented in Table 1 conclusively demonstrate that activation by molybdate and by the cofactor involve mutually exclusive populations of demolybdo sulfite oxidase. The extent of reconstitution in the simultaneous presence of molybdate and cofactor was equal to the sum of the activities generated by the two mechanisms independently. Furthermore, reconstitution with cofactor after initial reconstitution with molybdate also resulted in additivity of reconstitution. In contrast, when cofactor-dependent reconstitution was performed

Table 1.   Reconstitution of demolybdo sulfite oxidase by molybdate and by rat liver molybdenum cofactor

Source of molybdenum	Sulfite oxidase reconstituted (%)
Molybdate	34.7
Molybdenum cofactor	67.3
Molybdate + molybdenum cofactor[a]	98.7
Molybdate	34.9
+ molybdenum cofactor[b]	96.0
Molybdenum cofactor	65.2
+ molybdate[b]	62.8

[a] Simultaneous incubation.

[b] Sequential incubations.

before addition of molybdate, the latter failed to generate sulfite oxidase activity, in keeping with the expected effect of preincubation of the enzyme at 37°C on this process.

These studies have led to the formulation of the scheme presented in Figure 2 for the reconstitution of demolybdo sulfite oxidase. Crucial to the proposal is the existence of a cofactor molecule that is the molybdenum-binding prosthetic group present as the metal complex in native sulfite oxidase. In tungsten-containing sulfite oxidase molecules, the site is occupied by tungsten-substituted cofactor, whereas the metal-free enzyme molecules are devoid of the cofactor molecule as well. The latter population is readily activated by sources containing the preformed molybdenum-cofactor complex when incubated at 37°C. In the case of the tungsten-containing molecules, it is proposed that incubation at 37°C leads to dissociation of tungsten from the protein-bound cofactor and that, in the presence of molybdate, reconstitution is achieved through the binding of molybdenum soon after the release of tungsten. In the absence of molybdenum, the protein-bound empty cofactor undergoes a chemical modification that prevents subsequent activation by molybdate or by cofactor.

DISTRIBUTION OF MOLYBDENUM COFACTOR

Cofactor activity capable of reconstituting sulfite oxidase has been detected in various sources such as human liver, *N. crassa*, and *Escherichia coli* (Johnson, Jones, and Rajagopalan, 1977). All of these sources of cofactor, including rat liver, are also effective in the reconstitution of nitrate reductase activity in extracts of *N. crassa nit-1*, attesting to the identicalness of the cofactor from all sources tested.

RECONSTITUTION OF RAT LIVER SULFITE OXIDASE

W-Containing Molecules   Apo-Molecules

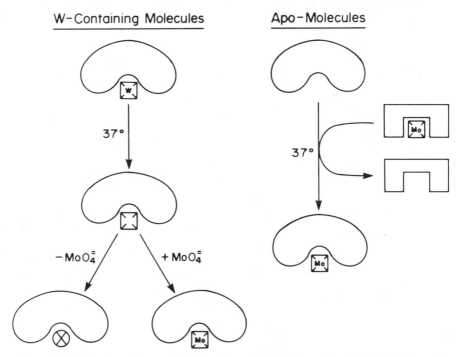

Figure 2. Mechanism of reconstitution of demolybdo sulfite oxidase by molybdate and by preformed molybdenum cofactor.

IN VITRO RECONSTITUTION BY
ACID-DENATURED XANTHINE OXIDASE

Acid-denatured preparations of bovine milk xanthine oxidase were found to be capable of activating purified demolybdo sulfite oxidase, as well as nitrate reductase in *N. crassa nit-1*. It is known that a cyanide-reactive sulfhydryl group, identified as "persulfide," serves an essential function at the active site of xanthine oxidase (Massey and Edmondson, 1970). Sulfite oxidase does not appear to have a similar group at its active site (Johnson and Rajagopalan, unpublished observations). The data in Table 2 show that cyanide-treated xanthine oxidase is functional as a source of cofactor for sulfite oxidase and nitrate reductase; thus, the persulfide group is not part of the transferable cofactor.

    In agreement with the findings of other researchers, we have found that the cofactor released from xanthine oxidase is extremely labile and loses its reconstitutive ability rapidly. However, the cofactor was somewhat stabi-lized in the presence of molybdate, and could be separated from the denatured protein by gel filtration (Figure 3). Under these conditions, cofactor activity eluted slightly ahead of FAD, and was found to coelute with a fluorescent material different from flavin.

CHARACTERIZATION OF A COMMON PROSTHETIC GROUP OF MOLYBDOENZYMES

The existence of a transferable, noncovalently bound prosthetic group that is involved in complexation of molybdenum in enzymes containing that metal has led to a search for a common, previously unidentified material in a group of molybdenum enzymes. The simplest of all molybdoproteins is the molybdenum domain of rat liver sulfite oxidase (Johnson and Rajagopalan, 1977), which contains only molybdenum as the prosthetic group. Figure 4 shows the absorption spectrum of the molybdenum center of the protein before and after the addition of sodium dodecyl sulfate (SDS). Addition of SDS quickly abolished the characteristic spectrum of the molybdenum center and generated a new band with a maximum at 380 nm. Prolonged incubation with SDS led to an alteration in this spectrum with gradual enhancement of the absorbance of the sample. These data could be explained on the basis of rapid release of the cofactor from the protein and subsequent modification to yield an altered species. Gel filtration of the mixture on Sephadex G-25 in SDS yielded three well-separated fluorescent fractions termed A, B, and C in order of elution. Fraction A eluted in the included column volume, whereas B and C were retarded, indicating an affinity for the Sephadex. Using a variety of denaturation techniques such as boiling, acid treatment and neutralization, and treatment with guanidine,

Table 2.   Effect of cyanide treatment on cofactor activity[a]

	Native xanthine oxidase	CN-treated xanthine oxidase
Sulfite oxidase reconstitution: units sulfite oxidase per μg xanthine oxidase	0.066	0.036
nit-1 reconstitution: nmoles NO_2^-/10 min per μg xanthine oxidase	0.111	0.122
Xanthine oxidase activity: units per mg	10.8	0
Xanthine oxidase molybdenum: μg per mg	0.32	0.30

[a] Native or cyanide-treated xanthine oxidase was acidified to pH 2.7 with HCl and maintained at 4°C for 10 min. Samples were neutralized with NaOH and incubated with purified demolybdo sulfite oxidase at 37°C for 60 min or extracts of *Neurospora nit-1* for 15 min at 25°C.

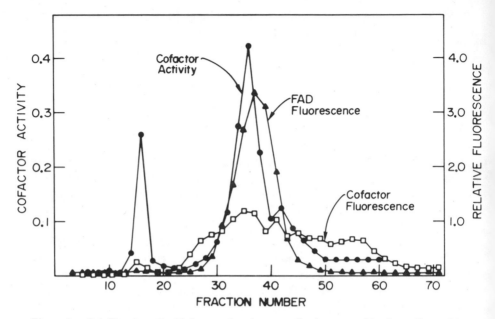

Figure 3.   Gel filtration of acid-denatured and reneutralized extract of bovine milk xanthine oxidase. Milk xanthine oxidase was acidified to pH 1.5 with 12 N HCl and incubated for 5 minutes at 0°C. The sample was neutralized with 3 N NaOH and centrifuged to remove denatured enzyme. Sodium molybdate was added to the supernatant to a final concentration of 0.01 M and the sample was fractionated on a Sephadex G-25 column using deaerated 0.05 M phosphate (pH 7.3) containing 0.01 M molybdate.

SDS, pyridine, or high salt, and filtration on Sephadex columns, we have obtained similar materials from several molybdoenzymes. The fluorescence spectra of two of the fractions (A and B) from various sources are shown in Figure 5. The third fraction (C) in each case showed a principal excitation peak at 350 nm and an emission peak at 420 nm. Thus, even though each enzyme yielded multiple forms of the molybdenum cofactor, the corresponding fraction from each enzyme had essentially similar fluorescent properties. The relationship of the three forms to one another is not yet clear; however, it is likely that they represent structural modifications that occur during isolation and subsequent exposure to light and oxygen. The absorption spectra of the cofactor from several sources are shown in Figure 6. They are quite similar to one another and to the spectrum of SDS-treated molybdenum fragment of sulfite oxidase (Figure 4). Because the samples used in Figure 6 were devoid of molybdenum, these spectra are the properties of the modified cofactor per se.

Although we have made no attempt at maintaining the reconstitutive activity of the cofactor during isolation, our approach has been to obtain

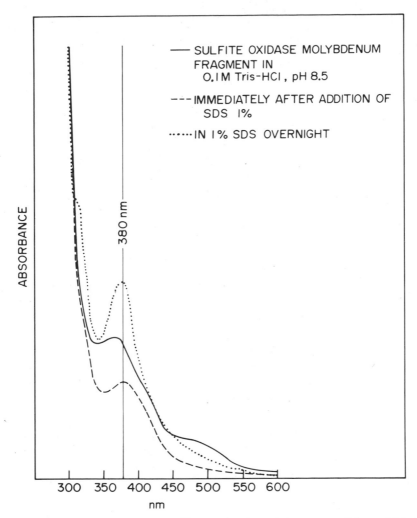

Figure 4. Effect of sodium dodecyl sulfate on the absorption spectrum of the molybdenum fragment of rat liver sulfite oxidase. The spectra represent: (———), native fragment in 0.1 M Tris HCl (pH 8.5); (------), immediately after addition of 1% SDS; (·········), after overnight treatment with SDS.

reasonable quantities of the material for structural studies. By using the criterion that the same material should be obtainable from several different molybdoenzymes, we have established the validity of the presumption that the isolated material is the molybdenum cofactor. Meaningful structural studies are now possible.

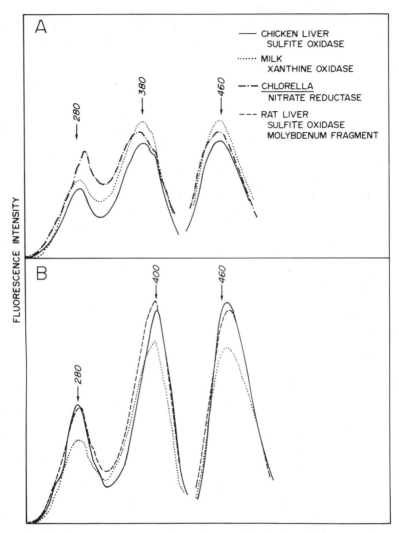

Figure 5. The fluorescence spectra of two forms of isolated molybdenum cofactor from various sources.

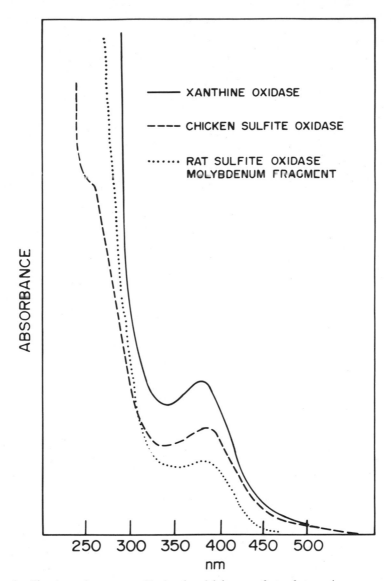

Figure 6.   The absorption spectra of isolated molybdenum cofactor from various sources.

REFERENCES

Johnson, J. L., H. J. Cohen, and K. V. Rajagopalan. 1974. Molecular basis of the biological function of molybdenum. Molybdenum-free sulfite oxidase from livers of tungsten-treated rats. J. Biol. Chem. 249:5046–5055.

Johnson, J. L., H. P. Jones, and K. V. Rajagopalan. 1977. *In vitro* reconstitution of demolybdo sulfite oxidase by a molybdenum cofactor from rat liver and other sources. J. Biol. Chem. 252:4994–5003.

Johnson, J. L., and K. V. Rajagopalan. 1976. Electron paramagnetic resonance of the tungsten derivative of rat liver sulfite oxidase. J. Biol. Chem. 251:5505–5511.

Johnson, J. L., and K. V. Rajagopalan. 1977. Tryptic cleavage of rat liver sulfite oxidase. Isolation and characterization of molybdenum and heme domains. J. Biol. Chem. 252:2017–2025.

Johnson, J. L., K. V. Rajagopalan, and H. J. Cohen. 1974. Molecular basis of the biological function of molybdenum. Effect of tungsten on xanthine oxidase and sulfite oxidase in the rat. J. Biol. Chem. 249:859–866.

Johnson, J. L., W. R. Waud, H. J. Cohen, and K. V. Rajagopalan. 1974. Molecular basis of the biological function of molybdenum. Molybdenum-free xanthine oxidase from livers of tungsten-treated rats. J. Biol. Chem. 249:5056–5061.

Jones, H. P., J. L. Johnson, and K. V. Rajagopalan. 1977. *In vitro* reconstitution of demolybdo sulfite oxidase by molybdate. J. Biol. Chem. 252:4988–4993.

Ketchum, P. A., and R. J. Downey. 1975. *In vitro* restoration of nitrate reductase: investigation of *Aspergillus nidulans* and *Neurospora crassa* nitrate reductase mutants. Biochim. Biophys. Acta 385:354–361.

Ketchum, P. A., and C. L. Sevilla. 1973. *In vitro* formation of nitrate reductase using extracts of the nitrate reductase mutant of *Neurospora crassa, nit-1*, and *Rhodospirillum rubrum*. J. Bacteriol. 116:600–609.

Ketchum, P. A., and R. S. Swarin. 1973. *In vitro* formation of assimilatory nitrate reductase: Presence of the constitutive component in bacteria. Biochem. Biophys. Res. Commun. 52:1450–1456.

Massey, V., and D. Edmondson. 1970. On the mechanism of inactivation of xanthine oxidase by cyanide. J. Biol. Chem. 245:6595–6598.

Nason, A., A. D. Antoine, P. A. Ketchum, W. A. Frazier III, and D. K. Lee. 1970. Formation of assimilatory nitrate reductase by *in vitro* intercistronic complementation in *Neurospora crassa*. Proc. Natl. Acad. Sci. U.S.A. 65:137–144.

Nason, A., K.-Y. Lee, S.-S. Pan, P. A. Ketchum, A. Lamberti, and J. De Vries. 1971. *In vitro* formation of assimilatory reduced nicotinamide adenine dinucleotide phosphate:nitrate reductase from a *Neurospora crassa* mutant and a component of molybdenum enzymes. Proc. Natl. Acad. Sci. U.S.A. 68:3242–3246.

Pateman, J. A., D. J. Cove, B. M. Rever, and D. B. Roberts. 1964. A common cofactor for nitrate reductase and xanthine dehydrogenase which also regulates the synthesis of nitrate reductase. Nature 201:58–60.

Pienkos, P. T., V. K. Shah, and W. J. Brill. 1977. Molybdenum cofactors for molybdoenzymes and *in vitro* reconstitution of nitrogenase and nitrate reductase. Proc. Natl. Acad. Sci. U.S.A. 74:5468–5471.

Shah, V. K., and W. J. Brill. 1977. Isolation of an iron-molybdenum cofactor from nitrogenase. Proc. Natl. Acad. Sci. U.S.A. 74:3249–3253.

Nitrogen Fixation, Volume I
Edited by W. E. Newton and W. H. Orme-Johnson
Copyright 1980   University Park Press   Baltimore

The Molybdenum Site in Nitrogenase—Structural Elucidation by X-Ray Absorption Spectroscopy

Keith O. Hodgson

Molybdenum is well known to be an essential constituent of nitrogenase. Approximately two molybdenum atoms and 24–32 iron atoms are contained in the molybdenum-iron (MoFe) protein. Despite numerous attempts to spectroscopically probe and chemically model the molybdenum site, there exist no experiments that as yet directly address the structure of the site. Furthermore, there are no experiments that implicate molybdenum in the catalytic cycle of dinitrogen fixation. It is obvious that there is a need for a spectroscopic method that can define the state of molybdenum under both resting and active conditions.

In principle, one such technique is x-ray absorption spectroscopy (XAS), which studies the influence of the electronic nature and structural environment of a given atom on its x-ray absorption behavior as a function of energy. The theoretical basis (Ashley and Doniach, 1975; Lytle et al., 1975; Stern et al., 1975; Lee and Beni, 1977) and practical aspects (Cramer et al., 1978c) of data collection and analysis have been discussed. We have recently used XAS and the study of the fine structure above the absorption edge (called extended x-ray absorption fine structure or EXAFS) to investigate the molybdenum environment in nitrogenase.

This paper provides a brief introduction to XAS and describes how the EXAFS can be used to elicit structural information. This is done by illus-

The EXAFS studies described herein were all performed at the Stanford Synchrotron Radiation Laboratory, which is a national facility supported by the National Science Foundation. This work was supported primarily through NSF grant PCM 17105.

trating how the EXAFS analysis of a structurally uncharacterized compound leads to determination of accurate coordination distances and numbers. The application of this method to the molybdenum in nitrogenase and related systems is then described.

STRUCTURAL ANALYSIS
USING X-RAY ABSORPTION SPECTROSCOPY

The x-ray absorption spectrum (Figure 1) exhibits decreasing absorption as the photon energy is increased. Superimposed on this smooth background is a steeply rising discontinuity in the absorption at an energy characteristic of the molybdenum in the sample. This abrupt increase in absorption occurs when the incident photon has just sufficient energy to promote a core electron to unoccupied valence levels or to the continuum. The molybdenum absorption edge at 20 keV (called the K edge) arises from excitation of the 1s electron into the continuum.

The absorption trend both near the edge and beyond it is not smoothly varying; rather, there exists a wealth of fine structure characteristic of the chemical environment of the x-ray absorbing atom. Structure at energies close to the absorption edge (not seen in Figure 1) typically consists of a series of approximately Lorentzian lines, superimposed on a steeply rising absorption step, that arise from transitions of core electrons to vacant optical levels below the continuum.

Although bound-state transitions can account for most of the structure within about 25 eV of the absorption edge, in virtually all cases additional structure is observed over several hundred electron volts past the edge. This long range oscillation, or EXAFS, results from interference between the photoelectron wave propagating from the x-ray absorbing atom and the wave backscattered by neighboring atoms. Depending on whether the scattered wave is in phase or out of phase with the outgoing photoelectron wave, there will be an increase or decrease in the absorption.

Edge structure and EXAFS clearly have quite different physical origins, and they contain complementary chemical information about the x-ray absorbing atom and its environment. Analysis of the positions and relative intensities of the absorption edge features reveals details about the metal site symmetry, its oxidation state, and the nature of the surrounding ligands. Interpretation of the phase, amplitude, and frequency of the EXAFS oscillations provides information about the type, number, and distances of atoms in the vicinity of the absorber. It is these latter effects with which this paper primarily deals.

Data Collection and Processing

Measurement of the XAS spectrum is quite easily accomplished by recording the ratio of incident beam flux to flux transmitted through the sample as

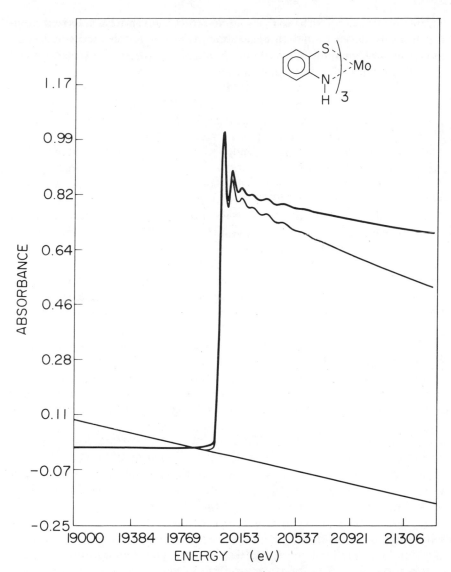

Figure 1.   X-ray absorption spectrum of tris(2-amidobenzenethiolato)molybdenum(VI). The K absorption edge of the molybdenum occurs at about 20,000 eV. The EXAFS is clearly observable to the high energy side of the absorption edge. The lighter line is the original data and the darker line shows the result of subtracting out a background determined by extrapolating the background absorption from before the edge.

a function of energy. The x-ray source for the experiment is synchrotron radiation produced by a high energy electron storage ring at the Stanford Synchrotron Radiation Laboratory. This source provides x-ray fluxes many orders of magnitude higher than those obtainable from conventional x-ray tubes (Bienenstock and Winick, 1978; Cramer and Hodgson, 1979). The spectrum of *tris*(2-amidobenzenethiolato)molybdenum(VI) shown in Figure 1 was recorded on a solid sample. The detailed procedures used for analysis of the data have been published (Cramer et al., 1978c). Briefly, the data are calibrated to a known energy standard and a background is removed to yield the EXAFS itself.

The EXAFS: Its Origin and Analysis

The factors that contribute to EXAFS are summarized as follows. Each coordination shell of surrounding ligands will result in the contribution of a term to the overall EXAFS pattern, which manifests itself as a damped, modulated sine wave. The frequency of the sine wave is determined by the distance to the coordination shell (commonly called the scatterer) and the phase shift, α. The longer the distance from absorber to scatterer, the higher the frequency of the wave. The absolute phase of the wave (its origin) is governed by the atomic numbers of the absorber and the scatterers in the shell. The overall magnitude of the wave is proportional to the atomic number of the scatterer and is further governed in an additive fashion by the number of atoms in the shell. Finally, the shape of the wave (its amplitude modulation) is again determined by the atomic number of the scatterer and by a Debye-Waller factor that accounts for an exponential falloff due to vibration effects.

Figure 2 illustrates these effects for three different Mo-to–single scatterer pairs in which the distances have been fixed at 2 Å. The Mo—Mo wave is seen to be substantially larger than the Mo—S wave and has a different shape. It would thus be impossible to describe Mo—S EXAFS with Mo—Mo parameters. In the case of the Mo—O and Mo—S, the waves have an origin that differs by almost π radians; thus, one could not describe Mo—O EXAFS with Mo—S parameters. These effects allow qualitative structural information to be extracted about the type of atoms in each coordination shell.

There are two basic approaches to the analysis of EXAFS data: Fourier transform and curve-fitting methods. I only consider curve-fitting analysis, which has been developed and tested in our laboratories for the structural analysis of molybdenum complexes (Cramer et al., 1978c). The essence of EXAFS curve-fitting analysis is to use a parameterized function that will model the observed EXAFS, and then to adjust the structure-dependent parameters in this theoretical EXAFS expression until the fit with the experimentally observed EXAFS is optimized. The final values of

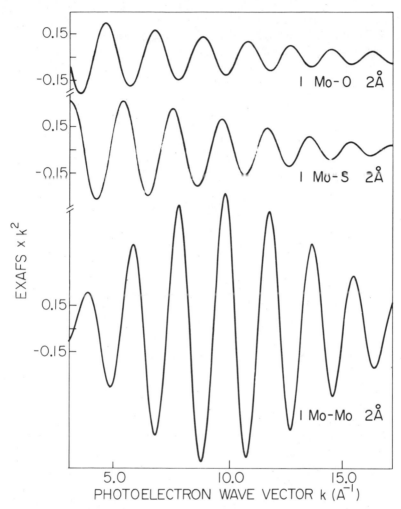

Figure 2.   The effects of phase and amplitude on waves from three different molybdenum-scatterer interactions each at 2.0-Å distance. Note that the difference in the phases of the oxygen and sulfur waves allows them to be distinguished on this basis alone (even though the atoms are at the same distance). Likewise, Mo—Mo and Mo—O have different phases. Although Mo—Mo and Mo—S have similar phases, their vastly different amplitude envelopes allow them to be distinguished as well. These plots were made using the semiempirical phase and amplitude parameters given in Cramer et al., 1978c.

the optimized parameters will then yield structural information about the compound under study.

This procedure is illustrated schematically in Figure 3 for a metal surrounded by a coordination shell of sulfur and of nitrogen atoms. The N and S shells each contribute a sine wave component to the overall EXAFS. Each

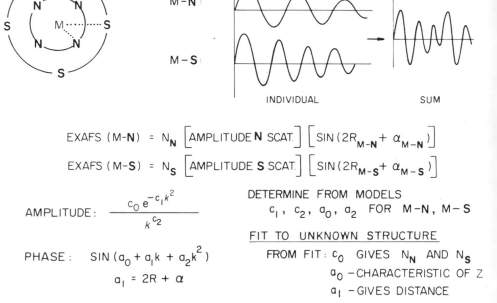

$$\text{EXAFS (M-N)} = N_N \left[\text{AMPLITUDE } N \text{ SCAT.} \right] \left[\sin(2R_{M-N} + \alpha_{M-N}) \right]$$

$$\text{EXAFS (M-S)} = N_S \left[\text{AMPLITUDE } S \text{ SCAT.} \right] \left[\sin(2R_{M-S} + \alpha_{M-S}) \right]$$

AMPLITUDE: $\dfrac{c_0\, e^{-c_1 k^2}}{k^{c_2}}$

DETERMINE FROM MODELS

c_1, c_2, a_0, a_2 FOR M-N, M-S

FIT TO UNKNOWN STRUCTURE

PHASE: $\sin(a_0 + a_1 k + a_2 k^2)$

$a_1 = 2R + \alpha$

FROM FIT: c_0 GIVES N_N AND N_S

a_0 - CHARACTERISTIC OF Z

a_1 - GIVES DISTANCE

Figure 3. Schematic representation of EXAFS and fitting parameters for a hypothetical metal complex with two coordination shells (S and N).

wave (M—N and M—S) has parameters that determine its amplitude shape and phase behavior. The phase is modeled by the function indicated in Figure 3, and the constants a_0 and a_2 are determined for each wave from curve-fitting data from model compounds of known structure. The amplitudes of the M—N and M—S waves are parameterized and the constants c_1 and c_2 determined from model data fits. Alternatively, the phase and amplitude parameters can be calculated and tabulated values are available for most elements (Lee, Teo, and Simons, 1977; Teo et al., 1977). Finally, the overall multiplier of each wave, N, is directly proportional to the number of atoms in the shell. Once determined, the constants $c_1, c_2, a_0,$ and a_2 (which describe the EXAFS for the S or N wave) can be used in fits to those EXAFS data of unknowns. From the fit to the data of the unknown structure, the EXAFS analysis provides the distance (from the a_1 term) and number of atoms (from c_0) in each coordination shell. In practice, up to four shells of ligands out to about 4.5 Å from the metal can be analyzed. The approximations and errors in this analysis have been discussed in detail (Cramer et al., 1978c; Shulman et al., 1978; Tullius, Frank, and Hodgson, 1978).

This approach to determining structure is illustrated by EXAFS analysis for *tris*(2-amidobenzenethiolato)molybdenum(VI), which at the time of the analysis had yet to be crystallized. Assuming that sulfur would ligate to molybdenum, a one-shell fit was first carried out. The results of fitting a Mo—S wave are shown on the left of Figure 4. It is apparent that only a single scatterer shell cannot account for the EXAFS. Note that the fit cannot reproduce the presence in the data of the beat region, which is characteristic of at least two different absorber-scatterer distances. Adding a second shell of atoms (in this case nitrogen atoms) dramatically improves the fit, as seen on the right of Figure 4. From the four variables in this two-shell fit, 3.2 sulfur atoms at 2.419 Å and 3.1 nitrogen atoms at 1.996 Å were calculated.

The uniqueness of the choice of atomic type can be tested by using phase and amplitude parameters characteristic of different Mo—scatterer pairs. For example, a reasonable fit with a second shell of sulfur atoms is impossible because the a_0 terms for Mo—N and Mo—S waves are so different. Likewise, the large and characteristic amplitude of a Mo—Mo pair (see Figure 2) makes it impossible to use these parameters to describe the Mo—N wave. Using such logic, in combination with knowledge of reasonable bond distances, one can generally arrive at a structure. It should be emphasized that it would be quite difficult to distinguish between scattering atoms that are adjacent in the periodic table (such as nitrogen and oxygen), and this introduces an ambiguity.

Subsequent to the EXAFS analysis, the structure of *tris*(2-amidobenzenethiolato)molybdenum (VI) was determined (Yamanouchi and Enemark, 1978) and is shown on the top in Figure 5 along with a comparison with the distances determined by EXAFS. The agreement is excellent. Figure 5 (bottom) shows another structure where EXAFS distances were determined prior to solution of the crystal structure (Berg et al., 1979). A comparison of 20 distances calculated by EXAFS on a variety of molybdenum complexes (a representative selection of which are given in Table 1) with crystallographic results yields an average deviation of 0.012 Å. Furthermore, coordination numbers were determined to an accuracy of around 1 atom in 5. Results on inorganic complexes and proteins containing copper (Tullius, Frank, and Hodgson, 1978) and iron (Bunker and Stern, 1977; Cramer et al., 1978d; Shulman et al., 1978) have revealed similar accuracy for determination of structure. As a coordination environment becomes more complex (for example, with a spread in coordination distance produced by Jahn-Teller distortions), theoretical formulation (Shulman et al., 1978) and practical experience (Cramer et al., 1978c; Tullius, Frank and Hodgson, 1978) reveals that for most data distances to the same ligand type cannot be resolved unless they differ by more than about 0.15 Å.

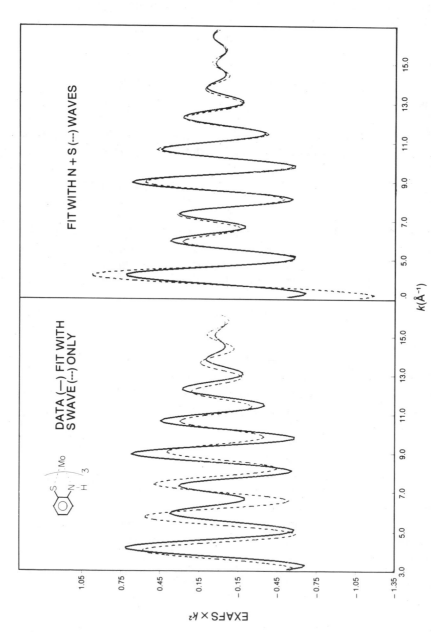

Figure 4.   Curve-fitting analysis of molybdenum EXAFS. For the structure shown on the left, the data (——) are fit (– – –) with only one sulfur wave and it can be clearly seen that this is inadequate to describe the phase and amplitude of the observed EXAFS. The fit on the right includes a second coordination shell (N) and gives an excellent fit. The numerical results of the fitting analysis are given in Table 1.

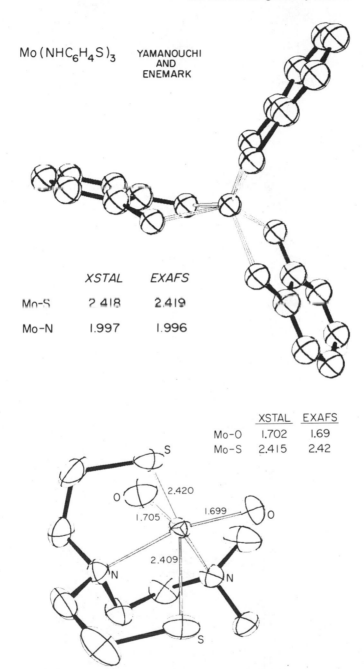

Mo(NHC_6H_4S)$_3$ YAMANOUCHI
                        AND
                        ENEMARK

	XSTAL	EXAFS
Mo-S	2.418	2.419
Mo-N	1.997	1.996

	XSTAL	EXAFS
Mo-O	1.702	1.69
Mo-S	2.415	2.42

Figure 5.   ORTEP plots from x-ray structure analysis of two molybdenum complexes. The distances determined from the crystal structures are shown along with the EXAFS results.

Table 1. Structure determination from Mo EXAFS

Compound	Mo—O [number]		Mo—S [number]		Mo—X [number]		Mo—Mo [number]	
	Fit	Crystal	Fit	Crystal	Fit	Crystal	Fit	Crystal
$MoS_4{}^{2-}$			2.179 [4.5]	2.180 [4]				
$Mo(S_2CNEt_2)_4$			2.532 [6.8]	2.529 [8]				
$MoO(S_2CNEt_2)_2$	1.661 [1.2]	1.664 [1]	2.426 [3.7]	2.414 [4]				
$Mo(C_6H_4NHS)_3$			2.419 [2.9]	2.418 [3]	1.996 [3.2]	1.997 [3]		
					2.328 [2.2]	2.326 [3]		
MoO_3dien	1.740 [3.3]	1.736 [3]						
$Mo_2O_4(S_2CNEt_2)_2$	1.670 [1.3]	1.679 [1]	2.445 [1.5]	2.455 [2]	1.939 [1.9]	1.941 [2]	2.578 [1.0]	2.580 [1]
$Mo_2O_2S_2(S_2CNEt_2)_2$	1.663 [1.0]	1.655 [1]	2.472 [1.7]	2.444 [2]	2.325 [2.5]	2.310 [2]	2.836 [0.5]	2.817 [1]
MoO_2	1.694 [2.4]	1.702 [2]	2.424 [1.4]	2.415 [2]				

APPLICATIONS OF EXAFS TO STUDY METAL IONS IN PROTEINS

Before discussing applications to molybdenum in nitrogenase, it is worthwhile to point out certain features of EXAFS analysis. The restricted nature of x-ray absorption data is often an advantage. In contrast to crystallographic studies, the information present in EXAFS is limited to the immediate vicinity of the absorber, and long-range features such as secondary and tertiary structure are not revealed. Thus, only a small number of variables need refining to obtain the local structure, whereas accuracy in diffraction results depends on refining all of the large number of atomic positions in the crystal. The distances obtained from careful EXAFS studies can be accurate to within a few hundredths of an angstrom, whereas most protein crystallographic results have accuracies an order of magnitude worse.

Some experimental advantages of x-ray absorption spectroscopy should be noted. First, this technique can be applied to any state of matter; therefore, crystals, transparent solutions, or frozen samples are not required. Second, the selection rules are such that an absorption edge and associated fine structure will always exist, so paramagnetic or isotopically enriched samples are unnecessary. The lack of isotopic sensitivity makes labeling experiments impossible, but metal and ligand substitution studies are conceivable. Finally, x-ray absorption spectra can be collected in minutes or hours, and, once obtained, the results may be available within days or weeks, in contrast with the years often required for high resolution single crystal protein diffraction data analysis and refinement. It is for all of these reasons that molybdenum in nitrogenase is an attractive candidate for EXAFS analysis.

EXAFS STUDY OF THE MOLYBDENUM SITE IN NITROGENASE

EXAFS studies were initially carried out with lyophilized MoFe component from *Clostridium pasteurianum*, and the results of these studies have recently appeared (Cramer et al., 1978b). Subsequently, the MoFe component from *Azotobacter vinelandii* and the FeMo cofactor isolated therefrom have been studied (Cramer et al., 1978a). Highlights of these results, along with more recent studies of model complexes, are given below and evidence is developed for the structure of the molybdenum site in the resting state of nitrogenase.

X-Ray Absorption Studies
of the MoFe Component from *Clostridium pasteurianum*

The first x-ray absorption spectrum recorded on lyophilized MoFe protein revealed a distinct beat pattern (see Figure 6). The presence of such a beat, reminiscent of that seen in model complexes, immediately suggested at least two shells of scatterers around the molybdenum. Careful curve fitting analysis of the EXAFS, using methods identical to those outlined in "Structural Analysis" above, identified the lower frequency component as sulfur at 2.35 Å from the molybdenum. The absolute phase of this wave is in excellent agreement with Mo—S ligation while being significantly different from Mo—O values. The amplitude of the wave indicated about three sulfurs. A second conclusion is that no Mo=O bonds exist in the semireduced state of nitrogenase. This result is based on the position and shape of the edge and the absence of a lower frequency component in the EXAFS, which would be present if there were Mo=O bonds (Cramer et al., 1978b).

The second set of scattering atoms is about 0.4 Å further out from the molybdenum than are the primary sulfur ligands. This shell of atoms, initially thought to be molybdenum, has a phase behavior that is significantly different from Mo—Mo or Mo—S parameters and is close to Mo—Fe parameters. Two to three iron atoms at a distance of 2.72 Å are calculated from the analysis.

Finally, a third set of neighboring atoms must be included to produce a model that completely fits the data. This shell is best fit by one to two sulfur atoms at 2.47 Å. These results are summarized in Table 2 and the best three wave fit is shown graphically in Figure 6.

These coordination distances and numbers are chemically reasonable and may be interpreted through comparison with known structures as representing three or four bridging sulfides, one or two terminal RS⁻ groups and two to three μ-sulfido–bridged iron atoms around the molybdenum. The numbers suggest the presence of a novel Mo-Fe-S cluster in nitrogenase. At least two specific models may be proposed:

I   II

It is also important to observe that these results dictate the absence of Mo=O, Mo=S, or Mo—Mo moieties at the molybdenum site.

Table 2.   Molybdenum bond distances and coordination numbers determined for nitrogenase MoFe component

	C. pasteurianum		*A. vinelandii*	
	Å	#	Å	#
Mo—S	2.35	3–4	2.36	3–4
Mo—Fe	2.72	2–3	2.72	2–3
Mo—S	2.48	1–2	2.51	1–2

Comparison of the Mo EXAFS in Nitrogenase with that of a $Mo_1Fe_3S_4$ Model Complex

The presence of a Mo-Fe-S cluster in nitrogenase suggested that synthetic attempts might lead to an inorganic analog of the molybdenum site. Noting the spontaneous self-assembly of the ferredoxin analog clusters $[Fe_4S_4(SR)_4]^{2-}$ from simple reagents, Holm and his coworkers initiated a similar approach to the synthesis of Mo-Fe-S clusters. Recently the first such complex, $[Mo_2Fe_6S_9(SEt)_8]^{3-}$, has been isolated and structurally characterized (Wolff et al., 1978). The molecular structure of the complex is shown in Figure 7 and consists of a dimer of two $Mo_1Fe_3S_4$ cubes.

A comparison of the EXAFS of this compound to that of molybdenum in nitrogenase (Figure 8) shows the data to be quite similar, especially the EXAFS phases. There are small differences in amplitude around $k = 9 \text{ Å}^{-1}$. Curve-fitting analysis reveals that the Mo—S (sulfide) and Mo—Fe interactions in the cluster and in nitrogenase are very similar (the numerical results are summarized in Table 3). The major difference between the model and the protein may reside in the nature of the external ligands to molybdenum. It must be emphasized that differences in stoichiometry and physical properties, and the absence of evidence in the EXAFS for the distant fourth cube sulfide, point out that the $[Mo_2Fe_6S_9(SEt)_8]^{3-}$ complex in its entirety should not be considered an analog of the molybdenum sites in nitrogenase.

An important question concerns the uniqueness of the $Mo_1Fe_3S_4$ model that current EXAFS results point to as the most likely candidate for the nitrogenase molybdenum environment. Besides the obvious structures ruled out, there are other variations on the Mo-Fe-S cube that should be considered. One possibility would be a molybdenum shared by two cubes, as shown schematically in Figure 9. The EXAFS that would be expected for such a structure is simulated and compared to the nitrogenase EXAFS. It is clear that such a structure is very unlikely. Less clearly ruled out is structure II, where one molybdenum is bridged by four sulfides to two irons. In this case, the fit residual χ^2 almost doubles when the simulated spectrum of this structure is compared to that of nitrogenase. Although this makes structure II appear to be unlikely, the evidence is not as great as for the situation

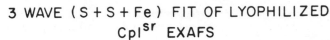

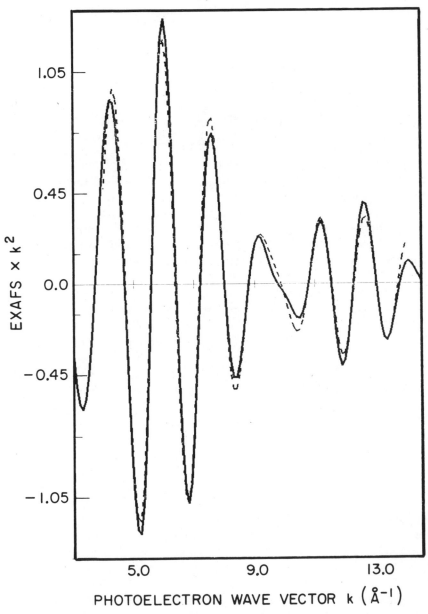

Figure 6.   Curve-fitting analysis of EXAFS data from lyophylized *Clostridium pasteurianum* molybdenum EXAFS. The data (———) are fit (– – – –) with three coordination shells (S, Fe, and S′). Numerical results of the fit are summarized in Table 2.

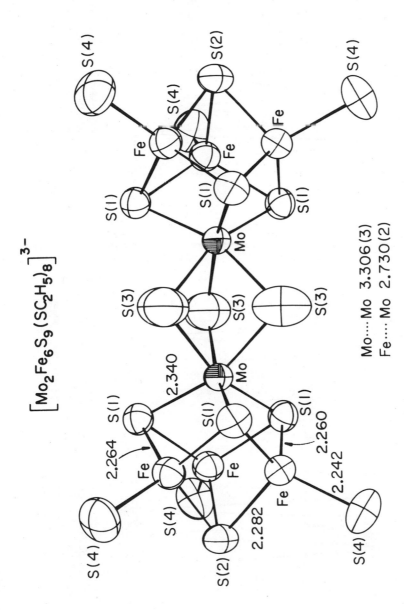

$$\left[Mo_2Fe_6S_9(SC_2H_5)_8 \right]^{3-}$$

Mo····Mo 3.306 (3)
Fe····Mo 2.730 (2)

Figure 7.   Structure of the complex containing two clusters of one molybdenum and three iron atoms each. The molybdenum atoms are shaded and the iron atoms are located diagonally across the three faces of each cube from the molybdenum.

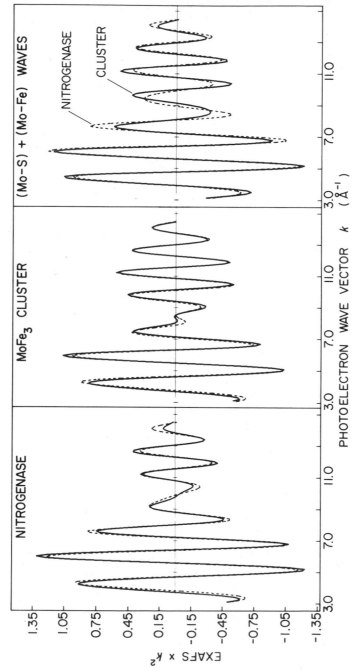

Figure 8.   Curve-fitting analysis of nitrogenase and the Mo-Fe cluster compound shown in Fig. 7. The nitrogenase data on the left and the cluster compound data in the middle are each shown fit to three waves—S, Fe, and S'. A comparison of the sum of the S + Fe waves for each fit is shown on the right. The close agreement between the two sums (and likewise the very close agreement between the parameters determined from the fits; see Table 3) establishes that the Mo_1Fe_3 cluster possesses a molybdenum structural fragment in common with nitrogenase. Further details are given in the text.

Table 3. Comparison of sulfur and iron wave parameters for nitrogenase and Mo_1Fe_3 cluster

	Nitrogenase	Cluster	Crystal (cluster)
Sulfur wave			
Overall amplitude	0.5657	0.5452	
Calculated number of sulfurs	3.79	3.65	
Linear phase	3.1068	3.1034	
Calculated distance	2.352 Å	2.351 Å	2.340 Å
Fe wave			
Overall amplitude	0.2534	0.2571	
Calculated number of irons	(3.04)	(3)	
Linear phase	4.1431	4.2268	
Calculated distance	2.722 Å	2.764 Å	2.73 Å

presented in Figure 9. Clearly, further work is needed both to elucidate the complete molybdenum environment and to characterize the iron environment in nitrogenase.

EXAFS Comparisons Between MoFe Components of Different Organisms

All of the initial studies described above were carried out with MoFe protein isolated from *Clostridium pasteurianum*. In order to confirm these results and to make a comparison between MoFe components from two different species, we have also studied the MoFe component from *Azotobacter vinelandii* (Cramer et al., 1978a). The EXAFS data recorded on the *Azotobacter* MoFe protein are compared (Figure 10) with the data for the lyophilized *Clostridium* protein. Detailed analysis (see Table 2 for summary) shows that the conclusions about primary sulfur ligation and the presence of a molybdenum-iron interaction are substantiated and that the molybdenum environment in both proteins is virtually identical.

EXAFS Study of the FeMo Cofactor

EXAFS studies of the FeMo cofactor isolated by Shah and Brill have been performed (Cramer et al., 1978a). EXAFS of the FeMo cofactor isolated from *Azotobacter* MoFe protein was recorded on a rigorously anaerobic solution of the material. Visual comparison of the data (Figure 10) with that of the intact MoFe proteins indicates that the basic features of the molybdenum environment in the cofactor are preserved during the extraction process and that the intact protein and the cofactor share a common molybdenum site. These data lend support to the idea of a common molybdenum site in the nitrogenase MoFe proteins.

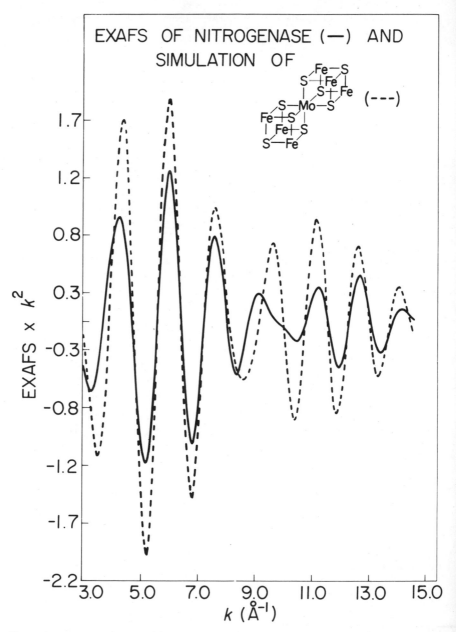

Figure 9.   Comparison of molybdenum EXAFS for nitrogenase (——) with data simulated for the molybdenum-bridged cube structure shown (————).

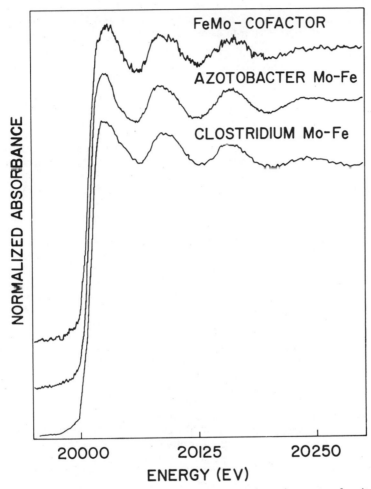

NORMALIZED ABSORBANCE

FeMo – COFACTOR

AZOTOBACTER Mo-Fe

CLOSTRIDIUM Mo-Fe

20000   20125   20250

ENERGY (EV)

Figure 10.   Comparison of the molybdenum K edge x-ray absorption spectra for nitrogenase MoFe protein from *Clostridium pasteurianum* and *Azotobacter vinelandii* with that of the FeMo cofactor extracted from *Azotobacter*.

ACKNOWLEDGMENTS

I wish to thank Steve Cramer, who personally carried out much of the research described in this paper as a part of his Ph.D. thesis research in my laboratories at Stanford. Bill Gillum and Len Mortenson made innumerable contributions to the protein preparations and data acquisition. Protein samples were also provided by Ed Stiefel and the cofactor was prepared by Winston Brill and Vinod Shah. Seb Doniach has continually provided numerical insights and support during the course of these studies. The MoFe$_3$S$_4$ complex was synthesized by Kathy Warrick and Tom Wolff in the laboratories of R. H. Holm, and the structural work was done by

Jeremy Berg, an undergraduate in my laboratories. I thank John Enemark for providing a structural preprint prior to its publication.

REFERENCES

Ashley, C. A., and S. Doniach. 1975. Theory of extended x-ray absorption edge fine structure (EXAFS) in crystalline solids. Phys. Rev. B 11:1279–1288.

Berg, J. M., K. O. Hodgson, S. P. Cramer, J. L. Corbin, A. Elsberry, N. Pariyadath, and E. I. Stiefel. 1979. Structural results relevant to the Mo sites in xanthine oxidase and sulfite oxidase. The crystal structures of MoO_2L, L = $(SCH_2CH_2)_2NCH_2CH_2X$ with X = SCH_3, $N(CH_3)_2$. J. Am. Chem. Soc. 101: 2774–2776.

Bienenstock, A., and H. Winick. 1978. Synchrotron radiation research. Annu. Rev. Nucl. Part. Sci. 28:33–113.

Bunker, B., and E. A. Stern. 1977. The iron-sulfur environment in rubredoxin. Biophys. J. 19:253–264.

Cramer, S. P., J. H. Dawson, K. O. Hodgson, and L. P. Hager. 1978d. Studies on the ferric forms of cytochrome p_{450} and chloroperoxidase by EXAFS. Characterization of the Fe-N and Fe-S distances. J. Am. Chem. Soc. 100:7282–7290.

Cramer, S. P., W. O. Gillum, K. O. Hodgson, L. E. Mortenson, E. I. Stiefel, J. R. Chisnell, W. J. Brill, and V. K. Shah. 1978a. The molybdenum site of nitrogenase. 2. A comparative study of Mo-Fe proteins and the Fe-Mo cofactor by x-ray absorption spectroscopy. J. Am. Chem. Soc. 100:3814–3819.

Cramer, S. P., and K. O. Hodgson. 1979. X-ray absorption spectroscopy. A new structural method and its applications to bioinorganic chemistry. Prog. Inorg. Chem. 25:1–39.

Cramer, S. P., K. O. Hodgson, W. O. Gillum, and L. E. Mortenson. 1978b. The Mo site of nitrogenase. Preliminary structural evidence from x-ray absorption spectroscopy. J. Am. Chem. Soc. 100:3398–3407.

Cramer, S. P., K. O. Hodgson, E. I. Stiefel, and W. E. Newton. 1978c. A systematic x-ray absorption study of Mo complexes. The accuracy of structural information from extended x-ray absorption fine structure. J. Am. Chem. Soc. 100:2748–2761.

Lee, P. A., and G. Beni. 1977. New method for the calculation of atomic phase shifts: Application to extended x-ray absorption fine structure (EXAFS) in molecules and crystals. Phys. Rev. B 15:2862–2883.

Lee, P. A., B.-K. Teo, and A. L. Simons. 1977. EXAFS: A new parameterization of phase shifts. J. Am. Chem. Soc. 99:3856–3859.

Lytle, F. W., D. E. Sayers, and E. A. Stern. 1975. Extended x-ray absorption fine structure technique. II. Experimental practice and selected results. Phys. Rev. B 11:4825–4835.

Shulman, R. G., P. Eisenberger, B.-K. Teo, B. M. Kincaid, and G. S. Brown. 1978. Fluorescence x-ray absorption studies of rubredoxin and its model compounds. J. Mol. Biol. 124:305–321.

Stern, E. A., D. E. Sayers, and F. W. Lytle. 1975. Extended x-ray absorption technique. III. Determination of physical parameters. Phys. Rev. B 11:4836–4846.

Teo, B.-K., P. A. Lee, A. L. Simons, P. Eisenberger, and B. M. Kincaid. 1977. EXAFS: Approximation, parameterization and chemical transferability of amplitude functions. J. Am. Chem. Soc. 99:3854–3856.

Tullius, T. D., P. Frank, and K. O. Hodgson. 1978. Characterization of the blue copper site in oxidized azurin by EXAFS. Determination of a short Cu-S distance. Proc. Natl. Acad. Sci. U.S.A. 75:4069–4073.

Wolff, T. E., J. M. Berg, C. Warrick, K. O. Hodgson, R. H. Holm, and R. B. Frankel. 1978. The Mo-Fe-S cluster complex $[Mo_2Fe_6S_9(SC_2H_5)_8]^{3-}$. A synthetic approach to the molybdenum site in nitrogenase. J. Am. Chem. Soc. 100:4630–4632.

Yamanouchi, K., and J. H. Enemark. 1978. Structure of Tris [2-aminobenzenethiolate(2-)-N,5] molybdenum (VI), $Mo(NHC_6H_4S)_3$. Inorg. Chem. 17:2911–2917.

Nitrogen Fixation, Volume I
Edited by W. E. Newton and W. H. Orme-Johnson
Copyright 1980   University Park Press   Baltimore

Investigation of Molybdenum-Substrate Interactions with X-ray Absorption Spectroscopy

J. P. Smith, J. A. Kirby, M. P. Klein, A. Robertson, and A. C. Thompson

Molybdenum is always found in active preparations of nitrogenase. This fact, along with the results of studies of inorganic models for nitrogenase, such as the molybdate-cysteine complex of Schrauzer and co-workers (Schrauzer and Doemeny, 1971), suggest that molybdenum plays an important role in the enzyme mechanism. In light of the many molybdenum-dinitrogen complexes that have been prepared, it is tempting to speculate that the molybdenum is somehow involved with the binding of nitrogenase substrates prior to or concomitant with reduction by the enzyme. To fully understand the function of molybdenum in nitrogenase, it is important to know whether, and under what conditions, molybdenum substrate interactions will occur. We have been using x-ray absorption spectroscopy to investigate this problem, and in this paper we describe our approach and the results we have obtained to date.

The intense, tunable x-ray source that is occasionally available at the Stanford Synchrotron Radiation Laboratory has made it practical to study samples with the low concentrations of the absorber of interest typical of metalloprotein systems. Several studies of metalloprotein systems using the

This work was supported by the U.S. Energy Research and Development Administration. Synchrotron radiation facilities were provided by the Stanford Synchrotron Radiation Laboratories, supported by National Science Foundation grant DMR 73-07692-AO2 in cooperation with the Stanford Linear Accelerator Center and the Energy Research and Development Administration.

synchrotron radiation x-ray source have been reported recently (Shulman et al., 1976; Hu, Chan, and Brown, 1977).

The theory and interpretation of x-ray absorption spectra have been discussed in the literature (Stern, 1974; Shulman et al., 1975) and only a brief synopsis is presented here. For purposes of discussion, the x-ray absorption spectrum may be divided into two parts—the edge region, located within 50 eV of the sharp increase in absorption at the absorption edge, and the extended x-ray absorption fine structure, or EXAFS, which extends from 50–1000 eV above an absorption edge.

In the edge region, we observe transitions from core levels (1s, in the present case) to low-lying unoccupied molecular orbitals. Edge structure is determined by the metal site symmetry, the identity of the surrounding ligands, and the strength of the metal-ligand interactions.

At higher energies, a continuous distribution of final states is available for occupation by the ejected photoelectron. The absorption coefficient in the energy range of 50–1000 eV above the edge is modulated because of backscattering of the ejected photoelectron off of atoms in the vicinity of the absorbing atom. These modulations are the EXAFS, which may be thought of as an interferogram whose structure is sensitive to the distances between the absorber and nearby scatterers, the nature of the scattering atom, and the degree of local order around the absorbing atom. The EXAFS is written as a normalized modulation of the absorption coefficient, μ, in photoelectron wave vector (k) space.

$$\chi(k) = \frac{\mu(k) - \mu_o(k)}{\mu_o(k)} \qquad (1)$$

In equation 1, $k = 2\pi/\lambda$, where λ is the DeBroglie wavelength of the photoelectron, μ_o is the absorption coefficient in the absence of any modulation. The dependence of EXAFS on local structure is given by equation 2 (Stern, 1974).

$$\chi(k) = \sum_j \frac{N_j}{kR_j^2} |f_j(\pi,k)| \; e^{-\sigma_j^2 k^2} \, e^{-R_j/\eta} \sin(2kR_j + \alpha_j(k)) \qquad (2)$$

Here, $\chi(k)$ has been written as a sum of damped sinusoids. Each term in the sum represents the contribution of a shell of N_j identical atoms located at a distance R_j from the absorber. The backscattering amplitude function, $f_j(\pi, k)$, describes the interaction between the photoelectron and the scatterer. The effects of thermal and structural disorder are represented by the $e^{-\sigma_j^2 k^2}$ factor. The effect of interatomic potentials on the photoelectron phase is given by the total phase shift $\alpha_j(k)$. Finally, a mean free path factor ($e^{-R_j/\eta}$) accounts for inelastic decay of the photoelectron. In equation 2, the backscattering amplitude function is characteristic of the type of scattering atom, and the total phase shift is a combined characteristic of the absorber-scatterer pair.

The above considerations indicate that the x-ray absorption spectrum is sensitive to the arrangement and nature of the ligands around a metal atom. In principle, if dinitrogen were to bind directly to molybdenum, or bind at some other enzyme site causing an indirect perturbation of the coordination environment of molybdenum, the x-ray absorption spectrum would be altered.

Our experimental approach to the use of x-ray absorption spectroscopy can be summarized as follows. Solutions of the molybdenum-iron component of nitrogenase are prepared and equilibrated with either argon or dinitrogen. The molybdenum absorption edge spectra and EXAFS of these samples are measured and compared, along with the spectra of selected model compounds.

EXPERIMENTAL METHODS

Enzyme Samples

The MoFe component of *Azotobacter vinelandii* was prepared according to the procedure of Shah and Brill (1973). In one experiment, the MoFe component of *Clostridium pasteurianum* nitrogenase, prepared by M. Henzl of the University of Wisconsin at Madison, was employed. The average specific activity of pooled *Azotobacter* preparations used in this work ranged from 1500 to 1650 nmoles of ethylene produced/min/mg of protein. For samples run as frozen solutions, the recovery of enzyme activity after x-ray experiments was 90% or better. For the experiment where spectra were taken on solutions held at 4°C, the recovery of the argon-equilibrated sample was 75% and that of the N_2-equilibrated sample was 88%.

Inorganic Model Compounds

Tris(3,4-dithiotoluene)molybdenum(VI) was prepared by the method of Gilbert and Sandell (1960). Other molybdenum compounds were commercially available reagent grade chemicals.

X-ray Spectra

X-ray spectra were recorded at the Stanford Synchrotron Radiation Laboratory. This facility has been described elsewhere (Kincaid, 1975). Model compound spectra were recorded in the absorption mode using ion chamber detectors. Enzyme spectra were recorded in both the absorption and fluorescence (Jaklevic et al., 1977) modes, using a three element silicon detector for fluorescence measurements. Because of severe absorption mode background problems, only fluorescence EXAFS data are presented here.

RESULTS

The absorption edge spectra of several molybdenum compounds are shown in Figure 1. In Figure 2, the absorption edge spectra of two nitrogenase samples at 4°C are presented.

In all of the edge spectra, the absorption of the molybdenum K shell was isolated by fitting a first- or second-order polynomial to the pre-edge region of the data, followed by extrapolation and subtraction of the fitted function from the entire data set.

Two types of EXAFS experiments were performed on nitrogenase samples. Three separate data collections were made on nitrogenase samples held at −90°C. Each data collection included samples equilibrated with argon and dinitrogen. A single large data collection was performed on samples held at 4°C. The average EXAFS of the nitrogenase samples held at −90°C are shown in Figure 3. In Figure 4, we show the EXAFS of nitrogenase samples at 4°C.

In an effort to measure the effects of dinitrogen coordination on molybdenum EXAFS, Timothy Walker of our laboratory has prepared samples of *trans*-bis(dinitrogen)bis[1,2-bis(diphenyl-phosphino)ethane]molyb-

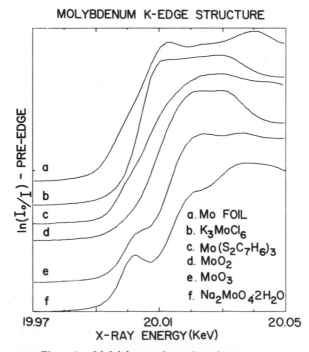

Figure 1.   Molybdenum absorption edge spectra.

MOLYBDENUM EDGE STRUCTURE
NITROGENASE Mo-Fe COMPONENT

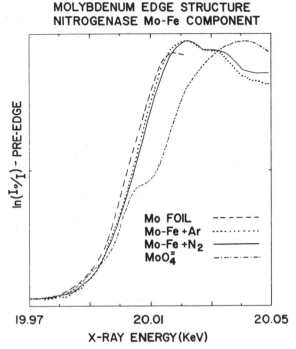

Figure 2. Molybdenum absorption edge spectra for nitrogenase samples. The spectra of molybdenum foil and sodium molybdate are shown as references.

denum(0) and its tetrahydride analog. In Figure 5, we present the EXAFS of these compounds.

DISCUSSION

Edge Spectra

The spectra of model compounds (Figure 1) illustrate the sensitivity of the edge region structure to the environment of the molybdenum. Progressing from the octahedral K_3MoCl_6 to more distorted hexacoordinate structures, tris(3,4-dithiotoluene)molybdenum(VI), MoO_2, and MoO_3, the absorption edge becomes broader and more complex. The spectrum of sodium molybdate dihydrate shows the edge shape characteristic of tetrahedrally coordinated species with formally d° configurations (Van Nordstrand, 1958).

The absorption edge of nitrogenase is broad and featureless on the low energy side of the first absorption maximum. On the high energy side of the first absorption maximum, there is a distinct shoulder. For hexacoordinate first row transition metal ions, such shoulders have been attributed to split-

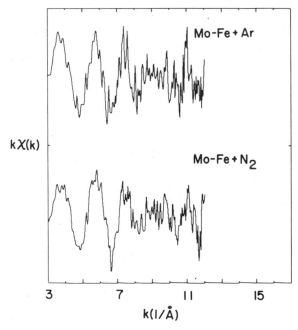

Figure 3.   EXAFS of nitrogenase samples at −90°C.

ting of the excited state due to asymmetric coordination (Glen and Dodd, 1968). Comparison of the two enzyme spectra shows little difference in the low energy side of the absorption edge. The energy of the inflection point of the absorption edge has been used as a qualitative indicator of molybdenum "coordination charge" (Cramer et al., 1976). The present results show that, within an error of 1 eV, there is no change in the inflection point energy due to the presence of dinitrogen.

When dinitrogen is added, the shoulder on the high energy side of the first absorption maximum becomes somewhat broader. Various schemes have been proposed for the interpretation of this sort of feature (Srivastava and Nigam, 1972–1973), but, in the absence of a detailed model of the molybdenum site in nitrogenase, a quantitative interpretation of the changes observed would be premature.

In summary, the presence of dinitrogen causes only minor changes in the molybdenum edge structure of nitrogenase, indicating that no drastic change in the coordination environment of molybdenum occurs on the addition of dinitrogen. At the present time, we cannot propose a detailed explanation of the molybdenum edge structure in nitrogenase. However, the extension of these experiments to nitrogenase samples equilibrated with other substrates or inhibitors might provide some qualitative information about the role of molybdenum in nitrogenase.

EXAFS Studies

The EXAFS of the nitrogenase samples have been analyzed by Fourier transformation (Stern, 1974). In Figure 6, the Fourier transforms of the $-90°C$ nitrogenase EXAFS are presented. In addition to the transforms of EXAFS of samples equilibrated with argon and dinitrogen, we show a "difference" Fourier transformation. The difference transform was calculated by Fourier transformation of the difference between the EXAFS of the MoFe $+ N_2$ sample and that of the MoFe $+$ argon sample and should highlight any features of the Fourier transforms due to the presence of dinitrogen. Examination of these transforms shows that both samples display a major peak at an apparent radius of 1.72 Å with a shoulder at 2.5 Å. There are no significant differences between the two transforms.

The results of the Fourier transform method depend on the weighting of the data in k space prior to Fourier transformation. By weighting different parts of the k space domain more heavily than others, information can be obtained about the dependence of the EXAFS amplitude on k. The EXAFS due to scatterers such as iron or molybdenum is typically stronger at large values of k than the EXAFS due to scatterers such as oxygen or sulfur. When the high k part of the data is given greater weight in the com-

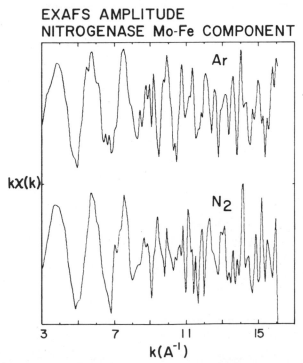

Figure 4.   EXAFS of nitrogenase samples at 4°C.

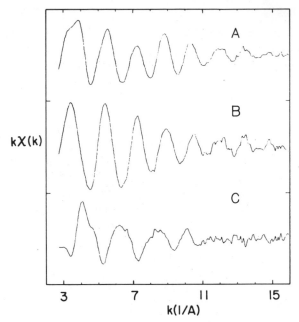

Figure 5. EXAFS of: a) *trans*-bis(dinitrogen)bis[1,2-bis(diphenylphosphino)ethane]-molybdenum(0); b) bis[1,2-bis(diphenylphosphino)ethane]molybdenum tetrahydride; and c) difference EXAFS, curve a minus curve b.

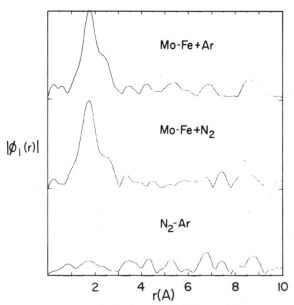

Figure 6. Fourier transforms of nitrogenase EXAFS. Data taken at −90°C. Data domain in k space: 3–12 ^{1}/Å. EXAFS weighted by k^1 before Fourier transformation.

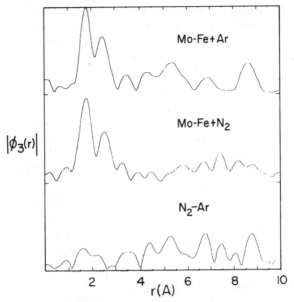

Figure 7.   Fourier transforms as shown in Figure 6, with EXAFS weighted by k^3.

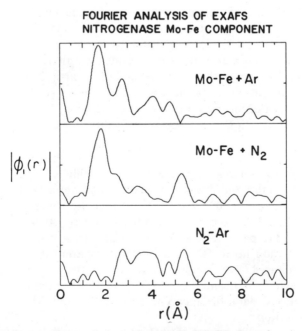

Figure 8.   Fourier transforms of nitrogenase (4°C) EXAFS. Data domain in k space, 3–12^1/Å. EXAFS weighted by k^1 before Fourier transformation.

putation of the Fourier transform, peaks due to scatterers with higher atomic numbers tend to grow with respect to peaks due to scatterers of lower atomic number. In Figure 7, we show the Fourier transforms of the $-90°C$ nitrogenase EXAFS data weighted by k^3. The growth of the peak at 2.5 Å relative to that at 1.72 Å suggests that the 2.5-Å component is due to a scatterer of higher atomic number than the scatterer(s) responsible for the peak at 1.72 Å.

In Figure 8, we show the transforms of nitrogenase data taken on samples at 4°C. There is a major peak at 1.72 Å, as was observed at $-90°C$. However, there are additional peaks at 3.62 and 5.32 Å (N_2) and 4.06 and 4.82 Å (argon) that are not present in the $-90°C$ data. There appear to be significant differences between the two transforms.

The interpretation of the data taken at 4°C presents several problems. The first is that of the peaks at distances greater than 4 Å. Experience with the analysis of EXAFS data from systems of known structure indicates that, unless there are multiple scatterers present, the detection of scatterers beyond 4 Å is not possible. It could be proposed, knowing that a number of iron atoms are present in the molybdenum-iron cofactor (Shah and Brill, 1977), that the molybdenum is present in a novel cluster structure with multiple molybdenum-iron distances. This proposal does not explain why the EXAFS of frozen and liquid solutions are different. To formulate an explanation consistent with both the $-90°C$ and the 4°C results, it would have to be assumed that: 1) on thawing, the molybdenum site undergoes a structural change; 2) the structure assumed on thawing is quite unusual by EXAFS standards; and 3) molybdenum-substrate interactions that occur at 4°C are somehow blocked in the frozen state. An alternative to the above speculations is to assume that some pathological artifact in the 4°C data has given rise to spurious peaks at large radii. These discrepancies could be resolved by repetition of the experiments in question; however, the present instrument time situation at the Stanford facility makes this unlikely in the near future.

In summary, the EXAFS results show that the presence of dinitrogen produces no significant change in the nearest neighbors of the molybdenum in nitrogenase. Both $-90°C$ and 4°C data are in agreement on this point. The transforms of the $-90°C$ data suggest that the scatterer(s) responsible for the peak at 2.5 Å have higher atomic number(s) than the scatterer(s) responsible for the peak at 1.72 Å. The transforms of data taken at 4°C show peaks at radii up to 5.3 Å, as well as apparent differences between samples equilibrated with argon and nitrogen. Because of the large differences between the 4°C and the $-90°C$ data, and the unusually large radii of peaks in the 4°C data, the apparent differences due to the presence of N_2 are open to question.

Finally, we consider the EXAFS of the molybdenum-dinitrogen and tetrahydride complexes (Figure 5). In these compounds, molybdenum is surrounded by four phosphorous atoms (at 2.45 Å), and in one case (at 2.01 Å) two nitrogen atoms as well. The beating of two slightly different frequencies is apparent in the low k region of the EXAFS of the dinitrogen complex. The tetrahydride EXAFS is composed of a single major frequency. These results show that the coordination of molybdenum by dinitrogen is detectable in some cases. It is clear, however, that the EXAFS is dominated by the scattering of the four phosphorous atoms. This has implications for the study of molybdenum-substrate interaction in nitrogenase by the EXAFS technique. If N_2 enters the coordination sphere of molybdenum, its presence could be hidden by the scattering due to the presence of several strong scatterers (X), such as sulfur or iron. This is particularly true if the Mo—N distance was close to the Mo—X distance. At the present time, we are continuing our EXAFS studies of molybdenum-dinitrogen complexes in an effort to better define the conditions under which Mo—N coordination could be detected by EXAFS.

ACKNOWLEDGMENTS

Timothy Walker contributed EXAFS data on the molybdenum dinitrogen complex and its tetrahydride analog and assisted in the collection of nitrogenase EXAFS data. Ninon Kafka assisted in the preparation of enzyme samples and in the collection of EXAFS data.

REFERENCES

Cramer, S. P., T. Eccles, F. Kutzler, K. Hodgson, and L. E. Mortenson. 1976. Molybdenum x-ray absorption edge spectra: The chemical state of molybdenum in nitrogenase. J. Am. Chem. Soc. 98:1286–1287.

Gilbert, T. W., and E. B. Sandell. 1960. Reaction of dithiol with molybdenum. J. Am. Chem. Soc. 82:1087–1091.

Glen, G. L., and C. G. Dodd. 1968. Use of molecular orbital theory to interpret x-ray K-absorption spectral data. J. Appl. Phys. 31:5372–5377.

Hu, V. N., S. I. Chan, and G. S. Brown. 1977. X-ray absorption edge studies on oxidized and reduced cytochrome c oxidase. Proc. Natl. Acad. Sci. U.S.A. 74:3821–3825.

Jaklevic, J., J. Kirby, M. Klein, A. Robertson, G. Brown, and P. Eisenberger. 1977. Fluorescence detection of EXAFS: Sensitivity enhancement for dilute species and thin films. Solid State Commun. 23:679–683.

Kincaid, B. M. 1975. Synchrotron radiation studies of K-edge x-ray photoabsorption spectra; Theory and experiment. Ph.D. thesis, Stanford University.

Schrauzer, G. N., and P. A. Doemeny. 1971. Chemical evolution of a nitrogenase model. II. Molybdate-cysteine and related catalysts in the reduction of acetylene to olefins and alkanes. J. Am. Chem. Soc. 93:1608–1618.

Shah, V. K., and W. J. Brill. 1973. Nitrogenase. IV. Simple method of purification to homogeneity of nitrogenase components from *Azotobacter vinelandii*. Biochim. Biophys. Acta 305:445–454.

Shah, V. K., and W. J. Brill. 1977. Isolation of an iron-molybdenum cofactor from nitrogenase. Proc. Natl. Acad. Sci. U.S.A. 74:3249–3253.

Shulman, R. G., P. Eisenberger, W. E. Blumberg, and N. A. Stombaugh. 1975. Determination of the iron-sulfur distances in rubredoxin by x-ray absorption spectroscopy. Proc. Natl. Acad. Sci. U.S.A. 72:4003–4007.

Shulman, R. G., Y. Yafet, P. Eisenberger, and W. E. Blumberg. 1976. The observation and interpretation of x-ray absorption edges in iron compounds and proteins. Proc. Natl. Acad. Sci. U.S.A. 73:1384–1388.

Srivastava, U. C., and H. L. Nigam. 1972–1973. X-ray absorption edge spectrometry (XAES) as applied to coordination chemistry. Coord. Chem. Rev. 9:275–310.

Stern, E. A. 1974. Theory of the extended X-ray absorption fine structure. Phys. Rev. B 10:3027–3037.

Van Nordstrand, R. A. 1958. X-ray absorption edge spectrometry. In: V. D. Frechette (ed.), Handbook of X-rays, pp. 168–198. John Wiley and Sons, Inc., New York.

Section IV
Chemical Models

Nitrogen Fixation, Volume I
Edited by W. E. Newton and W. H. Orme-Johnson
Copyright 1980   University Park Press   Baltimore

Chemical Aspects
of Nitrogen Fixation

J. H. Enemark

In beginning this discussion of nitrogen chemistry, it is useful to look briefly at the inorganic nitrogen cycle in the environment (Figure 1). The reductive fixation of dinitrogen (N_2) to ammonia (NH_3) is carried out naturally by the enzyme nitrogenase and industrially by the Haber process. About 25% of the total nitrogen fixed is produced by the Haber process and another 63% results from biological nitrogen reduction. Direct oxidative fixation of nitrogen occurs primarily as a by-product of combustion processes in power plants and internal combustion engines and accounts for about 9% of the total nitrogen fixed. Oxidation of dinitrogen in the atmosphere by lightning accounts for another 4%. Thus, the reductive fixation of dinitrogen to ammonia accounts for an estimated 87% of the nitrogen fixed, with oxidative fixation providing the remainder (Pratt, 1977).

Also shown in Figure 1 is the denitrification pathway whereby nitrate is reduced to nitrite, to N_2O, and ultimately to N_2. Finally, it should be emphasized that molybdenum plays a vital role in the biological conversions of inorganic nitrogen species. Not only is molybdenum present in nitrogenase, it also occurs in the assimilatory nitrate reductase of plants and fungi and in the dissimilatory nitrate reductase of denitrifying microorganisms (Payne, 1973).

ENERGETICS OF NITROGEN COMPOUNDS

It is well known that the free energy of formation of NH_3 from the elements is favorable (equation 1).

Financial support from the National Institutes of Environmental Health Sciences (Grant ES-00966) is gratefully acknowledged.

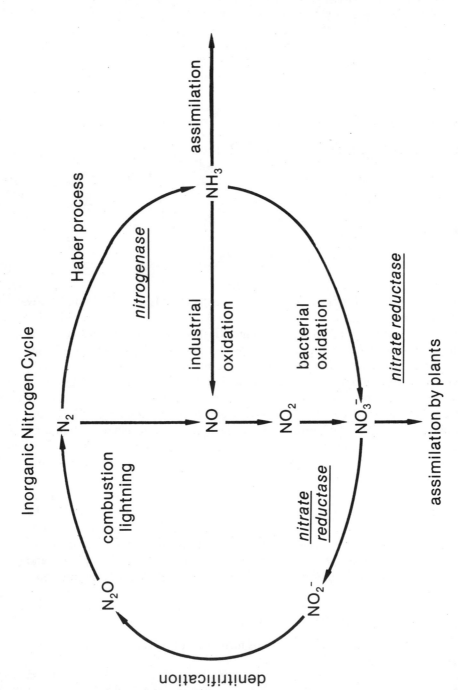

Figure 1. Simplified view of the inorganic nitrogen cycle in the environment. The molybdenum-containing enzymes in the cycle are underlined.

$$N_{2 (g)} + 3H_{2 (g)} \rightarrow 2NH_{3 (g)}$$
$$\Delta G° (298°) = -16.6 \text{ kJ/mole} \qquad (1)$$

However, equation 1 ignores the energetics of hydrogen production. The overall chemical reactions in a typical ammonia plant (Sweeney, 1976) are shown in equations 2–4.

$$\Delta G° (298°)$$

		$\Delta G° (298°)$	
$CH_{4 (g)} + H_2O_{(g)} \rightarrow CO_{(g)} + 3H_{2 (g)}$		142 kJ	(2)
$N_{2 (g)} + 3H_{2 (g)} \rightarrow 2NH_{3 (g)}$		−33 kJ	(3)
$N_{2 (g)} + CH_{4 (g)} + H_2O_{(g)} \rightarrow CO_{(g)} + 2NH_{3 (g)}$		+109 kJ	(4)

$$\Delta G° = 54.5 \text{ kJ/mole of } NH_3$$
$$\Delta H° = 57.5 \text{ kJ/mole of } NH_3$$

Thus, for the overall process both $\Delta G°$ and $\Delta H°$ are unfavorable. The microorganisms fixing nitrogen at ambient temperature and pressure in aqueous media while metabolizing carbon compounds will face similar energy demands (Watt and Bulen, 1976).

Nitrogen is fixed oxidatively when nitric oxide (NO) is formed as by-product in widely dispersed combustion processes. Nitric oxide has a large positive free energy of formation, but it is not an attractive precursor to ammonia. In fact, the first step in the commercial production of nitric acid (Matasa and Tonca, 1973) involves the very favorable reaction of ammonia with oxygen (equation 5).

$$2NH_{3 (g)} + \frac{5}{2} O_{2 (g)} \rightarrow 2NO_{(g)} + 3H_2O_{(g)}$$
$$\Delta G° (298°) = -478 \text{ kJ} \qquad (5)$$

TRANSITION METAL COMPOUNDS OF DINITROGEN

Since the first dinitrogen complex was isolated in 1965 by Allen and Senoff, a wide variety of dinitrogen complexes have been prepared or claimed. Earlier work on dinitrogen complexes and their derivatives has been reviewed in previous symposia (Newton and Nyman, 1976; Newton, Postgate, and Rodriguez-Barrueco, 1977) by several authors. Emphasis is given here to systems where the stoichiometry and stereochemistry of the various species are known, and to advances made since 1976. The possible structures (1–7) are shown below. For simplicity, only the metal atom(s) and the dinitrogen ligands are shown. Other ligands that are attached to the metal atoms are mentioned when specific compounds are discussed.

$$M\!-\!N\!\equiv\!N$$

$$M\!-\!N\!\equiv\!N\!-\!M$$

$$\textit{1} \qquad\qquad \textit{2} \qquad\qquad \textit{3} \qquad\qquad \textit{4}$$

$$M\!-\!N\!\equiv\!N\!-\!M$$

$$M\!-\!N\!\equiv\!N \qquad N\!\equiv\!N\!-\!M\!-\!N\!\equiv\!N$$

$$\textit{5} \qquad\qquad\qquad \textit{6} \qquad\qquad\qquad \textit{7}$$

Reactions of η^1-Mono-dinitrogen Compounds

Until recently, complexes with a single dinitrogen molecule attached "end-on" to a transition metal (structure *1*) have proved disappointingly unreactive. However, Sellmann and Weiss (1977) have now demonstrated nucleophilic attack on an M—N≡N complex (equation 6).

$$+ \ LiC_6H_5 \ \xrightarrow[-30°C]{THF} \ Li\left[C_5H_5(CO)_2Mn\!-\!N\!=\!N \atop \qquad\qquad\quad \mid \atop \qquad\qquad\quad C_6H_5\right]$$

$$\textit{8} \qquad\qquad\qquad\qquad\qquad\qquad \textit{9}$$

$$(6)$$

Attack occurs at the nitrogen adjacent to the metal atom (*endo*-N atom), consistent with the original assignment of charge distribution in coordinated dinitrogen from XPS studies (Leigh et al., 1970). Addition of acid to compound *9* gives a phenyldiazene complex that decomposes on warming (equation 7).

$$\left[C_5H_5(CO)_2Mn\!-\!N\!=\!N \atop \qquad\qquad\quad \mid \atop \qquad\qquad\quad C_6H_5\right]^- + \ H^+ \ \xrightarrow[0°C]{Et_2O/H_2O} \ C_5H_5(CO)_2Mn\!-\!N\!=\!NH \atop \qquad\qquad\qquad\qquad\qquad\quad \mid \atop \qquad\qquad\qquad\qquad\qquad\quad C_6H_5$$

$$\textit{9}$$

$$(7)$$

$$C_6H_6 \ + \ C_5H_5(CO)_2Mn\!-\!N\!\equiv\!N \ \xleftarrow[20°C]{THF}$$

$$\textit{8}$$

Sellmann and Weiss (1978) have also carried out sequential nucleophilic and electrophilic additions of CH_3 groups to compound 8 and have shown that the resulting coordinated dimethyldiazene can be displaced from compound 10 by N_2. Repeated cycling of equation 8 eventually results in destruction of structure 8 because of competing side reactions.

$$C_5H_5(CO)_2Mn-N\equiv N + CH_3^- \xrightarrow[-30°C]{THF} \left[C_5H_5(CO)_2Mn-N=N\right]^- $$

$$\underset{CH_3}{|}$$

$$8$$

$$(CH_3)_3OBF_4$$

$$\xrightarrow[-30°C]{THF} \qquad (8)$$

$$CH_3N=NCH_3 \xleftarrow[Et_2O,\ 20°C]{} C_5H_5(CO)_2Mn-N=N\overset{CH_3}{\diagup} \longleftarrow$$

$$\underset{CH_3}{|}$$

$$N_2(100\ bar)$$

$$10$$

Reactions of η^2-Mono-dinitrogen Complexes

Evidence for complexes with structure 2 is discussed in a later section.

Reactions of Linear M—N≡N—M Complexes

There appear to be no reactions of the coordinated dinitrogen molecule in complexes with structure 3.

Reactions of M_2N_2 Complexes with Structure 4

The structure of an example of 4 has been described in detail (Jonas et al., 1976). The reactions of the coordinated dinitrogen have not been extensively studied (Leigh, 1977).

Reactions of N_2M—N≡N—MN_2 Complexes

Dinitrogen complexes with structure 5 have been reviewed in a previous volume (Bercaw, 1977). Treatment of compound 11 with acid produces hydrazine (equation 9).

$$(Me_5C_5)_2Zr-N\equiv N-\overset{\overset{N_2}{|}}{Zr}(C_5Me_5)_2 \xrightarrow{HCl}$$

$$\underset{N_2}{|}$$

$$11 \qquad\qquad (9)$$

$$2(Me_5C_5)_2ZrCl_2 + N_2H_4 + 2N_2$$

The four-electron reduction of one dinitrogen molecule to hydrazine could result from the concerted action of the two d^2 metal ions [formally Zr(II)] on a single dinitrogen molecule, or from the self-reduction of diimide (diazene), N_2H_2, with one N_2H_2 molecule being formed at each zirconium center. Trapping experiments to demonstrate the presence of N_2H_2 are inconclusive and the mechanism of equation 9 remains unknown (Bercaw, unpublished data).

Reactions of $M(N_2)_2$ Complexes

Dinitrogen complexes with structures 6 and 7 have been extensively studied, especially the *cis* and *trans* bis-dinitrogen complexes of the formally d^6 transition metals Mo(0) and W(0), in which each metal atom is also coordinated by four tertiary phosphines. The early chemistry of $M(N_2)_2P_4$ complexes has been described (Chatt et al., 1976; George and Iske, 1976; Chatt, 1977). *cis*-$W(N_2)_2(PMe_2Ph)_4$ reacts readily with acid in methanol to form NH_3 and N_2 in a 2:1 ratio (equation 10).

$$cis\text{-}W(N_2)_2(PMe_2Ph)_4 \xrightarrow{H_2SO_4/MeOH} 2NH_3 + N_2 + W(VI) \qquad (10)$$

12

$$cis\text{-}Mo(N_2)_2(PMe_2Ph)_4 \xrightarrow{H_2SO_4/MeOH} \text{brown oil} + N_2$$

13

$$(11)$$

$$N_2 + NH_3 \xleftarrow{40\% \text{ KOH}}$$
$$(\sim 0.3 \text{ mol})(\sim 0.7 \text{ mol})$$

Reaction 10 is ~90% complete (Chatt, Pearman, and Richards, 1975; Chatt, Pearman, and Richards, 1977a). The analogous molybdenum complex, compound *13*, gives an unidentified intermediate from which NH_3 and N_2 are produced in a 2:1 ratio on treatment with strong base (equation 11). Reaction sequence 11 is ~30% complete (Chatt, Pearman, and Richards, 1977a).

It has recently been shown that *trans*-$W(N_2)_2(PMePh_2)_4$, compound *14*, also produces NH_3 and N_2 in acidic methanol (equation 12). In nonpolar solvents, compound *14* gives compound *15*, an N_2H_3 complex (equation 13); compound *15* yields NH_3 and N_2H_4 on hydrolysis (equations 14 and 15) (Chatt, Pearman, and Richards, 1977b).

$$\text{trans-W}(N_2)_2(PMePh_2)_4 \xrightarrow{\text{HCl/MeOH}} 2NH_3 + N_2 \qquad (12)$$

$$\mathbf{14}$$

$$\xrightarrow{\text{HCl/CH}_2\text{Cl}_2} \text{trans-WCl}_3(N_2H_3)(PMePh_2)_2 + N_2 \qquad (13)$$

$$\mathbf{15}$$

$$NH_3 + N_2H_4 \xleftarrow{\text{H}_2\text{SO}_4/\text{MeOH}}$$

$$(\sim 0.6 \text{ mol})(\sim 0.25 \text{ mol}) \qquad (14) \qquad (15) \Big\downarrow \text{40\% KOH}$$

$$NH_3 + N_2H_4$$

$$(0.76 \text{ mol}) \ (0.08 \text{ mol})$$

The mechanism(s) whereby coordinated dinitrogen in compounds *12–14* is converted to NH_3 remains unknown. The Chatt group currently favors the sequence in equation 16. Steps a through d and f are known, but step e, cleavage of the N—N bond of an η^1—N_2H_4 species, has not yet been demonstrated (Chatt, Pearman, and Richards, 1977a). The overall reaction involves the addition of six protons and the oxidation of M(0) to M(VI).

a) $M(N_2)_2 \rightarrow M(0)N_2 + N_2$

b) $MN_2 + H^+ \rightarrow M\text{—}NNH$

c) $M\text{—}NNH + H^+ \rightarrow M\text{—}NNH_2$

d) $M\text{—}NNH_2 + H^+ \rightarrow M\text{—}NHNH_2$ (16)

e) $M\text{—}NHNH_2 + H^+ \rightarrow M\text{—}NHNH_3 \rightarrow MNH + NH_3$

f) $MNH + 2H^+ \rightarrow M(VI) + NH_3$

A reaction that is quite different from that in equations 10–13 occurs between acids and $M(N_2)_2(dppe)_2$, where dppe is the chelating diphosphine $Ph_2PCH_2CH_2PPh_2$ (equation 17). Moreover, compound *17* does not undergo further reduction to liberate NH_3 or N_2H_4. The coordinated N_2H_2 group can be deprotonated to a coordinated N_2H group (Chatt, 1977). Reaction 17 is very fast and essentially diffusion controlled (Chatt et al., 1978).

$$\text{trans-W}(N_2)_2(dppe)_2 \xrightarrow{\text{HCl}} WCl_2(N_2H_2)(dppe)_2 + N_2 \qquad (17)$$

$$\mathbf{16} \qquad\qquad\qquad\qquad \mathbf{17}$$

Compound *16* and its molybdenum analog, compound *18*, undergo several other reactions in addition to that in equation 17. Representative examples are:

$$trans\text{-}M(N_2)_2(dppe)_2 \xrightarrow{\text{RCN}} M(N_2)(RCN)(dppe)_2 + N_2 \tag{18}$$

(Tatsumi, Hidai, and Uchida, 1975)

$$(M = Mo, W) \xrightarrow{\text{RX}} (dppe)_2XM\text{—}N\text{=}N\diagdown_R + N_2 \tag{19}$$

(Chatt et al., 1976; George and Iske, 1976)

$$\xrightarrow{\text{CH}_2\text{Br}_2} \left[(dppe)_2BrW\text{=}N\text{—}N\diagdown_{CH_2}\right]^+ + N_2 \tag{20}$$

(Ben-Shoshan et al., 1976)

$$\xrightarrow{\text{Br(CH}_2)_4\text{Br}} \left[(dppe)_2BrM\text{=}N\text{—}N\diagdown_{CH_2\text{—}CH_2}^{CH_2\text{—}CH_2}\right]^+ + N_2 \tag{21}$$

(Chatt et al., 1977)

$$\xrightarrow{\text{CH}_3\text{Br/THF}} [(dppe)_2BrM\text{=}N\text{—}N\text{=}CH(CH_2)_3OH]^+ + N_2 \tag{22}$$

(Bevan et al., 1977a)

In general, equations 18–22 occur thermally with M = molybdenum and with tungsten filament irradiation for M = tungsten. A distinctive feature of equations 18–22 is their slowness relative to equation 17. The kinetics of equation 18 has been studied independently by Chatt et al. (1978) and by Carter, Bercaw, and Gray (1978). Both studies find the reaction to be first order in compound *18*, with a rate constant of about $1 \times 10^{-4} \cdot \sec^{-1}$. Reaction 19 is also first order in compound *18*, with a similar rate constant (Chatt et al., 1978). The kinetics for equations 18 and 19 are consistent with loss of dinitrogen from compound *18* as the rate-determining step (equation 23).

$$trans\text{-}M(N_2)_2 (dppe)_2 \rightarrow M(N_2) (dppe)_2 + N_2 \tag{23}$$

$$(M = Mo, W)$$

Note, however, that if equation 23 is the slow step in equations 18–22, then some other step must presumably occur first in equation 17, which is a very fast reaction of compound *16*. Likewise, equation 16a may need to be reconsidered as being the first step in equations 10–13.

Cyclic voltammetry has shown that the $M(N_2)(RCN)(dppe)_2$ products of equation 18 undergo two successive reversible one-electron oxidations. Compounds in which RCN is covalently bound to an electrode surface exhibit at least one reversible electrochemical oxidation (Leigh and Pickett, 1977).

The mechanism of equation 19 has been investigated in detail (Chatt et al., 1978) and is described by equation 24.

$$M(N_2)_2(dppe)_2$$

$\downarrow$ Rate-determining step

$$M(N_2)(dppe)_2 + N_2$$

$\downarrow$ RX

$$M(N_2)(RX)(dppe)_2 \qquad (24)$$

$\downarrow$ Homolysis

$$M(N_2)X(dppe)_2 + R\cdot$$

Decomposition

$\downarrow$ Radical coupling

Alkylation

$$MX_2(dppe)_2 + R_2 \qquad\qquad MX_2(dppe)_2 + \text{olefin}$$

$$XM(N_2R)(dppe)_2$$

$$+ \qquad\qquad\qquad +$$

$$M(N_2)_2(dppe)_2 \qquad\qquad M(N_2)_2(dppe)_2$$

The final products depend on the pathways taken by the $M(N_2)X(dppe)_2$ and R· radicals after homolytic cleavage of the RX bond of the coordinated RX molecule. In equation 22, the $M(N_2)X(dppe)_2$ radical reacts with the THF solvent to give the final product.

The products of equation 19, diazenido (NNR) complexes, undergo further chemical reactions. Protonation occurs at the *exo*-N atom (equation 25) to give organohydrazido(2-)(NNHR) complexes (Chatt et al., 1976). Reaction 25 also weakens the N—N bond, as demonstrated by the structures of compounds *19* and *20*, where R = C_8H_{17} (Day et al., 1976).

$$M(N_2R)X(dppe)_2 + HX \rightarrow M(N_2HR)X(dppe)_2{}^+ \qquad (25)$$

$$I \xrightarrow{\;2.88\;} Mo \xrightarrow{\;1.85\;} N \xrightarrow{\;1.15\;} N \qquad\qquad I \xrightarrow{\;2.81\;} Mo \xrightarrow{\;1.80\;} N \xrightarrow{\;1.26\;} N{-}H$$

$$\underset{128^\circ \searrow C_8H_{17}}{} \qquad\qquad\qquad \underset{130^\circ \searrow C_8H_{17}}{}$$

19   *20*

Recently, George (unpublished data) has shown that diazenido ligands can be converted to primary amines and ammonia (equation 26).

$$MX(NNR)(dppe)_2 + NaBH_4 \xrightarrow[100^\circ C]{MeOH/C_6H_6} RNH_3 + NH_3 \qquad (26)$$

$$(\sim 60\%)$$

The conversion of organohydrazido(2-) ligands to secondary amines also takes place (equation 27), but the fate of the second nitrogen atom of the ligand is not known (Bevan et al., 1977b).

$$MX(NNR_2)(dppe)_2 \xrightarrow{\text{LiAlH}_4/\text{MeOH}/\text{HBr}/\text{H}_2\text{O}} HNR_2 \tag{27}$$

Summary of Reactions of Discrete Complexes of Dinitrogen

The above sections have provided several examples of the chemical fixation of nitrogen by reactions of discrete dinitrogen complexes. The "fixed" products include NH_3 (equations 10, 11, 12, 14, and 15), N_2H_4 (equations 9 and 14), primary amines (equation 26), secondary amines (equation 27), and diazenes (equation 8).

OTHER NITROGEN-FIXING SYSTEMS

Some provocative systems for nitrogen fixation have recently been described. Schrauzer and Guth (1977) reported that dinitrogen can be photoreduced to ammonia and traces of hydrazine over TiO_2 doped with iron. Shilov and co-workers report that the electrochemical reduction of dinitrogen to hydrazine takes place in methanol-guanidine buffer containing Ti(III) and Mo(III). Optimum N_2H_4 production occurs when small amounts of water (2% by volume) are also present. The electrolyses are carried out under N_2 pressure (30 atm) at -1.90 V (versus S.C.E.) (Kulakovskaya et al., 1977).

The reduction of dinitrogen to ammonia and hydrazine by reducing hydroxide gels (equation 28) was discovered in 1970 and has been reviewed by Shilov (1976, 1977) and by Nikanova and Shilov (1977).

$$V(OH)_2 + N_2 + H_2O \xrightarrow{\text{Mg(OH)}_2} V(OH)_3 + N_2H_4 + NH_3 \tag{28}$$

The relative amounts of hydrazine and ammonia formed depend on the Mg(II):V(II) ratio. It has been proposed that the initial step in equation 28 is the four-electron reduction of an N_2 molecule (Shilov, 1976). Recently, Schrauzer and co-workers have also investigated the $V(OH)_2$-$Mg(OH)_2$ system and have concluded that the reaction involves reduction to diimide, N_2H_2, which either disproportionates to N_2 and N_2H_4 or decomposes to N_2 and H_2 (equation 29) (Zones et al., 1978).

An η^2-mono-dinitrogen complex ("side-on" bonded; structure 2) is proposed for one of the steps in equation 29. Schrauzer has also proposed an η^2-intermediate for model molybdenum systems and for nitrogenase itself (Schrauzer, 1976, 1977). The evidence for discrete η^2-dinitrogen complexes (structure 2) is considered in the next section.

η^2-MONO-DINITROGEN COMPLEXES

Experimental Evidence

Stable dinitrogen complexes with one N_2 molecule bonded "side-on" to a single transition metal (structure 2) remain unusually elusive. Among the earliest evidence for a transitory η^1-dinitrogen complex is the infrared result of Armor and Taube (1970), which shows that η^2-$[(NH_3)_5Ru(^{15}N^{14}N)]^{2+}$ isomerizes to η^1-$[(NH_3)_5Ru(^{14}N^{15}N)]^{2+}$ in the solid state and in solution with no loss of N_2. Isomerization of the η^1-dinitrogen ligand also occurs for $[(das)_2Ru(^{15}N^{14}N)]^{2+}$ in the solid state (Quinby and Feltham, 1972). Early results on N_2 complexes of "$(C_5H_5)_2Ti$" were thought to indicate isolable η^2 species (Brintzinger, 1976). However, subsequent studies have shown these complexes to be binuclear (structure 3) (Bercaw, 1977). Recently, x-ray diffraction data from trans-$(i$-$Pr_3P)_2RhCl(N_2)$ have been interpreted in terms of η^2 coordination of dinitrogen (Busetto et al., 1977). Unfortunately, the compound crystallizes in a space group that requires that the N_2 molecule and chlorine atom be disordered. Consequently, the exact N_2 geometry is difficult to assess (see Note below). Moreover, the $N\equiv N$ stretching frequency (2100 cm^{-1}) is very similar to that observed in linear $M-N\equiv N$ systems such as the formally isoelectronic complex trans-$(t$-$Bu_3Ph)_2RhH(N_2)$, which has an η^1-dinitrogen ligand and $\nu_{N\equiv N} = 2155$ cm^{-1} (Hoffman et al., 1976). Another recent report claims that equation 30 leads to compound 21 (below), an η^2-mono-dinitrogen complex (Gynane, Jeffrey, and Lappert, 1978). Compound 21 is an unstable brown solid that produces low yields ($\sim20\%$) of hydrazine, traces of ammonia, and dinitrogen on reaction with aqueous hydrochloric acid. No $N\equiv N$ stretch is observed for compound 21, but it exhibits an ESR spectrum that shows hyperfine splitting from two equivalent nitrogen atoms. In a vacuum or in aromatic solvents, compound 21 is converted to a diamagnetic species formulated as compound 22 (below). The available evidence that compound 21 is an η^2-mono-dinitrogen complex is not conclusive, and additional physical studies of compound 21 are warranted.

$$[Zr(\eta\text{-}C_5H_5)_2Cl(R)] \xrightarrow[\text{THF 20°C}]{\text{Na/Hg, N}_2} [Zr(\eta\text{-}C_5H_5)_2(N_2)R] \qquad (30)$$

$$R = CH(SiMe_3)_2 \qquad \swarrow \qquad 21$$

$$(Zr(\eta\text{-}C_5H_5)_2R)_2N_2$$

$$22$$

Although there is only tenuous evidence for η^2-dinitrogen complexes, there are examples of stable η^2 complexes of RN_2 (R = fluorene) (compound *23*) (Nakamura et al., 1977) and PhN=NPh (compound *24*) (Dickson and Ibers, 1972).

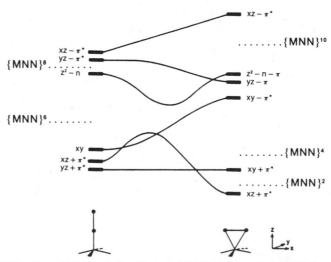

t-Bu$_{NC}$ N

Ni | 1.25 Å

t-Bu^{NC} N

        R

23

t-Bu$_{NC}$ NPh

Ni | 1.39 Å

t-Bu^{NC} NPh

24

Theoretical Considerations

Hoffmann, Chen, and Thorn (1977) have investigated the general problem of η^1-("end-on") versus η^2-("side-on") coordination for a diatomic molecule interacting with a transition metal. The results of their theoretical studies are shown in Figure 2. The η^2 geometry is preferred for $\{MNN\}^2$ and $\{MNN\}^4$ electron configurations, e.g., Mo(IV) and Mo(II), whereas η^1 coordination is more stable for $\{MNN\}^6$, e.g., Mo(0). [The superscript in $\{MNN\}^n$ denotes the number of electrons that are primarily located in metal d and/or $\pi^*(N_2)$ orbitals (Enemark and Feltham, 1974).]

The frontier orbitals involved in the bonding in the η^1 and η^2 geometries are shown in Figure 3 (Hoffmann, Chen, and Thorn, 1977). Of particular

Figure 2. Molecular orbital correlation diagram (after Hoffmann et al., 1977) showing the relationship between η^1-(left) and η^2-(right) complexes of dinitrogen. The superscript in $\{MNN\}^n$ denotes the number of electrons that are primarily associated with metal d and/or $\pi^*(NN)$ orbitals (notation of Enemark and Feltham, 1974).

interest is the η^2 diagram on the right-hand side of Figure 3. An $\{MNN\}^2$ complex will have a pair of electrons in the lowest orbital, which is primarily d_{xz} and $\pi^*(NN)$. This orbital has lobes on the nitrogen atoms that are spatially situated to bind one proton to each nitrogen atom. In an $\{MNN\}^4$ complex, the N—N bond will be further weakened by the δ-type interaction between d_{xy} and the other $\pi^*(NN)$ orbital. For an $\{MNN\}^6$ species, the highest occupied molecular orbital depends on the nature of the metal, because there are three orbitals of similar energy. Note, however, that the combination $d_z{}^2$-n-π should be strongly stabilized by binding one proton to each nitrogen atom.

Implications for Nitrogenase

Molybdenum is currently the favored coordination site of dinitrogen in nitrogenase. EXAFS studies on the enzyme indicate that the molybdenum atoms are well separated (Cramer et al., 1978). Thus, a MoN_2 moiety becomes the simplest chemical unit to consider. Schrauzer (1976) has proposed that the enzyme operates by Mo(IV), forming an η^2-dinitrogen complex (an $\{MoNN\}^2$ species) that accepts two protons and liberates diimide, N_2H_2. Diimide either decomposes or disproportionates to N_2H_4 and N_2 (equation 29). The enzyme then reduces N_2H_4 to two moles of NH_3 in a separate step. Chatt and co-workers have proposed that the enzyme may use equation 16, an η^1 nitriding mechanism. Figure 4 suggests a novel η^2 mechanism that contains some features of both equation 16 and equation 29 as well as an earlier η^2 scheme of Newton, Corbin, and McDonald (1976).

For N_2 being reduced at a single metal atom, it is convenient to use the $\{MNN\}^n$ formalism to count electrons while examining the alternative binding arrangements and protonation sequences that are possible for the six-electron reduction of one dinitrogen molecule to two of ammonia. For simplicity, the overall reduction is considered as a series of two-electron steps, i.e., $\{MNN\}^2 \rightarrow \{MNN\}^4 \rightarrow \{MNN\}^6$. However, the process could equally well be a series of one-electron steps, i.e., $\{MNN\}^2 \rightarrow \{MNN\}^3 \rightarrow \{MNN\}^4 \rightarrow \{MNN\}^5 \rightarrow \{MNN\}^6$. The first step in Figure 4 is the protonation of an η^2-$\{MNN\}^2$ complex to give structure 25. If no additional electrons are supplied, this species could presumably decompose to "N_2H_2" as in equation 29. Two more electrons reduce the complex to the $\{MNN\}^4$ level. Hydrolysis of structure 26 by strong acid or base should produce hydrazine, as observed by Thorneley, Eady, and Lowe (1978). Continued reduction of the diprotonated moiety to the $\{MNN\}^6$ level yields a dinitrene complex. Both of the limiting valence structures for coordinated nitrene ligands are shown for structure 27. Addition of four protons to the dinitrene species should produce two molecules of NH_3. It seems likely that in the enzyme a two-electron reduction accompanies the final protonations, consistent with Stiefel's hypothesis

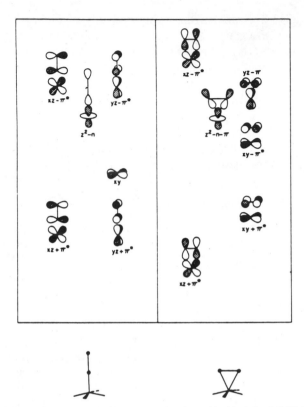

Figure 3.   Schematic drawing of the frontier molecular orbitals of the MNN fragment of an L_4MN_2 complex having η^1-(left) and η^2-(right) geometries (after Hoffmann et al., 1977).

(1973) of coupled proton and electron changes in reactions of molybdoenzymes.

An attractive feature of the mechanism of Figure 4 is the formation of multiple bonds between the metal and nitrogen atoms concomitant with the decrease in the bond order between the two nitrogen atoms. Chatt and Richards (1977) have previously pointed out that "the tendency of molybdenum to form multiple bonds to nitrogen . . . may be the foundation of its ability to effect the reductive cleavage of the dinitrogen molecule to ammonia." Another inherent feature of the mechanism is control of the proton affinity of the coordinated nitrogen atoms by the metal (Stiefel, 1973).

A key intermediate in Figure 4 is structure *27*, a dinitrene complex. Recently, bis(organonitrene) complexes of molybdenum and tungsten have been isolated and characterized by Haymore, Maatta, and Wentworth (1979). Interestingly, complex *28* has one nearly linear and one strongly bent nitrene ligand, as would be required by valence bond structure *27*.

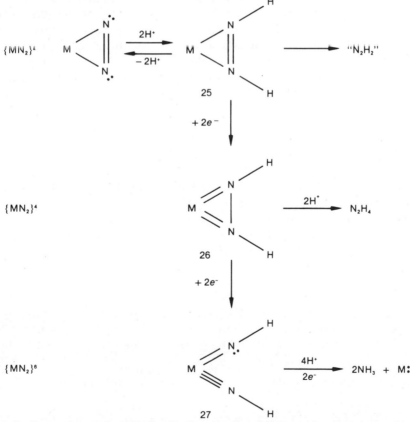

$$28$$

Mononitrene complexes of the type $MoX_2(NH)(dppe)_2$ with X = chlorine or bromine are known, and several nitrido complexes evolve ammonia in acid (Chatt and Dilworth, 1975, 1977).

The mechanism proposed in Figure 4 also explains the production of ammonia from cis-$M(N_2)_2(PMe_2Ph)_4$ compounds and acid (equations 10

Figure 4.   Proposed general mechanism for the reduction of dinitrogen by nitrogenase and by chemical systems. The mechanism involves: stabilization of η^2-geometry by protonation; formation of multiple bonds between the metal and nitrogen atoms; and mediation of the proton affinity of the nitrogen atoms by the metal.

and 11). For compounds *12* and *13*, protonation of one of the dinitrogen ligands may well precede the loss of one molecule of dinitrogen (Carter, Bercaw, and Gray, 1978; Manriquez et al., 1978). In any event, however, the loss of one molecule of N_2 from compound *12* or *13* will generate a η^1 species at the $\{MNN\}^6$ level of reduction with an adjacent vacant coordination site. Collapse of the η^1 complex to a protonated η^2 species and substitution of the PR_3 ligands by oxygen ligands (Chatt, 1977) should drive the reactions to completion. The differences observed between compounds *12* and *13* in their reactions with acid (equations 10 and 11) may reflect the relative stabilities of the nitrene complexes for M = W and M = Mo.

CONCLUSIONS

An extensive chemistry of η^1-dinitrogen complexes and their derivatives has developed since 1965. In contrast, relatively few definitive data are available for η^2 species. The relatively simple mechanism for the reduction of dinitrogen proposed in Figure 4 involves the formation and degradation of the MN_2H_2 ring system. The lack of information on this simple ring system presents a challenge to both the synthetic chemist and the theoretician. Once synthesized, such compounds should also provide a haven for spectroscopists of many frequencies.

NOTE ADDED IN PROOF

The x-ray structure of *trans*-$(i\text{-}Pr_3P)_2RhCl(N_2)$ has been redetermined at $-160°C$ by Thorn, Tulip, and Ibers (unpublished data). In addition, the ^{15}N and ^{31}P NMR spectra of the complex have been obtained in toluene-d_8 solution. The new x-ray data and the NMR spectra both show that the N_2 molecule adopts η^1 coordination rather than η^2 coordination as previously claimed (Busetto et al., 1977). Thus, there is as yet no definitive example of a stable dinitrogen compound with structure 2.

ACKNOWLEDGMENTS

I thank the organizing committee for their invitation to participate in this symposium. Special appreciation is accorded to Professors J. E. Bercaw, J. Chatt, T. A. George, B. Haymore, R. Hoffmann, J. A. Ibers, D. Sellman, A. E. Shilov, and G. N. Schrauzer for informing me of recent results from their laboratories. I thank Professor R. D. Feltham for numerous helpful discussions, Dr. V. K. Jones for translating a paper from Russian, and D. Spargur for technical assistance.

REFERENCES

Allen, A. D., and C. V. Senoff. 1965. Nitrogenopentammineruthenium(II) complexes. J. C. S. Chem. Comm., pp. 621–622.

Armor, J. N., and H. Taube. 1970. Linkage isomerism in nitrogen-labeled [Ru(NH₃)₅N₂]Br₂. J. Am. Chem. Soc. 92:2560–2562.

Ben-Shoshan, R., J. Chatt, W. Hussain, and G. J. Leigh. 1976. A diazomethane complex of tungsten. J. Organometallic Chem. 112:C9–C10.

Bercaw, J. E. 1977. Reduction of molecular nitrogen to hydrazine at titanium and zirconium. In: W. Newton, J. R. Postgate, and C. Rodriguez-Barrueco (eds.), Recent Developments in Nitrogen Fixation, pp. 25–40. Academic Press, Inc., New York.

Bevan, P. C., J. Chatt, A. A. Diamantis, R. A. Head, G. A. Heath, and G. J. Leigh. 1977a. Diazobutanol complexes of molybdenum and tungsten from the reaction of bis[1,2-bis(diphenylphosphino)-ethane]bis(dinitrogen)molybdenum or -tungsten with tetrahydrofuran in the presence of alkyl bromides. J. C. S. Dalton, pp. 1711–1715.

Bevan, P. C., J. Chatt, G. J. Leigh, and E. G. Leelamani. 1977b. The conversion of ligating dinitrogen into amines. J. Organometallic Chem. 139:C59–C62.

Brintzinger, H. H. 1976. Coordination requirements for the reduction of a dinitrogen ligand molecule. In: W. E. Newton and C. J. Nyman (eds.), Proceedings of the First International Symposium on Nitrogen Fixation, Vol 1, pp. 33–41. Washington State University Press, Pullman.

Busetto, C., A. D'Alfonso, F. Maspero, G. Perego, and A. Zazzeta. 1977. Side-on bonded dinitrogen and dioxygen complexes of rhodium(I) Synthesis and crystal structures of trans-chloro(dinitrogen)-, chloro(dioxygen)-, and chloro(ethylene)-bis(tri-isopropylphosphine)-rhodium(I). J. C. S. Dalton, pp. 1828–1834.

Carter, B. J., J. E. Bercaw, and H. B. Gray. 1978. Substitution kinetics of trans-M(N₂)₂[(C₆H₅)₂PCH₂CH₂P(C₆H₅)₂]₂ and cis-M(N₂)₂[P(CH₃)₂(C₆H₅)]₄, M = Mo,W. Presented at the 175th National Meeting of the American Chemical Society, March 13–17, Anaheim, Calif. Paper INOR-203.

Chatt, J. 1977. The activation of molecular nitrogen. In: A. A. Addison, W. R. Cullen, D. Dolphin, and B. R. James (eds.), Biological Aspects of Inorganic Chemistry, pp. 229–243. Wiley-Interscience, New York.

Chatt, J., A. A. Diamantis, G. A. Heath, N. E. Hooper, and G. J. Leigh. 1977. Reactions of ligating dinitrogen to form carbon-nitrogen bonds. J. C. S. Dalton, pp. 688–697.

Chatt, J., A. A. Diamantis, G. A. Heath, G. J. Leigh, and R. L. Richards. 1976. Some reactions of ligating dinitrogen in stable mononuclear complexes. In: W. E. Newton and C. J. Nyman (eds.), Proceedings of the First International Symposium on Nitrogen Fixation, Vol. 1. pp. 17–26. Washington State University Press, Pullman.

Chatt, J., and J. R. Dilworth. 1975. Preparation of stable imido (nitrene)-complexes of molybdenum and their conversion into nitrido-complexes. J. C. S. Chem. Comm., pp. 983–984.

Chatt, J., and J. R. Dilworth. 1977. Nitrido- and imido- or nitrene complexes of molybdenum. J. Indian Chem. Soc. 54:13–18.

Chatt, J., R. A. Head, G. J. Leigh, and C. J. Pickett. 1978. The mechanism of alkylation and acylation of dinitrogen coordinated to molybdenum and tungsten. J. C. S. Dalton, pp. 1638–1647.

Chatt, J., A. J. Pearman, and R. L. Richards. 1975. The reduction of mono-co-ordinated molecular nitrogen to ammonia in a protic environment. Nature 253:39–40.

Chatt, J., A. J. Pearman, and R. L. Richards. 1977a. Conversion of dinitrogen in its molybdenum and tungsten complexes into ammonia and possible relevance to the nitrogenase reaction. J. C. S. Dalton, pp. 1852–1860.

Chatt, J., A. J. Pearman, and R. L. Richards. 1977b. The preparation and oxidation, substitution, and protonation reactions of *trans*-bis(dinitrogen)tetrakis-
(methyldiphenylphosphine)tungsten. J. C. S. Dalton, pp. 2139–2142.

Chatt, J., and R. L. Richards. 1977. The binding of dinitrogen and dinitrogen
hydrides to molybdenum. J. Less-Common Metals, 54:477–484.

Cramer, S. P., K. O. Hodgson, W. D. Gillum, and L. W. Mortenson. 1978. The
molybdenum site of nitrogenase—Preliminary structural evidence from x-ray
absorption spectroscopy. J. Am. Chem. Soc. 100:3398–3407.

Day, V. W., T. A. George, S. D. A. Iske, and S. D. Wagner. 1976. The protonation
of alkyldiazenido derivatives of molybdenum. The crystal and molecular structure
of the benzene solvate of iodo-*N*-octylhydrazidobis[1,2-bis(diphenylphosphino)-
ethane]-molybdenum iodide. J. Organometallic Chem. 112:C55–C58.

Dickson, R. S., and J. A. Ibers. 1972. A π-bonded azo-transition metal complex.
The structure of bis(*tert*-butyl isocyanide)(azobenzene)nickel(0). J. Am. Chem.
Soc. 94:2988–2993.

Enemark, J. H., and R. D. Feltham. 1974. Principles of structure bonding and
reactivity for metal nitrosyl complexes. Coord. Chem. Rev. 13:339–406.

George, T. A., and S. D. A. Iske, Jr. 1976. Reactions of coordinated dinitrogen.
Potential application of photochemistry. In: W. E. Newton and C. J. Nyman
(eds.) Proceedings of the First International Symposium on Nitrogen Fixation,
Vol 1, pp. 27–32. Washington State University Press, Pullman.

Gynane, M. J. S., J. Jeffrey, and M. F. Lappert. 1978. Organozirconium(III)-
dinitrogen complexes: evidence for (η^2-N$_2$)—metal bonding in [Zr(η-C$_5$H$_5$)$_2$-
(N$_2$)(R)][R = (Me$_3$Si)$_2$CH]. J. C. S. Chem. Comm. pp. 34–36.

Haymore, B., E., A. Maatta, and R. A. D. Wentworth. 1979. A biphenylnitrene complex of molybdenum with a bent nitrene ligand. Preparation and structure of *cis*-
Mo(NC$_6$H$_5$)$_2$(S$_2$CN(C$_2$N$_5$)$_2$)$_2$. J. Am. Chem. Soc. 101:2063–2068.

Hoffman, P. R., T. Yoshida, T. Okano, S. Otsuka, and J. A. Ibers. 1976. Crystal
and molecular structure of hydrido(dinitrogen)bis-[phenyl(di-*tert*-butyl)phos-
phine]rhodium(I). Inorg. Chem. 15:2462–2466.

Hoffmann, R., M. M.-L. Chen, and D. L. Thorn. 1977. Qualitative discussion of
alternative coordination modes of diatomic ligands in transition metal complexes.
Inorg. Chem. 16:503–511.

Jonas K., D. J. Brauer, C. Krüger, P. J. Roberts, and Y.-H. Tsay. 1976. "Side-on"
dinitrogen-transition metal complexes. The molecular structure of {C$_6$H$_5$[Na·O-
(C$_2$H$_5$)$_2$]$_2$[(C$_6$H$_5$)$_2$Ni]$_2$N$_2$NaLi$_6$(OC$_2$H$_5$)$_4$·O(C$_2$H$_5$)$_2$}$_2$. J. Am. Chem. Soc. 98:74–81.

Kulakovskaya, S. I., V. N. Tsarev, O. N. Efimov, and A. E. Shilov. 1977. Electrolytic reduction of nitrogen in a homogeneous system including Ti(III) and
Mo(III). Kinetika i Kataliz. 18:1045–1047.

Leigh, G. J. 1977. Chemistry of dinitrogen. In: W. Newton, J. R. Postgate, and C.
Rodriguez-Barrueco (eds.), Recent Developments in Nitrogen Fixation, pp. 1–24.
Academic Press, Inc., New York.

Leigh, G. J., J. N. Murrell, W. Bremser, and W. G. Proctor. 1970. On the state of
dinitrogen bound to rhenium. J. C. S. Chem. Comm., p. 1661.

Leigh, G. J., and C. J. Pickett. 1977. Electrochemical behavior of organonitrile
dinitrogen complexes of molybdenum(0) and tungsten(0) and the anchoring of a
dinitrogen complex to an electrode surface. J. C. S. Dalton, pp. 1797–1800.

Manriquez, J. M., D. R. McAlister, E. Rosenberg, A. M. Shiller, K. L. Williamson,
S. I. Chan, and J. E. Bercaw. 1978. Solution structure and dynamics of binuclear nitrogen complexes of bis(pentamethylcyclopentadienyl)titanium(II) and bis
(pentamethylcyclopentadienyl)zirconium(II). J. Am. Chem. Soc. 100:3078–
3083.

Matasa, C., and E. Tonca. 1973. Basic Nitrogen Compounds, 3rd ed., pp. 393–420. Translated by S. Marcus. Chemical Publishing Co., New York.

Nakamura, A., T. Yoshida, M. Cowie, S. Otsuka, and J. A. Ibers. 1977. Cyclopropanation reactions of diazoalkanes with nickel(0) and palladium(0) catalysts. The structure of (diazofluorene)bis(*tert*-butylisocyanide)nickel(0); a complex containing a π-bonded diazofluorene molecule. J. Am. Chem. Soc. 99: 2108–2117.

Newton, W. E., J. L. Corbin, and J. W. McDonald. 1976. Nitrogenase: Mechanism and models. In: W. E. Newton and C. J. Nyman (eds.), Proceedings of the First International Symposium on Nitrogen Fixation, Vol. 1, pp. 53–74. Washington State University Press, Pullman.

Newton, W. E., and C. J. Nyman (eds.) 1976. Proceedings of the First International Symposium on Nitrogen Fixation, Vol. 1. Washington State University Press, Pullman.

Newton, W., J. R. Postgate and C. Rodriguez-Barrueco (eds.). 1977. Recent Developments in Nitrogen Fixation. Academic Press, Inc., New York.

Nikanova, L. A., and A. E. Shilov. 1977. Dinitrogen fixation in homogeneous protic media. In: W. E. Newton, J. R. Postgate, and C. Rodriguez-Barrueco (eds.), Recent Developments in Nitrogen Fixation, pp. 41–52. Academic Press, Inc., New York.

Payne, W. J. 1973. Reduction of nitrogenous oxides by microorganisms. Bacteriol. Rev. 37:409–452.

Pratt, P. F. 1977. Effect of increased nitrogen fixation on stratospheric ozone. Climactic Change 1:109–135.

Quinby, M. S., and R. D. Feltham. 1972. The infrared spectra of ruthenium derivatives of nitrogen, nitric oxide and carbon monoxide. Experimental evidence regarding dπ-pπ bonding. Inorg. Chem. 11:2468–2476.

Schrauzer, G. N. 1976. Biological nitrogen fixation: Model studies and mechanism. In: W. E. Newton and C. J. Nyman (eds.), Proceedings of the First International Symposium on Nitrogen Fixation, Vol. 1, pp. 79–116. Washington State University Press, Pullman.

Schrauzer, G. N. 1977. Nitrogenase model system and the mechanism of biological nitrogen reduction: Advances since 1974. In: W. Newton, J. R. Postgate, and C. Rodriguez-Barrueco (eds.), Recent Developments in Nitrogen Fixation, pp. 109–118. Academic Press, Inc., New York.

Schrauzer, G. N., and T. D. Guth. 1977. Photolysis of water and photoreduction of nitrogen on titanium dioxide. J. Am. Chem. Soc. 99:7189–7193.

Sellmann, D., and W. Weiss. 1977. The first reaction of the N_2 ligand with bases; reduction of coordinated dinitrogen by nucleophilic attack. Angew. Chem. Int. Ed. Engl. 16:880–881.

Sellmann, D., and W. Weiss. 1978. Consecutive nucleophilic and electrophilic attack on N_2 ligands: synthesis of azo compounds from molecular nitogen. Angew. Chem. Int. Ed. Engl. 17:269–270.

Shilov, A. E. 1976. Dinitrogen reduction in protic media. In: W. E. Newton and C. J. Nyman (eds.), Proceedings of the First International Symposium on Nitrogen Fixation, Vol. 1, pp. 42–52. Washington State University Press, Pullman.

Shilov, A. E. 1977. Dinitrogen reduction in protic media: A comparison of biological dinitrogen fixation with its chemical analogues. In: A. A. Addison, W. R. Cullen, D. Dolphin, and B. R. James (eds.), Biological Aspects of Inorganic Chemistry, pp. 197–228. Wiley-Interscience, New York.

Stiefel, E. I. 1973. Proposed molecular mechanism for the action of molybdenum in enzymes: Coupled proton and electron transfer. Proc. Natl. Acad. Sci. U.S.A. 70:988–992.

Sweeney, G. C. 1976. Technology and economics of ammonia production. In: W. E. Newton and C. J. Nyman (eds.), Proceedings of the First International Symposium on Nitrogen Fixation, Vol. 2, pp. 648–673. Washington State University Press, Pullman.

Tatsumi, T., M. Hidai, and Y. Uchida. 1975. Preparation and properties of dinitrogen-molybdenum complexes. II. Dinitrogen(organonitrile) complexes of molybdenum. Inorg. Chem. 14:2530–2534.

Thorneley, R. N. F., R. R. Eady, and D. J. Lowe. 1978. Biological nitrogen fixation by way of an enzyme-bound dinitrogen-hydride intermediate. Nature 272: 557–558.

Watt, G. D., and W. A. Bulen. 1976. Calorimetric and electrochemical studies on nitrogenase. In: W. E. Newton and C. J. Nyman (eds.), Proceedings of the First International Symposium on Nitrogen Fixation, Vol. 1, pp. 248–256. Washington State University Press, Pullman.

Zones, S. I., M. R. Palmer, J. G. Palmer, J. M. Doemeny, and G. N. Schrauzer. 1978. Hydrogen evolving systems. 3. Further observations on the reduction of molecular nitrogen and of other substrates in the $V(OH)_2$-$Mg(OH)_2$ system. J. Am. Chem. Soc. 100:2113–2121.

Nitrogen Fixation, Volume I
Edited by W. E. Newton and W. H. Orme-Johnson
Copyright 1980   University Park Press   Baltimore

A Quantum-Chemical Theory of Transition Metal–Dinitrogen Complexes

Hsu Chi-Ching

We have previously discussed the chemical bonding in mononuclear and binuclear dinitrogen complexes possessing only end-on coordination structures (Nitrogen Fixation Research Group, 1974). Using HMO and graphical methods (Tang and Kiang, 1976, 1977), we have extended our theoretical studies to the chemical bonding in mono-, bi-, and trinuclear dinitrogen complexes with a variety of reasonable coordination structures.

MONONUCLEAR DINITROGEN COMPLEXES

Molecules with Single End-on
Coordination Structure of the M-3—N-2—N-1 Type

We use a right-handed coordinate system with the z axis oriented in the direction of the molecular axis of N_2. For this coordination type, two equivalent sets of π bonding orbitals can be formed from the two different sets of atomic orbitals (d_{xz}, p_x^2, p_x^1) and (d_{yz}, p_y^2, p_y^1). These molecular orbitals (MOs) are twofold degenerate. Now consider the MO formed from the atomic orbital (AO) set (d_{xz}, p_x^2, p_x^1) and described by the linear combination of atomic orbitals (LCAO):

$$\Psi = C_1\phi_1 + C_2\phi_2 + C_3\phi_3$$

where ϕ_1, ϕ_2, and ϕ_3 represent the AOs p_x^1, p_x^2, and d_{xz}, respectively. Using MO graphical theory, we obtained the following results. First, the MO graph can be written as:

This paper is part of the research effort of the Nitrogen Fixation Research Group, Department of Chemistry, Kirin University.

$$\overset{-\eta}{\underset{x-\delta}{\odot}}\!\!-\!\!\!\overset{-1}{\underset{x}{\bigcirc}}\!\!-\!\!\!\underset{x}{\bigcirc}$$

$$3 \qquad 2 \qquad 1$$

with $x = \dfrac{\epsilon - \alpha}{\beta}$, $\delta = \dfrac{\alpha' - \alpha}{\beta}$, and $\eta = \dfrac{\beta'}{\beta}$; where $\alpha(\alpha')$, $\beta(\beta')$, and ϵ denote, respectively, the AO energy of $p_x{}^1$ or $p_x{}^2$ (d_{xz}), the interaction energy between ϕ_1 and ϕ_2 (ϕ_2 and ϕ_3), and the MO energy.

Second, the eigen equation corresponding to this MO graph is:

$$(x - \delta)(x^2 - 1) - \eta^2 x = 0$$

Third, the LCAO coefficients in this MO can be obtained explicitly through the equations:

$$C_1(x) = \frac{1}{N(x)} x(x - \delta) - \eta^2$$

$$C_2(x) = \frac{1}{N(x)} (x - \delta)$$

$$C_3(x) = \frac{1}{N(x)} \eta$$

$$N(x) = \{[x(x - \delta) - \eta^2]^2 + (x - \delta)^2 + \eta^2\}^{1/2}$$

where $N(x)$ is the normalization factor. Substituting the roots x into $\epsilon = \alpha + \beta x$, we have been able to evaluate the stabilization energy ΔE (or $\Delta E/2\beta$), which is very useful for discussion of the stability of such a dinitrogen complex. From the explicit form of $C_j(x)$, the charge density on the jth atom will be given by

$$q_j = \sum_i \eta_i C_j(x_i) C_j(x_i)$$

The two quantum-chemical parameters, $\Delta E/2\beta$ and q_1, are plotted against δ, as shown in curves I of Figures 1 and 2, respectively. For the d^6 configuration, these plots show that the stability of the end-on complex increases monotonously as δ decreases. The cases $\delta = 1$ and $\delta = -1$ correspond, respectively, to the π and π^* levels of the N_2 molecule. Hence, when δ tends toward -1 and even becomes more negative, i.e., the energy level α' of d_{xz} approaches the π^* level of N_2 from below and even gets beyond it, the complex tends to become more stable. With α' now lying rather high up, the electron will tend to migrate from the d orbital to the AO of the nitrogen atoms, resulting in an increase in q_1. This situation favors attack by electrophilic groups such as H^+, and thus facilitates the reduction of N_2. For the d^0 configuration, in contrast, the complex becomes more stable only when α' approaches the π level of N_2 from above and then goes further down. With α' now lying rather low, the N-1 atom will become more positive and thus the N_2 molecule will be reduced through attack by nucleophilic agents, such as H^-.

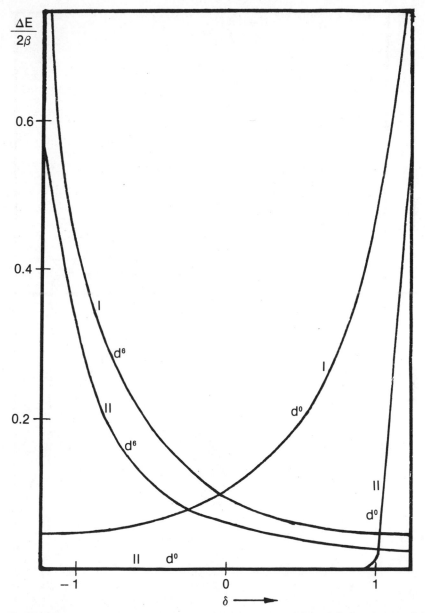

Figure 1.   Plots of the stabilization energy ($\Delta E/2\beta$) versus δ, which is a relative measure of the energy of the metal d orbital involved in π bonding with N_2, for mononuclear complexes. I represents the coordination type, M—N-2—N-1, and II the coordination type $M\overset{\displaystyle N\text{-}1}{\underset{\displaystyle N\text{-}1'}{\big|}}$ ($\eta^2 = 0.1$; $\eta'^1 = 0.025$; $\eta''^2 = 0.00625$). d^n represents the occupation (n) of the metal d orbitals.

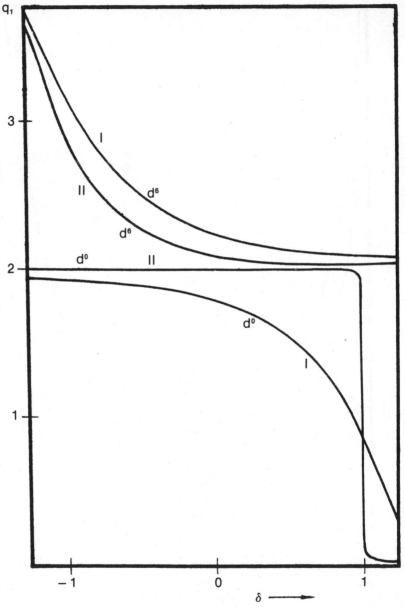

Figure 2.   Plots of charge density, q_1, on the *exo*-nitrogen (N-1) atom versus δ for mononu-clear complexes. I = M—N-2—N-1 and II = M$\diagdown$ (N-1, N-1') ($\eta^2 = 0.1$; $\eta'^2 = 0.025$; $\eta''^2 = 0.00625$). d^n is the number (n) of d electrons available.

Molecules with
Single Side-on Coordination Structure of the M-2 $\overset{\displaystyle N\text{-}1}{\underset{\displaystyle N\text{-}1'}{\diagdown\ \mid\ }}$ **Type**

Here, the MO graph can be written down as:

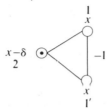

For such a complex, two nonequivalent sets of π bonding orbitals can be formed from the two different AO sets $(d_{xz}, p_x{}^{1'}, p_x{}^{1'})$ and $(d_{yz}, p_y{}^{1}, p_y{}^{1'})$. It is obvious from the accompanying diagram that the molecular parameters, η' and η'', for this case cannot be equal to each other, nor can they be equal to the parameter η of the previous case. In order to discuss the two different LCAOs for this side-on coordination structure on a basis comparable to the end-on case, we assume that the distance between the metal and N-2 in the first case is equal to that between the metal and the midpoint of the N—N bond in the present case, and the three parameters may be approximately interrelated by $\eta' = \frac{1}{2}\eta$ and $\eta'' = \frac{1}{4}\eta$.

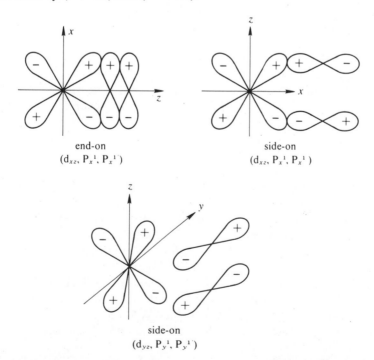

end-on
$(d_{xz}, P_x{}^1, P_x{}^{1'})$

side-on
$(d_{xz}, P_x{}^1, P_x{}^1)$

side-on
$(d_{yz}, P_y{}^1, P_y{}^1)$

Because there is a symmetry plane through the transition metal ion (M) and perpendicular to the N—N bond, the MO graph for the MO (d_{xz}, p_x^1, $p_x^{1'}$) can be reduced to a symmetric plus an antisymmetric part.

For the symmetric graph $x^0 - 1$, the eigen equation is $x - 1 = 0$, and the symmetric MO is given as

$$\Psi^{(S)} = \frac{1}{\sqrt{2}} (\phi^1 + \phi^{1'})$$

For the antisymmetric graph,

$$\underset{x + 1}{\overset{1}{\bullet}} \quad \overset{-\sqrt{2\eta'}}{\rule{3cm}{0.4pt}} \quad \underset{x - \delta}{\overset{2}{\bullet}}$$

the eigen equation is $(x - \delta)(x + 1) - 2\eta'^2 = 0$, and thus

$$\Psi^{(A)} = C_1^{(A)} \frac{1}{\sqrt{2}} (\phi^1 - \phi^{1'}) + C_2^{(A)} \phi_2$$

where

$$C_1^{(A)} = \frac{1}{N(x)} (x - \delta)$$

$$C_2^{(A)} = \frac{1}{N(x)} \sqrt{2\eta'}$$

$$N(x) = [(x - \delta)^2 + 2\eta'^2]^{1/2}$$

It is obvious that these solutions hold good for the MO of (d_{yz}, p_y^1, $p_y^{1'}$), provided that the parameter η' is replaced by η''.

The stabilization energy $\Delta E/2\beta$ and the charge density q_1 (on the N-1 atom) can then be plotted against δ for the side-on coordination structure (curves II of Figures 1 and 2). Figure 1 shows that, for the d^6 configuration, curve II is quite similar to curve I but is always lower in position, indicating that the side-on coordination structure is certainly less stable. For the d^0 configuration, however, the stabilization energy vanishes for $\delta < 1$ and thus no stable dinitrogen complexes can be formed at all. It is only when $\delta > 1$ (i.e., the energy level of the d orbital is equal to or lower than the π level of N_2) that we can obtain stable dinitrogen complexes. This result helps explain why no stable mononuclear side-on–coordinated dinitrogen complex was synthesized prior to 1977 (Busetto et al., 1977).

A comparison of curves I and II in Figure 2 for the d^6 configuration shows that, since q_1 is always larger for end-on coordination, the end-on complex will be more easily reduced. For the d^0 configuration, only where $\delta > 1$ (i.e., the energy level of the d orbital is lower than the π level of N_2) will the N_2 molecule be activated by a transition metal ion. Becoming more positive under this circumstance, the nitrogen atom is now in a more advantageous position of being successfully attacked by a nucleophilic group, such as H^-, and being reduced accordingly. In contrast, when $\delta < 1$, it is the end-on–coordinated N_2 molecule that is more easily activated.

Molecules with *trans*-Double–End-on Coordination Structures of the N-1—N-2—M-3—N-2′—N-1′ Type

The MO graph can be written as:

$$
\begin{array}{ccccc}
\overset{-1}{\underset{\displaystyle x}{\bigcirc}} & \overset{-\eta}{\underset{\displaystyle x}{\bigcirc}} & \overset{-\eta}{\underset{\displaystyle x-\delta}{\bullet}} & \overset{-\eta}{\underset{\displaystyle x}{\bigcirc}} & \overset{-1}{\underset{\displaystyle x}{\bigcirc}} \\
1 & 2 & 3 & 2' & 1'
\end{array}
$$

Thus, two equivalent sets of π bonding MO orbitals can be formed from the two different AO sets (p_x^1, p_x^2, d_{xz}, $p_x^{2'}$) and (p_y^1, p_y^2, d_{yz}, $p_y^{2'}$, $p_y^{1'}$), which are obviously twofold degenerate. The charge densities q_1 and q_2 can now be plotted against δ (Figure 3). Because q_1 is always larger than q_2 for all values of δ (i.e., N-1 is always more negative than N-2), the N-1 atom is the first to be attacked by electrophilic groups, including H^+. The higher activity of the *exo*-N atom is entirely analogous to the single end-on–coordinated N_2 complexes, in agreement with the x-ray photoelectron spectroscopic studies (Brant and Feltham, 1977).

Molecules with *cis*-Double–End-on Coordination Structures of the M-3—N-2—N-1 Type

$$
\begin{array}{c}
\mid \\
\text{N-2}' \\
\mid \\
\text{N-1}'
\end{array}
$$

Three nonequivalent sets of π bonding MOs can be formed from the AO sets (p_x^1, p_x^2, d_{xz}, $p_z^{2'}$, $p_z^{1'}$), (p_y^1, p_y^2, d_{yz}), and (d_{xy}, $p_y^{2'}$, $p_y^{1'}$), respectively. The π bonding corresponding to the first AO set is analogous to that of the third case, whereas the other two sets are analogs of the first case.

For a comparison of the three end-on coordination types, q_1 is plotted against δ for the d^6 configuration as curves I (single N_2), II (*trans*-$(N_2)_2$), and III (*cis*-$(N_2)_2$) of Figure 4. It is found that $q_1(I) > q_1(III) > q_1(II)$ for all values of δ. This result explains the fact that, in the reduction of bisdinitrogen complexes of molybdenum and tungsten, one of the N_2 molecules is easily replaced by a π-donating ligand, such as SO_4H^- or OH^- (Chatt, Pearman, and Richards, 1975, 1976), producing an intermediate simple dinitrogen complex. Obviously, such an intermediate complex with a more negative *exo*-N atom can be easily reduced.

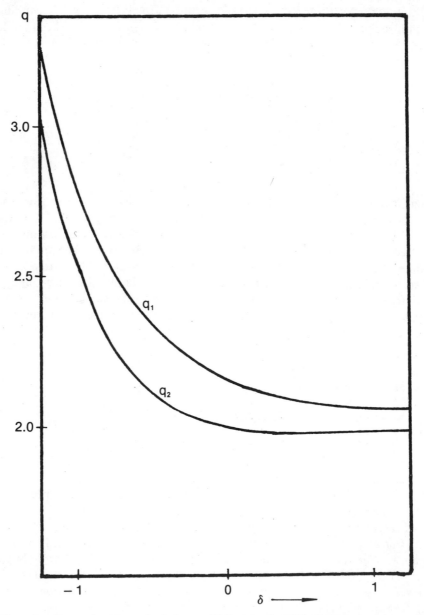

Figure 3.   Charge densities on both the *exo*-(N-1) and endo-(N-2) nitrogen atoms in mononuclear *trans*-double–end-on coordination compound (N-1—N-2—M-3—N-2′—N-1′) of a d^6 configuration as a function of δ. q_1 is the charge density on N-1 and q_2 is for N-2. $\eta^2 = 0.1$.

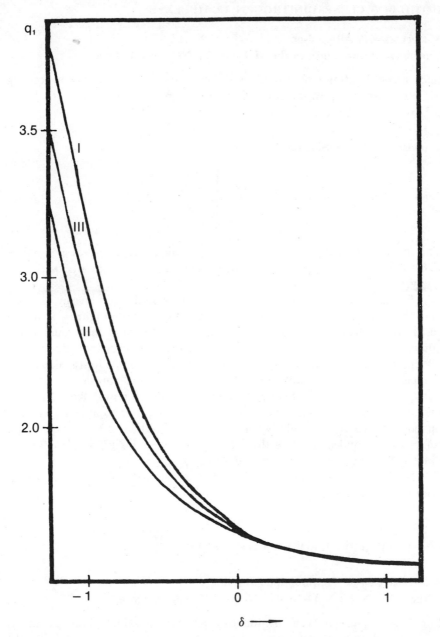

Figure 4. A comparison of charge density (q_1) on the *exo*-nitrogen (N-1) atom for the single–end-on [M—N—N-1; I], *trans*-double–end-on [N-1—N—M—N—N-1′; II], and *cis*-double–end-on [M$\leftarrow$N—N-1′)₂; III] coordination types. d⁶ Configuration and $\eta^2 = 0.1$.

HOMO-BINUCLEAR DINITROGEN COMPLEXES

Molecules with All–End-on
Coordination Structures of the M-1—N-2—N-2′—M-1′ Type

Such a coordination structure gives rise to two equivalent sets of π bonding MOs corresponding to the two AO sets $(d_{xz}^1, p_x^2, p_x^{2'}, d_{xz}^{1'})$ and $(d_{yz}^1, p_y^2, p_y^{2'}, d_{yz}^{1'})$.

Molecules with All–Side-on

Coordination Structures of the M-1 [diagram: N-2 above, N-2′ below, with M-1 and M-1′ bonded] M-1′ Type

This coordination type gives rise to two nonequivalent sets of π bonding MOs arising from $(d_{xz}^1, p_x^2, p_x^{2'}, d_{xz}^{1'})$ and $(d_{yz}^1, p_y^2, p_y^{2'}, d_{yz}^{1'})$, respectively.

For molecules of both structure types, the stabilization energy, $\Delta E/2\beta$, and the charge density, q_2, can be plotted against δ (Figures 5 and 6, respectively). The $\Delta E/2\beta$ curves are analogous to those for mononuclear complexes (Figure 1), just as expected. Thus these curves indicate that the all–end-on coordination type is more stable, which accounts for the paucity of binuclear dinitrogen complexes with all–side-on coordination (Krüger and Tsay, 1973; Jonas et al., 1976). Moreover, it is worth noting from Figure 6 that when $\delta > -1$ for d^{12} and $\delta < 1$ for d^0 it is easier to activate an end-on–coordinated N_2 molecule, whereas when $\delta < -1$ for d^{12} (the superscript refers to the total number of d electrons) or $\delta > 1$ for d^0, a side-on–coordinated N_2 molecule will be activated more easily instead.

Similar considerations can be extended immediately to hetero-binuclear dinitrogen complexes with essentially similar results.

TRINUCLEAR DINITROGEN COMPLEXES

All-*trans*-End-on Coordination Structure
of the M-1—N-2—N-3—M-4′—N-3′—N-2′—M-1′ Type

For this coordination type, the quantum-chemical parameters $\Delta E/2\beta$ and q_2 are again plotted against δ_2, as shown in Figures 7 and 8, respectively (the superscript n in the figures refers to the total number of d electrons available in the three transition metal atoms). Figure 7 shows that, when $\delta_1 = 0.5$ and $-1.25 < \delta_2 < 0.5$, the dinitrogen complexes become more stable as n

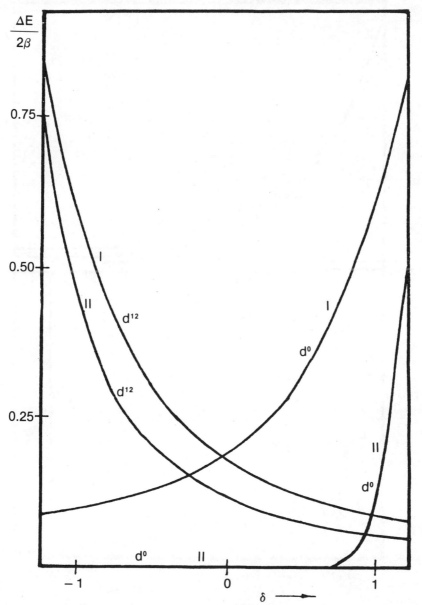

Figure 5.   Stabilization energy ($\Delta E/2\beta$) as a function of δ for homo-binuclear dinitrogen complexes of the types M—N-2—N-2'—M' (I) and M⟨N-2 / N-2'⟩M' (II), $\eta^2 = 0.1$, $\eta'^2 = 0.025$; $\eta''^2 = 0.00625$), with d⁰ and d¹² electron configurations.

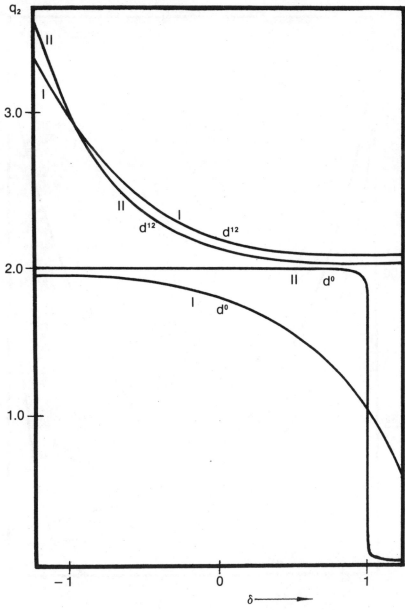

Figure 6. Charge density (q_2) on the nitrogen atom as a function of δ for M—N-2—N-2′—M′

(I) and M
```
        N-2
       /   \
      M     M′
       \   /
        N-2′
```
(II, $\eta^2 = 0.1$; $\eta'^2 = 0.025$; $\eta''^2 = 0.00625$) for both d^0 and d^{12} configura-

tions.

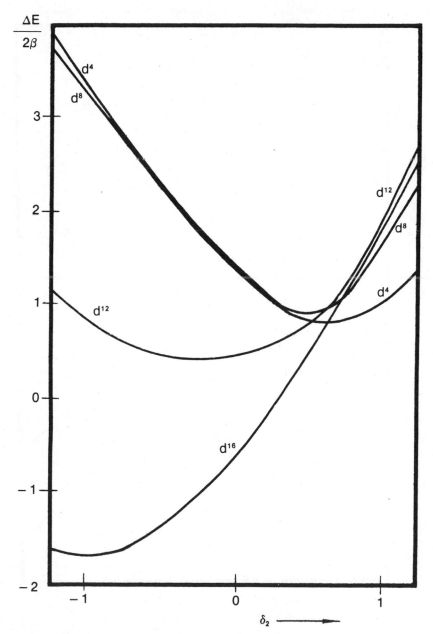

Figure 7.   Stabilization energy ($\Delta E/2\beta$) for the trinuclear complexes M—N2′—N-3′—M′—N-3—N-2—M as a function of δ_2 (the relative energy of the d orbital on the M′ metal atom) for the total electron populations of d^4, d^8, d^{12}, and d^{16} (with $\delta_1 = 0.5$ and $\eta_1{}^2 = \eta_2{}^2 = 0.1$).

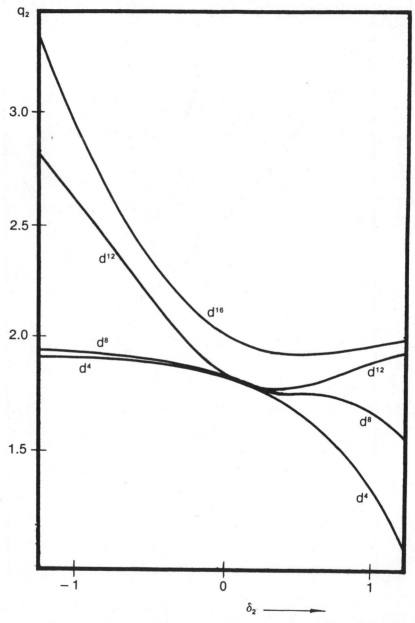

Figure 8.   Charge density (q_2) of the nitrogen (N-2) atoms in the complexes M—N-2′—N-3′—M′—N-3—N-2—M as a function of δ_2 and the total electron configurations d^4, d^8, d^{12}, and d^{16} (with $\delta_1 = 0.5$ and $\eta_1^2 = \eta_2^2 = 0.1$).

decreases; whereas in the case $\delta_1 = 0.5$ and $0.5 < \delta_2 < 1.25$, any decrease in n will result in decreased stability of these complexes. Thus, for the d^{16} configuration with $\delta_2 < 0.5$, no dinitrogen complex can be formed because $\Delta E/2\beta$ would then become negative. When $\delta_2 > 0.5$, however, the stability of these complexes will increase as δ_2 increases. This result is contrary to that expected from the results obtained for mononuclear and binuclear dinitrogen complexes.

In contrast, Figure 8 shows that the q_2 curves for these trinuclear complexes do bear similarities to those found for mononuclear and binuclear complexes. For example, when n is rather large (say, 16) and the energy level of the d orbital of the central atom M' approaches the π^* level of N₂ from below and even gets beyond it, d electrons will tend to migrate from M' toward the nitrogen atom, giving rise to an increase in q_2 and thus facilitating an electrophilic reaction. When n is small enough (say, 4) and the energy level of the d orbital of M' approaches the π level of N₂ from above and gets even further down, the N-2 atom will become somewhat positive, thus favoring a nucleophilic reaction.

Double–End-on Plus Single–Side-on Coordination Structure of the M-1—N-2—N-2'—M-1' Type
$$\diagdown \diagup$$
$$\mathbf{M'}$$

The plots of $\Delta E/2\beta$ and q_2 against δ_2 for such a coordination type are shown in Figure 9 and 10, respectively. Again, we find that the variation of the stabilization energy is, in general, quite analogous to the cases of all–trans–end-on coordination. For example, when $\delta_2 < 0.5$, the stability of the dinitrogen complexes increases as n decreases, so that complexes with d^{16} configuration would be most unstable, yet they are more stable than the d^{16} complexes with all–trans-end-on coordination. On the other hand, Figure 10 shows that when n is rather large (say, 16) and the energy level of the d orbital of M' lies rather high, the nitrogen atoms tend to carry more negative charge. However, if n becomes small enough (say, 4) and the energy level of the d orbital of M' lies rather low, the nitrogen atoms would tend to become somewhat positive instead.

Single–End-on Plus Double–

M-4

|

N-3

Side-on Coordination Structure of the M-1   M-1' Type

N-2

The $\Delta E/2\beta$ versus δ_2 and q_2 versus δ_2 plots (Figures 11 and 12) again show that the variation of both the stabilization energy and the charge density for

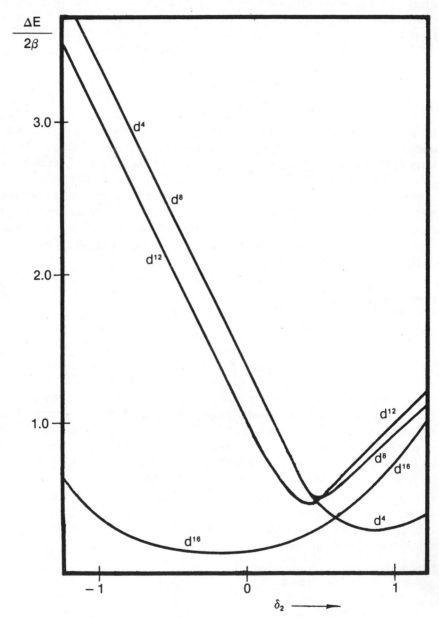

Figure 9.   Plots of $\Delta E/2\beta$ against δ_2 for the total electron populations d^4, d^8, d^{12}, and d^{16} for the complex M—N-2—N-2'—M (with $\eta_1 = 0.5$; $\eta_1{}^2 = \eta_2{}^2 = 0.1$; $\eta_2{}'^2 = 0.025$; $\eta_2{}''^2 = 0.00625$).

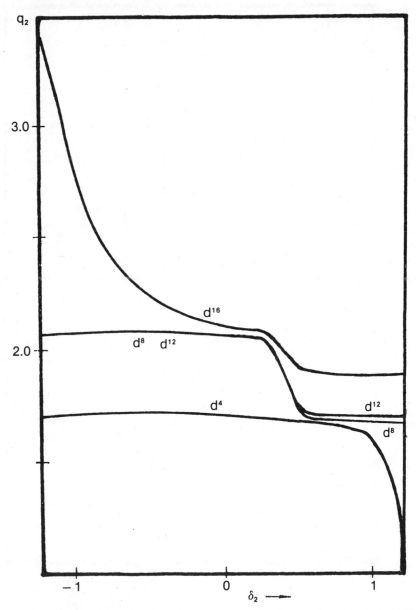

Figure 10.   Charge density (q_2) on the nitrogen atoms for the complexes M—N-2—N-2′—M

for d^4, d^8, d^{12}, and d^{16} electron populations (with $\eta_1 = 0.5$; $\eta_1{}^2 = \eta_2{}^2 = 0.1$; $\eta_2{}'^2 = 0.025$; $\eta_2{}''^2 = 0.00625$).

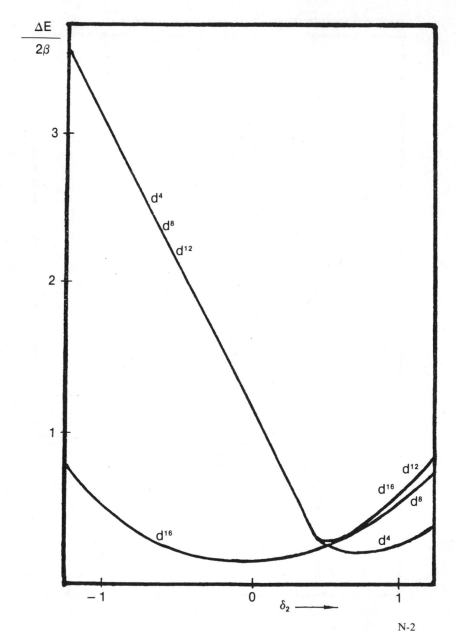

Figure 11. Plots of stabilization energy ($\Delta E/2\beta$) against δ_2 for the complexes $M\underset{N\text{-}3}{\overset{N\text{-}2}{\diamond}}M$ as a function of total electron populations of d^4, d^8, d^{12}, and d^{16} (with $\delta_1 = 0.5$; $\eta_1^2 = \eta_2^2 = 0.1$; $\eta_1'^2 = 0.025$; $\eta_1''^2 = 0.00625$).

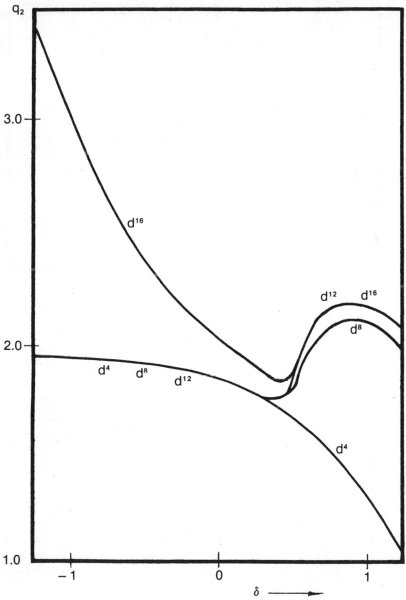

Figure 12.   Plots of charge density (q_2) on nitrogen atom (N-2) against δ_2 as a function of total electron populations d^4, d^8, d^{12}, and d^{16} for the complexes M⟨N-2 / N-3⟩M (with $\delta_1 = 0.5$; $\eta_1^2 =$ $\eta_2^2 = 0.1$; $\eta_1'^2 = 0.025$; $\eta_1''^2 = 0.00625$).

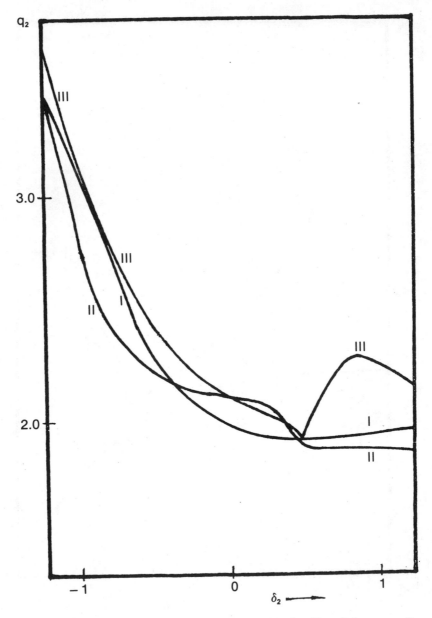

Figure 13.   Comparison of the charge density (q_2) curves for the d^{16} total electron configurations of the three trinuclear structures, M—N—N—M'—N—N—M (I), M—N—N—M (II), with M' below, and M (III) (with $\eta_1{}^2 = \eta_2{}^2 = 0.1$).

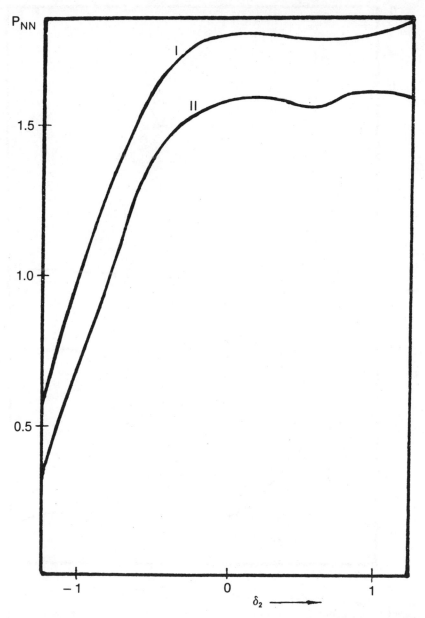

Figure 14.   Comparisons of plots of bond order (P_{NN}) versus δ_2 for the d^{16} configurations in the trinuclear complexes M—N—N—M'—N—N—M (I) and M\[structure\]M (II).

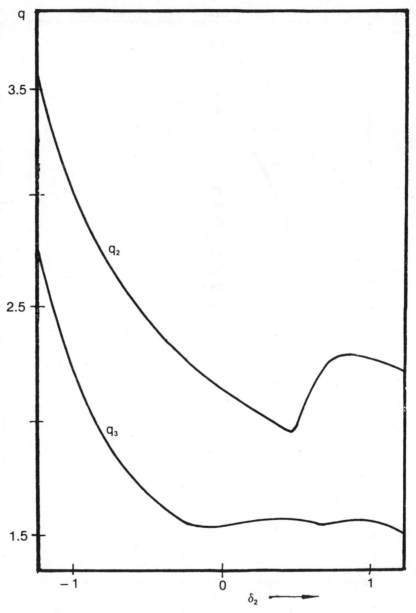

Figure 15.   A plot of the charge density (q) on the *exo* (q_2) and *endo* (q_3) nitrogen atoms in the complexes M⟨N-2 / N-3⟩M as a function of δ_2 (with $\delta_1 = 0.5$; $\eta_2^2 = 0.1$; $\eta_1^2 = 0.025$; $\eta_2'^2 = 0.00625$).

this coordination structure follow patterns similar to the double–end-on plus single–side-on coordination type.

On comparison of the charge density curves for these three trinuclear coordination structures for the d^{16} configuration (Figure 13), q_2 is in general somewhat larger for the third case, in conformity with the result of bond order (P_{NN}) calculations shown in Figure 14. Again, the N-2 atom, which is *exo* with respect to M', carries more negative charge, in general, than the other *endo*-N atom (cf. Figure 15).

SUMMARY

We are now in a position to give a comparative discussion on dinitrogen complexes, particularly trinuclear versus mononuclear and binuclear, on the basis of our quantum-chemical calculations. There are some essential similarities. First, when the total number of d electrons is quite large and the energy level of the d orbital of the "donor" transition metal atom lies quite high up, the nitrogen atoms tend to become more negative; such complexes will favor an electrophilic reaction. On the other hand, if the total number of d electrons is small and the energy level of the d orbital of the "acceptor" transition metal atom lies rather low, the nitrogen atoms tend to become somewhat positive; such complexes may then be preferentially reduced via a nucleophilic reaction. Second, the *exo*-N atom is in general more negative than the *endo*-N atom.

However, there are also some striking differences worthy of note. First, for both mononuclear and binuclear dinitrogen complexes, the end-on coordination type is in general more stable. For trinuclear dinitrogen complexes, however, Figure 16 shows that, when $-1.25 < \delta_2 < 0.5$, coordination structures involving both end-on and side-on coordinations to transition metal atoms, such as the second and third cases presented, turn out to be more stable than the all–end-on coordination type. This conclusion is true as long as the total number of d electrons is not less than 12. We therefore predict that the active center for catalytic nitrogen fixation in nitrogenase must be at least a trinuclear transition metal cluster involving both end-on and side-on coordinations of N_2.

Second, mononuclear and binuclear dinitrogen complexes are stable provided that either the energy level of the d orbital lies rather high up when the total number of d electrons is large or the energy level of the d orbital lies rather low when the total number of d electrons is sufficiently small. On the contrary, trinuclear dinitrogen complexes are unstable under such circumstances. So far, very few trinuclear dinitrogen complexes have been successfully synthesized. Examples are $[MoCl_4\{(N_2)ReCl(PMe_2Ph)_4\}_2]$ with a d^{14} configuration and $[Mg(THF)_4\{(N_2)Co(PMe_3)_3\}_2]$ with a d^{18} configuration, both of which possess an all–*trans*-end-on coordination structure (Cradwick et al., 1975; Hammer et al., 1976). For these complexes, the total number of

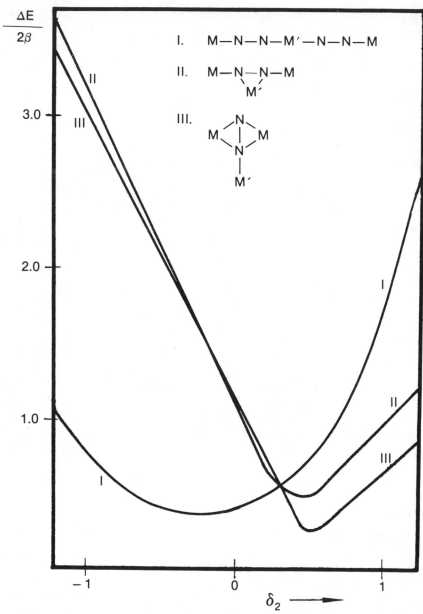

Figure 16.   Plots of stabilization energy ($\Delta E/2\beta$) against δ_2 for the d^{12} electron configurations in the trinuclear complexes M—N—N—M'—N—N—M (I), M—N—N—M (II), and M (III).

d electrons is quite large. Thus, it is inferred that the energy level of the d orbital involved must lie in the range of low energies.

ACKNOWLEDGMENTS

The author wishes to express his gratitude to Professor Tang Au-Ching for his suggestion of this problem as well as for his continued interest and untiring guidance in the course of this work.

REFERENCES

Brant, P., and R. D. Feltham. 1977. X-ray photoelectron spectra of molybdenum dinitrogen complexes and their derivatives. J. Less-Common Metals 54:81–87.

Busetto, C., A. D'Alfonso, F. Maspero, G. Perego, and A. Zazzetta. 1977. Side-on bonded dinitrogen and dioxygen complexes of rhodium(I). Synthesis and crystal structures of *trans*-chloro(dinitrogen)-, chloro(dioxygen)-, and chloro(ethylene-bis(tri-isopropylphosphine)rhodium(I). J. C. S. Dalton, pp. 1828–1834.

Chatt, J., A. J. Pearman, and R. L. Richards. 1975. Reduction of mono-coordinated molecular nitrogen to ammonia in a protic environment. Nature 253:39–40.

Chatt, J., A. J. Pearman, and R. L. Richards. 1976. Relevance of oxygen ligands to reduction of ligating dinitrogen. Nature 259:204.

Cradwick, P. D., J. Chatt, R. H. Crabtree, and R. L. Richards. 1975. Preparation and x-ray structure of a trinuclear dinitrogen-bridged complex, *trans*-[$MoCl_4\{(M_2)ReCl(PMe_2Ph)_4\}_2$]. J. C. S. Chem. Comm., pp. 351–352.

Hammer, R., H.-F. Klein, U. Schubert, A. Frank, and G. Hüttner. 1976. A novel hetero-bimetallic dinitrogen complex. Angew. Chem. Int. Ed. Engl. 15:612–613.

Jonas, K., D. J. Brauer, C. Krüger, P. J. Roberts, and Tsay, Y.-H. 1976. "Side-on" dinitrogen-transition metal complexes. Molecular structure of $\{C_6H_5[Na \cdot O-(C_2H_5)_2]_2[(C_6H_5)_2Ni]_2N_2NaLi_6(OC_2H_5)_4 \cdot O(C_2H_5)_2\}_2$. J. Am. Chem. Soc. 98:74–81.

Krüger, C., and Y.-H. Tsay. 1973. Molecular structure of a π-dinitrogen-nickel-lithium complex. Angew. Chem. Int. Ed. Engl. 12:998–999.

Nitrogen Fixation Research Group, Department of Chemistry, Kirin University. 1974. Theory of chemical bonding in dinitrogen complexes. Scientia Sinica 17:193–208.

Tang, A.-C., and Y.-S. Kiang. 1976. Graph theory of molecular orbitals. Scientia Sinica 19:207–226.

Tang, A.-C., and Y.-S. Kiang. 1977. Graph theory of molecular orbitals. II. Symmetrical analysis and calculations of MO coefficients. Scientia Sinica 20:585–612.

Nitrogen Fixation, Volume I
Edited by W. E. Newton and W. H. Orme-Johnson
Copyright 1980   University Park Press   Baltimore

Composite "String Bag" Cluster Model for the Active Center of Nitrogenase

Lu Jiaxi

In recent years, the problem of biological nitrogen fixation and its chemical simulation has attracted the attention of many biochemists and chemists. This problem of nitrogen fixation under much milder conditions than those heretofore possible with the Haber-Bosch or any other industrialized process is, of course, a rather hard "nut" to crack, yet it is very interesting and meaningful in itself, both theoretically and experimentally. Indeed, such a research problem has particular significance in China (with a population of over 800 million and an area of 9.6 million square kilometers) and other developing countries. Any prospect of developing, in the not too distant future, an industrializable nitrogen-fixation process on this basis is certainly worth an organized nationwide research effort. This paper presents a brief account of some phases of our research work together with their implications.

The rate-determining step in biological nitrogen fixation must be the activation of dinitrogen, as in the Haber-Bosch process. Temkin has developed his well-known rate equation with this as its first assumption. We started from the first principles of structural chemistry and proceeded to a theoretical analysis of the structural requirement for such an activation process. From this viewpoint, the biochemical studies of nitrogen fixation by nitrogenase can be rather satisfactorily interpreted in terms of a composite, multinuclear, transition-metal, "string bag" cluster structural model. The simplest example is a tetranuclear cluster of molybdenum and iron atoms, four in total, with three sulfur atoms bonded to them. The recent characterization of both the FeMo cofactor from nitrogenase (Shah and

Brill, 1977) and the environment of the molybdenum atom in the MoFe protein (and FeMo cofactor) in the semi-reduced state derived from the MoKα edge and EXAFS studies (Cramer et al., 1978) are very much compatible with our proposed model.

The enzymatic activation and reduction of dinitrogen do not seem to be very specific, since they are affected by a number of chemical inhibitors. Moreover, the turnover value is relatively very low. These observations indicate that the active N_2-fixing center of the enzyme cannot be structurally simple. Several interesting features deserve mention here. First, all reducible substrate molecules contain one multiple bond of order 2 or greater and carry at most only one terminal group of appreciable size, such as CH_3. This multiple bond is easily chemisorbed and sometimes exhibits noncompetitive or even competitive adsorption. Second, except for C_2H_2 (or a monosubstituted acetylene) cyclopropene, and NO (which is not reducible by nitrogenase), each reducible substrate has only one multiple bond ruptured. An alkyne or a cycloalkene is hydrogenated in the *cis* manner only (e.g., to C_2H_4) without rupture of the multiple bond. Also, in N_2O, the lower-ordered N—O bond is cleaved, rather than the higher-ordered N—N bond. Third, in comparison with RCN, the substrates CN^- and RNC exhibit much higher catalytic reduction rates by several orders of magnitude. These data indicate that nitrogenase activates substrate molecules via coordination to transition metal atoms in order to effect catalysis under mild conditions.

THE UNUSUAL INERTNESS OF THE N≡N BOND IN DINITROGEN: PRINCIPAL CONTRADICTION IN THE PROBLEM OF CHEMICAL ACTIVATION OF DINITROGEN MOLECULE

The N≡N bond in the dinitrogen molecule is unusually inert, as manifested in the following experimental facts. First, the N≡N bond length of 1.0976 Å is the shortest ever observed in molecules not containing any hydrogen atom and shorter than the C—C bond in C_2H_2 (1.205 Å) by as much as 0.107 Å. The stretching frequency is 2.331 cm^{-1}, again the highest among molecules not containing any hydrogen atom. Second, Tables 1 and 2 list all of the important molecular orbital energy levels of N_2 as well as the isoelectronic molecules NO^+, CO, C_2H_2, and CN^- (Figure 1), together with their "frontier" molecular orbitals, LV (lowest vacant), HO (highest occupied), and NHO (next to highest occupied) and their respective energy levels (ϵ_{LV}, ϵ_{HO}, ϵ_{NHO}) with the energy level separations $\Delta_1 \equiv \epsilon_{LV} - \epsilon_{HO}$, $\Delta_2 \equiv \epsilon_{HO} - \epsilon_{NHO}$. In the case of Δ_1, N_2 is unusually large, whereas C_2H_2 is much smaller, implying that N_2 is exceptionally inert to excitation. In the case of Δ_2, on the other hand, N_2 is unusually small, with an additional notable feature that the $1\pi_u$ and the $3\sigma_g$ levels are reversed, whereas C_2H_2 is

Table 1.   Important molecular orbitals with their designations and energies: the molecular orbital energy levels of N_2 and its isoelectronic molecules C_2H_2, NO^+, CO, and CN^-

Orbital designation for centrosymmetric molecules	Molecular orbital energy level (eV)[a]					Orbital designation for noncentrosymmetric molecules
	C_2H_2[b]	N_2[c]	NO^{-d}	CO^e	CN^{-f}	
$2\sigma_g$	(−28.34)	(−39.52)	(−40.34)	(−41.70)	(−24.50)	3σ
$2\sigma_u$	−18.42 (−21.11)	−18.78 (−19.88)	−18.24 (−23.28)	−41.65 (−20.68)	(−7.82)	4σ
$1\pi_u$	−11.40 (−12.01)	−16.73 (−15.77)	−15.65 (−15.22)	−16.58 (−16.66)	(−4.62)	1π
$3\sigma_g$	−16.44 (−18.57)	−15.59 (−14.82)	−16.52 (−14.62)	−14.00 (−13.80)	−3.82 (−3.31)	5σ
$1\pi_g$	(6.83)	(7.425)	−9.23 (−9.14)	(6.03)	(16.29)	2π
$3\sigma_u$	(9.60)	(30.00)	(29.55)	(23.40)	(36.91)	6σ

[a] Numerical data not in parenthese are experimental values taken from the literature as noted; those in parentheses are calculated values taken from literature as noted.

[b] Experimental values from Berry (1969); calculated values from McLean et al. (1960).

[c] Experimental values from Berry (1969); calculated values from Scherr (1955).

[d] Experimental values from Berry (1969); calculated values from Brion et al. (1959).

[e] Experimental values from Berry (1969); calculated values from Ransil (1960).

[f] Experimental values from Berkowitz et al. (1969); calculated values from Doggett and McKendrick (1970).

Table 2.   "Frontier" molecular orbitals with orbital energies and their differences[a]

Molecule		Molecular orbitals			Orbital energies and their differences				
		LV	HO	NHO	ϵ_{LV}	Δ_1	ϵ_{HO}	Δ_2	ϵ_{NHO}
Centrosymmetrical	C_2H_2	$1\pi_g$	$1\pi_u$	$3\sigma_g$	(6.83)	18.23	−11.40	5.04	−16.44
	N_2	$1\pi_g$	$3\sigma_g$	$1\pi_u$	(7.425)	23.02	−15.59	1.14	−16.73
Noncentrosymmetrical	NO^+	2π	1π	5σ	−9.23	6.42	−15.65	0.87	−16.52
	CO	2π	5σ	1π	(6.03)	20.03	−14.00	2.58	−16.58
	CN^-	2π	5σ	1π	(16.29)	20.11	−3.82	0.80	(−4.62)
Regularity noted	In general in the following order: $1\pi_g$ or 2π; $3\sigma_g$ or 5σ; $1\pi_u$ or 1π; with the following exceptions: C_2H_2—, $1\pi_g$, $1\pi_u$, $3\sigma_g$; NO^+—, 2π, 1π (and then 5σ).				$\epsilon_{LV}^\pi : NO^+ \ll CO < C_2H_2 < N_2 \ll CN^-$; $\epsilon_{LV}^\sigma : C_2H_2 \ll CO < NO^+ \sim N_2 < CN^-$; $\epsilon_{HO}^\sigma : CN^- \gg CO > N_2 > C_2H_2 \sim NO^+$ $\epsilon_{HO}^\pi : CN^- \gg C_2H_2 \gg NO^+ > CO \sim N_2$ $\Delta_1 : N_2 > CN^- \sim CO > C_2H_2 \gg NO^+$; $\Delta_2 : C_2H_2 \gg CO > N_2 \sim NO^+ \sim CN^-$				

[a] The numerical data not in parentheses are experimental values; those in parentheses are calculated values. It must be noted that, in quantum-chemical calculations, the molecule under consideration is assumed to be in the normal state configuration. Thus, for the HO molecular orbitals as well as those of lower energy, the computed values can be compared at once with the experimental data. However, for the LV molecular orbitals as well as those of higher energy, the computed values, too high in general, should not be compared with the experimental data from observed excited levels.

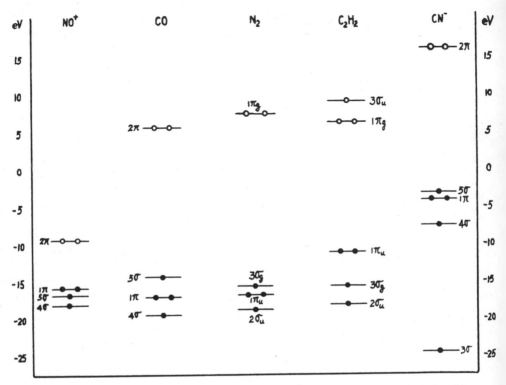

Figure 1.   The molecular orbital energy levels of N_2, and its isoelectronic molecules NO^+, CO, C_2H_2, and CN^-.

extremely large. This implies that the much closer internuclear approach in the case of dinitrogen molecule gives rise to a pair of much bigger "bananas" in the $1\pi_u$ positive overlap of the $2p_x$ (or $2p_y$) atomic orbitals but a much larger void in the $3\sigma_u$ negative overlap of the $2p_z$ atomic orbitals, whereas the $3\sigma_g$ positive overlap and the $1\pi_g$ negative overlap of the relevant atomic orbitals are not much affected. Thus, much more energy must be expended in the rupture of the first bond in N_2. A reasonable measure of the energy expenditure required is the difference between the triple-bond energy and the corresponding double-bond energy, which is 5.4 eV for N_2 as compared with 2.0 eV for C_2H_2. This difference indicates that the first "line of defense" cannot be the pair of $1\pi_u$ molecular orbitals alone, as in the case of C_2H_2; indeed, it is worth noting that this pair of $1\pi_u$ molecular orbitals comes somewhat lower in energy than the $3\sigma_g$ level in the case of N_2, whereas in C_2H_2 they are quite a way up.

    A third experimental fact is that, in the photoelectron spectra of N_2 (Figure 2), the $2\sigma_u$ and $3\sigma_g$ bands are very sharp, whereas the $2\sigma_g$ band is practically as diffuse as the $1\pi_u$ band. Brundle (1971) is correct, but not

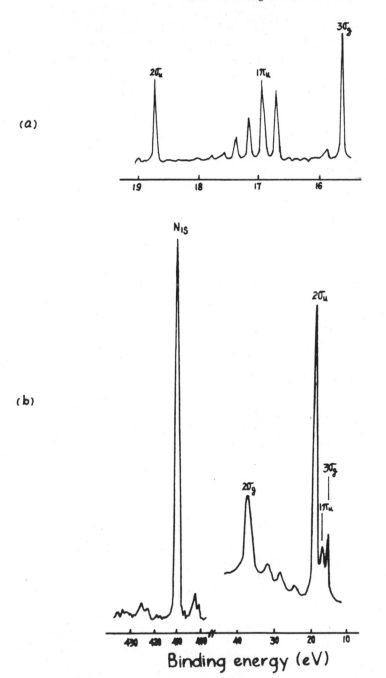

Figure 2.   The photoelectron spectra of N_2: (a) excited by HeI vacuum ultraviolet radiation; (b) excited by MgK$\alpha_{1,2}$ x-radiation.

entirely so, in his explanation of this anomalous behavior. It is certainly true that, because of the extremely short N—N internuclear distance, the $(2p_z\text{-}2p_z)_\alpha$ positive overlap is practically unaffected, whereas the $(2s\text{-}2s)_\alpha$ positive overlap becomes more prominent in the $3\sigma_g$ orbital. In other words, it is the $[(2s + 2p_z) - (2s + 2p_z)]_\alpha$ positive interaction that counts in this case. Hence this molecular orbital not only describes the σ bond within the triple bond, but takes account of the lone electron pair as well to a certain extent. Similar analysis applies also to the $2\sigma_u$ orbital. The extraordinary inertness of the N_2 molecule can therefore be satisfactorily accounted for. It is noteworthy that a pair of "normal" (mutually orthogonal) π orbitals, such as those found in the acetylenes, can be sufficiently activated by withdrawing π electrons from the completely occupied π orbitals and back-donating them into the completely vacant π orbitals, via two mutually perpendicular side-on coordinations to a pair of transition metal atoms, together with the synergic action of a pseudo (bent) end-on coordination to a third metal atom, which stabilizes these side-on modes of coordination.

We can thus presume that the "abnormal" triple bond in dinitrogen can be sufficiently activated by first pulling the N—N bond to a "normal" triple-bond length (about 1.20 Å) via a sufficiently strong end-on coordination to a transition metal atom (which suppresses internuclear repulsion and restores the $1\pi_u$ orbital level to its "normal" position) and then activating this "normalized" triple bond in the aforementioned manner (see also Newton, Corbin, and McDonald, 1976). This additional "normalizing" activation is a new, challenging problem to tackle in the formulation of a chemical model for catalytic nitrogen fixation. However, the absence of terminal atoms or groups on the dinitrogen molecule provides a possible straightforward (although by no means simple) solution to this problem by allowing true end-on coordination.

STRUCTURAL CRITERIA FOR THE ACTIVATION OF DINITROGEN MOLECULE

In a qualitative quantum-chemical discussion of mononuclear dinitrogen complexes involving d electrons under typical octahedral coordination conditions, we have been led to the following conclusions:

1. The mononuclear complexes with the N_2 ligand coordinated (either true or pseudo) end-on are in general much more stable than the corresponding complexes coordinated in the side-on fashion, particularly complexes with transition metal atoms (or ions) containing two to eight electrons.

2. The end-on mode of coordination leads to a lowering of the N—N bond stretching frequency and a lengthening of this triple bond, and thus sup-

presses the extraordinary inertness of this bond to a certain extent. Provided that this interaction stretches the bond to about 1.20 Å, sufficient activation can then be achieved by introducing additional coordination in the side-on fashion.

3. It is occasionally possible to stabilize the side-on mode of coordination in transition metal atoms or ions containing either about 0–2 or 8–10 d electrons. However, these mononuclear N_2 complexes are usually only of limited stability or even metastable. These qualitative conclusions are in satisfactory agreement with the semiquantitative quantum-chemical results obtained by Yatsimirskii (1971).

A similar qualitative quantum-chemical discussion has also been made for binuclear dinitrogen complexes, with the following conclusions:

1. The binuclear complexes with the N_2 ligand coordinated in the double-(either true or pseudo) end-on fashion (i.e., M—N—N—M' or

$$\overset{N-N}{\underset{M \qquad M'}{\diagup \qquad \diagdown}}$$

) are in general much more stable than the corresponding complexes coordinated in the double–side-on or single–end-on plus

single–side-on manner (i.e., $M\overset{N}{\underset{N}{\diagup\!\!\!\diagdown\big|\diagup\!\!\!\diagdown}}M'$ or $M{-}\overset{N}{\underset{M'}{N\diagdown\diagup}}$); this is particu-

larly true for complexes with the pair of transition metal atoms or ions containing a total of about 2–12 d electrons.

2. It is perhaps possible to stabilize other (double–side-on or single–end-on plus single–side on) modes of coordination in the case of transition metal atoms or ions containing either about 0–3 or 11–20 d electrons. However, these binuclear N_2 complexes are usually only of limited stability. We therefore conclude that binuclear N_2 complexes will never lead to a satisfactory chemical model for the active center of nitrogenase.

This qualitative discussion leads immediately to the following structural criteria for dinitrogen activation via coordination. First, end-on coordination is absolutely necessary to stretch the "abnormal" triple bond to restore its "normal" chemical behavior. One such end-on coordination bond may suffice, if it is strong enough. If it proves not to be so, then enough space must be provided for a second end-on coordination to the substrate. Second, side-on coordination is also absolutely necessary. In general, one such side-on bond will not suffice for dinitrogen with its two mutually orthogonal π orbitals. It is therefore desirable to have several such bonds in synergic action; i.e., double-side-on coordination. Third, it is absolutely necessary to protect such side-on coordination against possible isomerization into coordination of the true or pseudo end-on type. the syntheses and crystal structures of the stable lithium-nickel dinitrogen complexes, $(PhLi)_6Ni_2N_2(Et_2O)_2$ and $[Ph(Na(OEt_2))_2(Ph_2Ni_2NaLi_6(OEt)_4 \cdot OEt_2]_2$, by Jonas (1973), Krüger and Tsay (1973), and Jonas et al. (1976), give unequivocal support for the three proposed structural criteria. The lithium atoms working in sychronization provide metal atoms for double-(pseudo) end-on coordination to the N_2 ligand, and the pair of nickel atoms provides two centers for side-on coordination. Taken together, these suffice to protect the side-on coordination against isomerization. The N—N bond of this complex is stretched to 1.35 Å, the stretching frequency is below 1.550 cm^{-1}, and the N_2 ligand can be quantitatively displaced by C_2H_4. Such a complex liberates ammonia readily in the presence of water (Fischler and von Gustorf, 1975). Thus, a chemical mode for the active center of nitrogenase must be a polynuclear transition metal cluster containing at least four metal atoms.

These proposed structural criteria may also help in the mechanistic elucidation of the activation of dinitrogen by the industrial Haber-Bosch iron catalysts.

A MULTINUCLEAR COMPOSITE "STRING BAG" CLUSTER MODEL FOR THE ACTIVE CENTER OF CATALYTIC NITROGEN FIXATION IN NITROGENASE

The active center of catalytic nitrogen fixation in nitrogenase is undoubtedly the main constituent of the prosthetic group in the MoFe protein, which must be a kind of iron-sulfur protein. Since 1973, we have been developing a tetranuclear "string bag" cluster structural model as our first-stage working model (Figure 3), which we now describe. The essential part is composed of one "basal" iron atom, three "neck" M (transition metal) atoms (one molybdenum and two iron), and three inorganic labile sulfur atoms, seven atoms altogether, forming an incomplete cubane-like cluster. The three neck M atoms and the three sulfur atoms are situated on the vertices of two parallel (essentially) equilateral triangles arranged in a staggered configuration, with the basal iron atom serving as an interface with a 4Fe-4S* cubane-like iron-sulfur cluster in the MoFe protein for electron transport. Reliable estimates of the structural parameters for this "string bag" model are based on the structural data (Herskovitz et al., 1972; Averill et al., 1973; Mayerle et al., 1973; Carter et al., 1972) for the iron-sulfur proteins and their analogs, $[(PhCH_2S)\tfrac{1}{3}Fe_4S_4]^{2-}$ and $\{[(C_6H_4)(CH_2S)_2]_2Fe_2S_2\}^{2-}$. Our structural model has an Fe—S bond length of about 2.3 Å, whereas the M—S bond has a length of about 2.4 Å, but the bond angles S—M—S and M—S—M differ so much that the internuclear distances S···S and M···M become 3.6–3.7 Å and 3.2–3.3 Å, respectively, thus making the

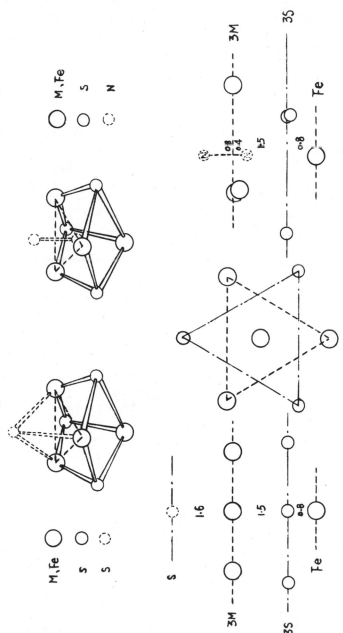

Figure 3.   A tetranuclear h-type string bag cluster structural model for the catalytic active center of nitrogenase (in Å).

string bag structure somewhat narrow necked. Each of the three sulfur atoms is bonded to the two adjacent neck M atoms (as well as to the basal iron atom) but not to the other sulfur atoms. Each of the three neck M atoms is not only bonded to the two adjacent sulfur atoms but also to the other neck M atoms. Thus, these three neck M atoms, together with the basal iron atom, form a tetranuclear cluster with sufficient chemical reactivity for the chemisorption of substrate molecules. With S—M—S about 100° and M—S—M only about 75°, the interlayer distances are about 0.8 Å and 1.5 Å, respectively, from the basal iron atom up, with a separation of about 1.6 Å between the "hole" (left by the missing sulfur atom) and the plane of neck M atoms. When N_2 plunges into this string bag, it is stopped by the retarding action of the basal iron atom. The resulting multiple side-on coordination prevents isomerization to other unfavorable configurations. If this h-type (h for hepta) string bag is not large enough to accommodate all substrates, the alternative n-type (n for nona) string bag (Figure 4), with a basal metal atom, a square of four sulfur atoms, and a similar square layer of four neck M atoms staggered with respect to the sulfur layer (with the same structural parameters as the h-type string bag), will certainly do so.

As a reasonable second-stage working model, more compatible with the Mo:Fe:S* atom ratios of 1:8:6 of MoFe cofactors and the MoKα edge and EXAFS data, it is proposed that two identical h-type $Fe(S_3)(MoFe_2)$ string bags, h1 and h2, are oriented essentially parallel to each other and form a "Siamese twin." They share a common neck molybdenum atom with two additional "purely basal" iron atoms joining the two pairs of non-connected adjacent sulfur atoms of the MoS_4 rectangular pyramid. Thus, a very much distorted n-type $Mo(S_4)Fe_4$ string bag in an inverted orientation is formed with molybdenum as its basal atom and two pairs of neck iron atoms (i.e., one pair of "purely basal" iron atoms plus one pair of "dual-role" iron atoms) forming an almost planar rectangle (Figure 5). Such a composite "three string bags–in–one" multinuclear cluster structure therefore consists of two coupled (molybdenum atom-shared) h-type string bags joined onto an n-type string bag with a pair of shared faces.

The chemisorption of substrate molecules on this composite string bag structure and their subsequent reduction can be of four different types. First, there is the net plunging mode of coordination (Figure 6a) in which the "abnormal" triple bond of N_2 is plunged into an h-type string bag (say, h1) and is reduced and ruptured there. As this substrate dives into the string bag, the neck narrows to tighten its grip on its "prey," which also feels the retarding action of the basal iron atom. The triple bond will be stretched to instill "normal" character resulting in side-on coordination by the neck M atoms. Hence, with electrons supplied via the dual-role iron atoms of the n-type string bag and hydronium ions supplied (preferentially to the exo-N

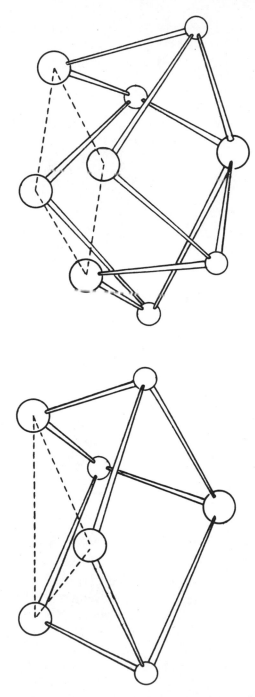

Figure 4.   An n-type string bag cluster structure (right) versus an h-type string bag cluster structure (left).

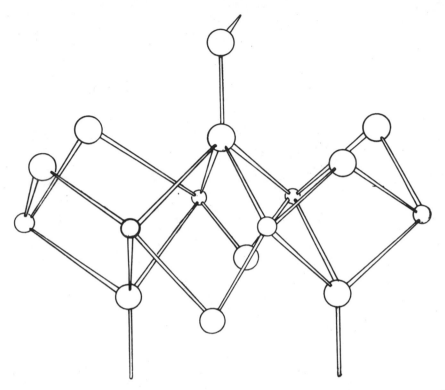

Figure 5. A composite string bag cluster structural model for the catalytic active center of nitrogenase.

atom) from the environment, the nitrogen atoms will be hydrogenated to give, successively, Fe—N—N—H, Fe—N—N$\diagdown$_{H}^{H}, Fe—N—H(+NH$_3$↑), Fe—N$\diagdown$_{H}^{H}, and Fe$\cdots$(+NH$_3$↑), and to restore h1 to its initial straight-necked configuration ready for renewal of the cycle. The neck narrowing of h1 will induce a deformation (and even deactivation) of h2, which will not be restored before the completion of the catalytic cycle in h1.

The net plunging mode of coordination will also work with small, nonpolar, neutral, e.g., N$_2$O (Figure 6b), or presumably cationic, e.g., X—Ph—N$_2^+$ (X≡H or p—NO$_2$), substrate molecules that contain a high-ordered bond and carry a small terminal group at one end only. Even H$_2$ can be adsorbed onto this composite structure with its simple σ bond

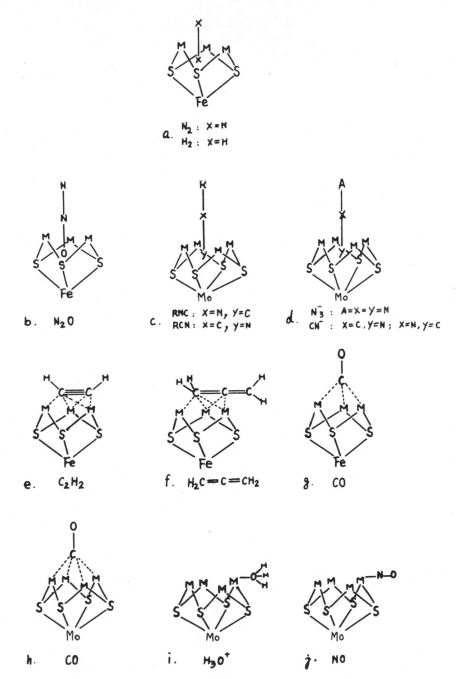

Figure 6.   A schematic diagram of the different modes of coordination of substrate molecules of various types in an h-type or n-type transition metal cluster.

weakened under the action of the apical iron atom and the basal M atoms (Figure 6a). Thus, the observed competitive inhibition of N_2 reduction by N_2O and H_2 and of N_2O reduction by N_2 as well as the observed noncompetitive inhibition of N_2O reduction by H_2 reduction by H_2 can be satisfactorily explained by our working model. However, the highly polar neutral, e.g., RNC, RCN (Figure 6c), or anionic, e.g., N_3^-, CN_3^- (Figure 6d) substrates are bulkier and incapable of plunging into an h-type string bag (since the neck is too small); but an n-type string bag may serve as an ideal plunging site for these molecules (Figure 6d) and also explains the observed mutual competitive inhibition.

This net plunging mode of coordination has been subjected to partial experimental justification. Miwa and Iwasawa (1976) have recently studied the reaction between the black Roussinate anion $[Fe_4S_3(NO)_7]^-$ and the diazonium cation $(p—NO_2—Ph—N_2)^+$ and obtained the paramagnetic product $(p—NO_2—Ph—N_2)[Fe_4S_3(NO)_4]$. In our laboratory, we have reproduced this reaction using an unsubstituted diazonium salt. The black Roussinate anion $\{[(ON)Fe](S_3)[Fe(NO)_2]_3\}^-$ is known to possess a straight-necked string bag structure (Johansson and Lipscomb, 1958; Chu and Dahl, 1977). We have found that the ESR signal of the product exhibits axial symmetry and a quintuplet hyperfine structure. Thus, the unpaired electron is close to the N—N bond in the diazonium radical in a local environment of no less than threefold symmetry, implying that the diazonium cation has plunged into the string bag following the expulsion of three NO ligands and possibly accompanied by a narrowing of the neck. Further experimental support comes from the appearance of new spectral bands and their systematic variation in a preliminary infrared spectroscopy investigation of a black Roussinate with N_2, acrylonitrile, and $(X—Ph—N_2)^+$ $(X \equiv p—NO_2,$ H) separately adsorbed onto it at very low pressure.

With substrate molecules such as C_2H_2 and its monosubstituted derivatives, this net plunging mode of coordination is impossible. Hence, the high-ordered bond may take the alternative, essentially lateral, side-on coordination across one of the M—S—M gaps in an h-type string bag plus a

pseudo end-on coordination ($\begin{smallmatrix} H \\ \diagdown \\ \diagup \\ M \end{smallmatrix} C{\equiv}C—H$) to the third neck M atom at

an angle of about 30°. The midpoint of the high-ordered bond would be above the neck plane, and the terminal groups both bent outward to practically the same extent, with the terminal hydrogen atom lying closer too the third M atom (Figure 6e), as observed with the alkyne complexes $[(OC)_3Fe]_3(C_2Ph_2)$ and $[(CH_3)_3CNC]_4Ni_4(C_2Ph_2)$ (Blount et al., 1966; Thomas et al., 1976). The hydrogenation of such a substrate molecule, therefore, takes place only in the *cis* manner. Moreover, the reduced substrate molecule is desorbed immediately on hydrogenation of the first π bond; thus, alkenes are formed and the high-ordered bond does not rupture. The other h-type string bag will undergo sympathetic deformation and even deactivation. With three available mesh gaps in an h-type string bag, there will be three different available coordination sites, which is why our working model is compatible with the much higher reduction rate of C_2H_2. The chemisorption of allene (Figure 6f) and cyclopropene must belong to this category, although the reduction rate is definitely much lower. Their reduction products can only be propylene for allene and cyclopropane plus propylene (possibly via isomerization to complexed allene) for cyclopropene.

Carbon monoxide is known to be an essentially noncompetitive inhibitor for all reducible substrates, so its chemisorption must involve a new mode of coordination (Figure 6g). When a CO molecule comes within 2 Å of a string bag neck, it is symmetrically coordinated pseudo end-on to each of the neck M atoms (three for h1 or h2; four for n). A "cap" is so formed that shuts off this string bag from both net plunging and side-on coordination. Each string bag so capped is poisoned. However, it is very unlikely that both h-type string bags could be simultaneously capped because of dipole-dipole repulsions, and the inductive effect that one $\sigma\pi$-coordinated CO ligand would exert on the other because of competition for the back-donation of electrons from the same neck molybdenum atom (K. R. Tsai, private communication). On the other hand, the much larger nitrogen-type string bag with a nonplanar rectangle of four iron atoms cannot be efficiently capped by the $\sigma\pi$-coordinated CO ligand alone. Moreover, the pair of dual-role iron atoms will also give rise to an inductive effect, resulting in noncompetitive inhibition (by this CO ligand) of substrates adsorbed thereon in the net plunging mode.

The H_2 evolution process is in fact the coordination (onto the transition metal atoms) and subsequent reduction (by the activated electrons) of hydronium ions. Experimentally, it has been shown that H_2 evolution is noncompetitively inhibited by all other reducible substrates, including NO, which is not reduced by nitrogenase, although it is not at all poisoned by CO. NO reacts readily with Fe(II), but nitrosyl complexes of molybdenum are not so common. It is believed that both the H_2 evolution from H_3O^+ and the inhibition reaction of NO are similar and can only take place from the extra-bag side on the purely basal Fe(II) atom of the n-type string bag (Figure 6i, j). This fourth mode of coordination satisfactorily accounts for the inhibition properties of H_2 evolution in our working model. The four different modes of coordination taken together seem to conform rather well to the substrate-inhibitor interactions of nitrogenase (Rivera-Ortiz and Burris, 1975; Burris and Orme-Johnson, 1976).

RNC is rather peculiar in its reduction process. This substrate plunges into an n-type string bag (with the negative end of the N—C dipole heading toward the basal molybdenum atom). On completion of the bond fission process, the products RNH_2 and CH_4 are released, and the possible fission fragments, the methyl radical ($\cdot CH_3$) and the methylene radical ($:CH_2$) will migrate toward one of the neck iron atoms or one of the $\overset{M \qquad M}{\underset{S}{\diagdown \diagup}}$ mesh gaps and anchor themselves there. When a second RNC molecule dives into this string bag, these free radicals will move over and add to RNC to form $RN{=}\overset{CH_3}{\underset{\vert}{C}}\cdot$ and $RN{=}C{=}CH_2$. The former is then reduced to yield $R\overset{\cdot}{N}H_2$ and $H_3C{-}CH_3$ (or the fission fragments $\cdots CH_2{-}CH_3$ and $\overset{\cdot}{C}H{-}CH_3$, which anchor themselves as above). The latter is reduced to yield RNH_2 and also $H_2C{=}CH_2$, or similarly the fragment $\cdots HC{=}CH_2$. These unstable intermediate methyl radicals may unite with one another to form $H_3C{-}CH_3$, or with $\cdot CH{=}CH_2$ to form $H_3C{-}CH{=}CH_2$, or with $\cdot CH_2CH_3$ to form $CH_3{-}CH_2CH_3$. $H_2C{:}$ either unites with another $:CH_2$ to form $H_2C{=}CH_2$, or with $:CHCH_3$ to form $H_2C{=}CH{-}CH_3$. Space limitation within the n-type string bag sets the upper limit of trimerization for the "polymerized" hydrogenation product. Of course, whether the reduction will go beyond the simple CH_4 stage or even the intermediate C_2H_4 and C_2H_6 stages depends mainly on the relative rates of RNH_2 release versus union of the free radicals with an entering RNC molecule, and this is in turn determined to a certain extent by the size of R. Obviously, this phenomenon of polymerized hydrogenation products will never occur for RCN.

The role of the interdependent ATP hydrolysis reaction and the structural-chemical implications of the ATP:$2e$ ratio are important parameters in understanding biological nitrogen fixation. Tsai has proposed a novel two-step, ATP-driven mechanism for electron activation and transport with some inevitable electron back-flow to account for the part of the ATP:$2e$ ratio that exceeds 4, which has not been satisfactorily explained so far (Nitrogen Fixation Research Group, Amoy University, 1976). Jiang and Tang (unpublished data from Kirin University) have suggested that: 1) ATP plays the dual role of providing a "lead wire" for electrons flowing from a molecular system of lower reduction potential to one of higher reduction potential by inducing conformational changes in both the Fe protein (A) and the MoFe protein (M); and 2) MgADP is an inhibitor of nitrogenase catalysis because its presence will inhibit the reduction of A^{ox} back to A as well as the electron transport from A to M. Figure 7 is a scheme of the changes in reduction potential of the nitrogenase constituents in the overall catalytic process and Figure 8 shows the two-step electron activation

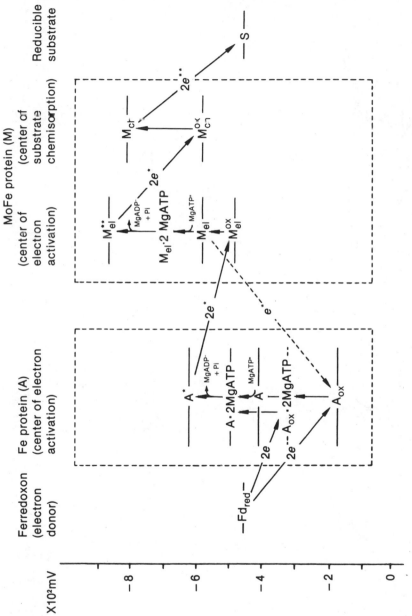

Figure 7.   Changes in reduction potential of various main constituents of nitrogenase.

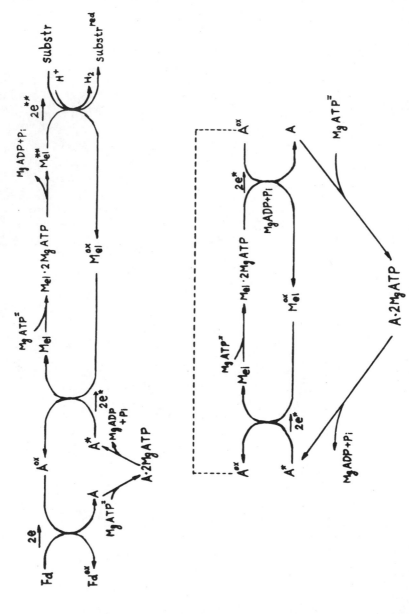

Figure 8.   The two-step electron activation and transport process and the electron back-flow cycle in relation to the ATP hydrolysis reaction. (a) The normal electron activation and transport process (ATP:$2e$ = 4). (b) The electron backflow cycle (ATP:$2e$ > 4).

and electron transport process and the electron backflow cycle in relation to ATP hydrolysis. We suggest that the Fe protein consists of 4Fe-4S* cubane-like clusters, which serve as centers for electron activation and transport to the MoFe protein. This dimeric protein has two monomeric moieties consisting of a pair of 4Fe-4S* cubane-like clusters (M_{el}), which act as centers of electron activation and transport, in association with a MoFe cofactor (M_{ch}) of the composite string bag cluster structure, which acts as a center of substrate chemisorption and coordination (Figure 9).

Using dinitrogen as substrate, we have carried out a qualitative MO analysis of the h-type string bag cluster by Dahl's treatment (Foust and Dahl, 1970; Trinh-Toan, Fehlhammer, and Dahl, 1972; Simon and Dahl, 1973a, 1973b; Trinh-Toan et al., 1977). Table 3 shows the qualitative molecular orbital (MO) energy level scheme (based on idealized C_s local symmetry) for dinitrogen in the net plunging mode of coordination. The four lowest MOs—(d_{z^2}) $(d_{xz}(+++))$ $(2\sigma_u)$:a', (d_{xz}) $(d_{x^2}(+-+))$ $(1\pi_x)$:a', (d_{yz}) $(d_{y^2}(+0-))$ $(1\pi_y)$:a', and $(d_{x^2}(+++))$ $(3\sigma_g)$:a'—are essentially the N_2 MOs, and are thus completely occupied. Next come the eleven bonding MOs up to (d_{xy}) $(d_{yz}(+-+))$ $\vdots(1\pi_y^*)$:a'' and $(d_{x^2-y^2})$ $(d_{yz}(+0-))$ $\vdots(1\pi_x^*)$:a'. In order to attain the maximum weakening of the pair of N_2 1π MOs, each is filled with two d electrons up to (d_{z^2}) $(d_{xz}(+++))\vdots(2\sigma_u)$:$a'$, with only one d electron for the aforementioned pair of highest-energy MOs, making a total of twenty d electrons altogether. Thus, this tetranuclear transition metal cluster must consist of one $Mo(IV)(d^2)$ plus three $Fe(II)(d^6)$ atoms,

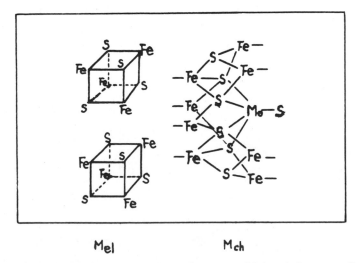

M_{el} M_{ch}

Figure 9.   The center of electron activation and transport (M_{el}) and the center of substrate chemisorption and coordination (M_{ch}) in MoFe protein.

Table 3. Qualitative MO energy level scheme (based on C_8 idealized local symmetry) for the N_2 substrate molecule h-type Fe(S$_3$)(MoFe$_2$) string bag coordination complex in the net plunging mode of coordination

Symmetry behavior symbol	Fe			(M′−M−M′)		N_2	
	Orbital symbol	Orbital diagram		Orbital symbol	Orbital diagram (end view)	Orbital symbol	Orbital diagram
		Side view	End view				End view Side view
a′	$d_{x^2-y^2}$			d_{yz} $(+\,0\,-)$			
a″	d_{xy}			d_{yz} $(+\,-\,+)$			
a′	d_{xz}			d_{y^2} $(+\,-\,+)$			

(continued)

Table 3. *(Continued)*

Symmetry behavior symbol	Fe Orbital symbol	Fe Orbital diagram Side view	Fe Orbital diagram End view	(M'−M−M') Orbital symbol	(M'−M−M') Orbital diagram (end view)	N$_2$ Orbital symbol	N$_2$ Orbital diagram End view	N$_2$ Orbital diagram Side view
a"				d_{xz} $(+-+)$				
a'				d_{xz} $(+0-)$				
a'	d_{z^2}			d_{xz} $(+++)$				
a'	$d_{x^2-y^2}$			d_{yz} $(+0-)$		$1\pi_x^*$		

$1\pi^*_y$

$2\sigma_u$

$1\pi_x$

$1\pi_y$

$3\sigma_g$

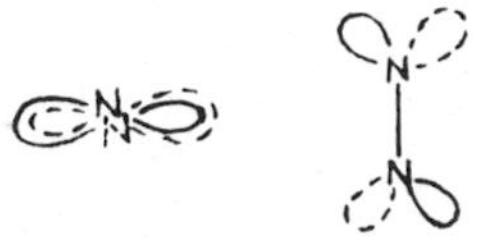

d_{yz}
$(+ - +)$

d_{xz}
$(+ + +)$

d_{x^2}
$(+ - +)$

d_{x^2}
$(+ 0 -)$

d_{x^2}
$(+ + +)$

d_{xy}

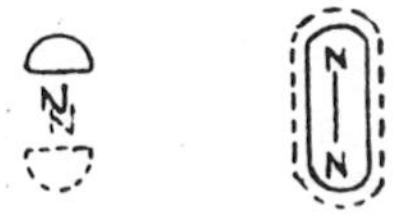

d_{z^2}

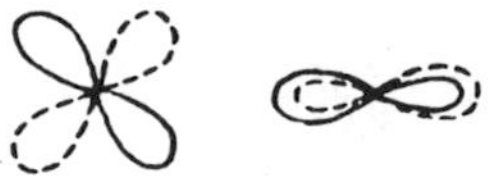

d_{xz}

d_{yz}

a''

a'

a'

a''

a'

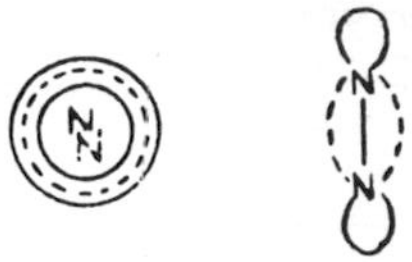

(continued)

Table 3. (Continued)

Symmetry behavior symbol	Fe Orbital diagram Orbital symbol	Fe Side view	Fe End view	(M′−M−M′) Orbital symbol	(M′−M−M′) Orbital diagram (end view)	N₂ Orbital symbol	N₂ End view	N₂ Side view
a″	d_{xy}			d_{yz} $(+\,-\,+)$		$1\pi^*_y$		
a′	$d_{x^2-y^2}$			d_{yz} $(+\,0\,-)$		$1\pi^*_x$		
a′				$d_{y^2-z^2}$ $(+\,+\,+)$				
a″				d_{xy} $(+\,-\,+)$				

a′			d_{xy} (+0−)				
a′			d_{x^2} (+ + +)		$3\sigma_g$		
a″	d_{yz}		d_{x^2} (+0−)	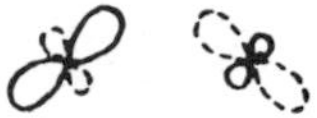	$1\pi_y$	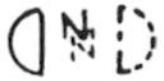	
a′	d_{xz}		d_{x^2} (+ − +)		$1\pi_x$		
a′	d_{z^2}		d_{xz} (+ + +)		$2\sigma_u$		

with one of these iron atoms as the basal atom to provide the best interface for electron transport. Hence, the sharing of a molybdenum atom in the formation of the "Siamese twin" h-type string bag structural model is not accidental. Interestingly, the black Roussinate anion, $[Fe_4S_3(NO)_7]^-$, has 34 d electrons, 14 of which form seven π bonds between the seven NO ligands and the four M atoms (one apical plus three basal), with the remaining 20 d electrons being responsible for the stabilization of this straight-necked h-type string bag structure. In the case of $(X—Ph—N_2)[Fe_4S_3(NO)_4]$, the expulsion of three NO ligands from the anion reduces the number of d electrons to only 31. With only four π bonds to NO, there remain three d electrons in excess of the number required for the stability of an h-type string bag and some nonbonding MOs must be occupied, which may explain the lower stability of this product. When the six electrons are fed into the N_2 "trapped" in an h-type string bag via the basal iron atom, the weakly bonding a', (d_{xz}) $(d_{x^2} (+ - +))$ $(1\pi_x)$, a'', (d_{yz}) $(d_{x^2}(+0-))$ $(1\pi_y)$, and the antibonding MOs $(d_{y_{-x}} (+0-))$, $(d_{y_{-z}} (+ - +))$ will be completely filled. The resulting instability will prove fatal to the integrity of the N_2 ligand unless it readjusts by undergoing successive bond fission and hydrogenation.

Our composite string bag cluster model has matured sufficiently so that we are considering synthesizing model compounds with an h-type string bag structure. For some years, we have taken the black Roussinate anion, $\{[(ON)Fe](S_3^*)[Fe(NO)_2]_3\}^-$, together with the red Roussinate anion $\{[(ON)_2Fe]_2S_2^*\}^{2-}$ (Thomas, Robertson, and Cox, 1958) and the "Dahl-Roussinate" anion $\{[(ON)Fe]_4S_4^*\}$ (Gall, Chu, and Dahl, 1974), as the first-step models for a mixed h-type string bag $Fe(S_3)(MoFe_2)$ structure and for 2Fe-2S* and 4Fe-4S* ferredoxins, respectively. As early as 1858, Roussin synthesized the black Roussinates (Mellor, 1958). Black Roussinate decomposes irreversibly in alkaline solutions to yield the corresponding red Roussinate, which resembles the dissociation of a 4Fe-4S* ferredoxin into the corresponding 2Fe-2S* ferredoxin. Conversely, a red Roussinate could also be converted into the corresponding black Roussinate either on weak acidification or on addition of a ferrous nitrosyl complex, $[Fe(NO)]^{2+}SO_4^{2-}$, or even a ferrous salt. This reconstitution reaction may give an important clue in the attempted synthesis of the "pure" or even "mixed" h-type string bag clusters. Gall, Chu, and Dahl (1974) synthesized $Fe_4S_4(NO)_4$ with a narrow-necked cubane-like cluster structure, and even reduced it with sodium amalgam to obtain "unexpectedly" a black Roussinate (Chu and Dahl, 1977). This interesting discovery might reveal a certain underlying principle in the chemistry of ferredoxins and related Fe-S proteins. It is therefore not surprising that a red-to-black conversion can be accomplished by a mild sulfide precipitant. Perhaps the black Roussinate should be looked on as a "condensation dimer" of the red Roussinate produced by a simultaneous expulsion of one NO from one red Roussinate anion and one S^{2-} from the other. Mild oxidation of a black Roussinate in the presence of S^{2-} might be a simple synthetic route to the "Dahl-Roussinate." Attempted synthesis of "mixed" black Roussinate has been under contemplation. If we continue to "sum up experience and go on discovering, inventing, creating and advancing," we believe that before too long we could succeed in synthesizing a composite string bag cluster and effecting a real breakthrough in the field of chemical nitrogen fixation.

ACKNOWLEDGMENTS

The author is indebted to all the past and the present members of the Nitrogen Fixation Research Group of this Institute for their contributions in the formulating of this composite string bag model. Helpful discussions with Professors A. C. Tang of Kirin University and K. R. Tsai of Amoy University are also gratefully acknowledged.

REFERENCES

Averill, B. A., T. Herskovitz, R. H. Holm, and J. A. Ibers. 1973. Synthetic analogs of the active sites of iron-sulfur proteins. II. Synthesis and structure of the tetra[mercapto-μ_3-sulfido-iron] clusters, $[Fe_4S_4(SR)_4]^{2-}$. J. Am. Chem. Soc. 95:3523–3534.

Berkowitz, J., 1969. Photoionization of hydrogen cyanide: electron affinity and heat of formation of CN. J. Chem. Phys. 50:1497–1500.

Berry, R. S. 1969. Electronic spectroscopy by electron spectroscopy. Annu. Rep. Phys. Chem. 20:357–406.

Blount, J. F., L. F. Dahl, C. Hoogzand, and W. Hübel. 1966. Structure of and bonding in an alkyne-nonacarbonyltriiron complex. A new type of iron-acetylene interaction. J. Am. Chem. Soc. 88:292–301.

Brion, H., 1959. Electron structure of nitric oxide. J. Chem. Phys. 30:673–681.

Brundle, C. R. 1971. Some recent advances in photoelectron spectroscopy. Appl. Spectroscopy 25:8–23.

Burris, R. H., and W. H. Orme-Johnson. 1976. Mechanism of biological nitrogen fixation. In: W. E. Newton and C. J. Nyman (eds.), Proceedings of the First International Symposium on Nitrogen Fixation, Vol. 1, pp. 208–233. Washington State University Press, Pullman.

Carter, Jr., C. W., J. Kraut, S. T. Freer, R. A. Alden, L. C. Sieker, E. Adman, and L. H. Jensen. 1972. A comparison of $Fe_4S_4^*$ clusters in high-potential iron protein and in ferredoxin. Proc. Natl. Acad. Sci. U.S.A. 69:3526–3529.

Chu, C. T.-W., and L. F. Dahl. 1977. Structural characterization of $[AsPh_4]^+$ $[Fe_4(NO)_7(\mu_3-S)_3]^-$. Stereochemical and bonding relationship of the Roussin black monoanion with the red ethyl ester, $Fe_2(NO)_4(\mu_2-SC_2H_5)_2$ and $Fe_4(NO)_4(\mu_3-S)_4$. Inorg. Chem. 16:3245–3251.

Cramer, S. P., W. O. Gillum, K. O. Hodgson, L. E. Mortenson, E. I. Stiefel, J. R. Chisnell, W. J. Brill, and V. K. Shah. 1978. The molybdenum site in nitrogenase. 2. A comparative study of Mo-Fe proteins and the iron-molybdenum cofactor by X-ray absorption spectroscopy. J. Am. Chem. Soc. 100:3814–3819.

Doggett, G., and A. McKendrick. 1970. Electron structure of the cyanide anion. J. Chem. Soc. A:825–827.

Fischler, I., and E. K. von Gustorf. 1975. Chemische und biologische Aspekte der fixierung und reduktion molekularen Stickstoffs. Naturwiss. 62:63–70.

Foust, A. S., and L. F. Dahl. 1970. Organometallic pricogen complexes. V. Preparation, structure, and bonding of the tetrameric antimony-cobalt cluster system, $Co_4(CO)_{12}Sb_4$: the first known (main group element)-(metal carbonyl) cubane-type structure. J. Am. Chem. Soc. 92:7337–7341.

Gall, R. S., C. T.-W. Chu, and L. F. Dahl. 1974. Preparation, structure, and bonding of two cubane-like iron-nitrosyl complexes, $Fe_4(NO)_4(\mu_3—S)_4$ and Fe_4 $(NO)_4(\mu_3—S)_2(\mu_3—NC(CH_3)_2)_2$. Stereochemical consequences of bridging ligand substitution on a completely bonding tetrametal cluster unit and of different terminal ligands on the cubane-like Fe_4S_4 core. J. A. Chem. Soc. 96:4019–4023.

Herskovitz, T., B. A. Averill, R. H. Holm, J. A. Ibers, W. D. Phillips, and J. F. Weiher. 1972. Structure and properties of a synthetic analogue of bacterial iron-sulfur proteins. Proc. Natl. Acad. Sci. U.S.A. 69:2437–2441.

Johansson, G., and W. N. Lipscomb. 1958. The structures of Roussin's black salt; $Cs[Fe_4S_3(NO)_7] \cdot H_2O$. Acta Crystallog. 11:594–598.

Jonas, K. 1973. π-Bonded nitrogen in a crystalline nickel-lithium complex. Angew. Chem. Internat. Edit. Engl. 12:997–998.

Jonas, K., D. J. Brauer, C. Krüger, P. J. Roberts, and Y.-H. Tsay. 1976. "Side-on" dinitrogen-transition metal complexes. Molecular structure of $\{C_6H_5[Na \cdot O$ $(C_2H_5)_2]_2[(C_6H_5)_2Ni]_2N_2NaLi_6(OC_2H_5)_4 \cdot O(C_2H_5)_2\}_2$. J. Am. Chem. Soc. 98:74–81.

Krüger, C., and Y.-H. Tsay. 1973. Molecular structure of a π-dinitrogen-nickel-lithium complex. Angew. Chem. Internat. Edit. Engl. 12:998–999.

Mayerle, J. J., R. B. Frankel, R. H. Holm, J. A. Ibers, W. D. Phillips, and J. F. Weiher. 1973. Synthetic analogs of the active sites of iron-sulfur proteins. Structure and properties of bis[o-xylyldithiolato-μ_2-sulfidoferrate(III)], an analog of the 2Fe-2S proteins. Proc. Natl. Acad. Sci. U.S.A. 70:2429–2433.

McLean, A. D., 1960. LCAO-MO-SCF ground state calculations on C_2H_2 and CO_2. J. Chem. Phys. 32:1595–1597.

Mellor, J. W. 1958. A Comprehensive Treatise on Inorganic and Theoretical Chemistry, Vol. 8, pp. 439–440. Longmans, London.

Miwa, M., and K. Iwasawa. 1976. The reaction product of an iron-sulfur complex with an arylazo compound (in Japanese). Seikei Daigaku Kogakubu Kogaku Hokoku 22:1571–1572.

Newton, W. E., J. L. Corbin, and J. W. McDonald. 1976. Nitrogenase: Mechanism and models. In: W. E. Newton, and C. J. Nyman (eds.), Proceedings of the First International Symposium on Nitrogen Fixation, Vol. 1, pp. 53–74. Washington State University Press, Pullman.

Nitrogen Fixation Research Group, Department of Chemistry, Amoy University. 1976. A model of the nitrogenase active center and the mechanism of nitrogenase catalysis. Scientia Sinica 19:460–474.

Ransil, B. J. 1960. Rev. Mod. Phys. 32:245.

Rivera-Ortiz, J. M., and R. H. Burris. 1975. Interactions among substrates and inhibitors of nitrogenase. J. Bacteriol. 123:537–545.

Scherr, C. W. 1955. J. Chem. Phys. 23:569.

Shah, V. K., and W. J. Brill. 1977. Isolation of an iron-molybdenum cofactor from nitrogenase. Proc. Natl. Acad. Sci. U.S.A. 74:3249–3253.

Simon, G. L., and L. F. Dahl. 1973a. Organometallic chalcogen complexes. XXVI. Synthesis, structure, and bonding of the cubane-like $[Co_4(h^5—C_5H_5)_4S_4]^n$ tetramers ($n = 0, +1$). Stereochemical influence due to oxidation of a completely nonbonding tetrahedral metal system. J. Am. Chem. Soc. 95:2164–2174.

Simon, G. L., and L. F. Dahl. 1973b. Organometallic pricogen complexes. VII. Synthesis, structure, and bonding of the cubane-like metal cluster [Co$_4$ (h^5—C$_5$H$_5$)$_4$P$_4$]. The first reported organometallic complex containing a naked phosphorus atom as a ligand. J. Am. Chem. Soc. 95:2175–2183.

Thomas, J. T., J. H. Robertson, and E. G. Cox. 1958. The crystal structure of Roussin's red ethyl ester. Acta Crystallog. 11:599–604.

Thomas, M. G., E. L. Muetterties, R. O. Day, and V. W. Day. 1976. Metal clusters in catalysis. 5. Four-electron η^2-ligand bonding in clusters and catalytic intermediates. J. Am. Chem. Soc. 98:4645–4646.

Trinh-Toan, W. P. Fehlhammer, and L. F. Dahl. 1972. Structure and bonding of the tetrameric cyclopentadienyliron carbonyl monocation, [Fe$_4$(h^5—C$_5$H$_5$)$_4$(CO)$_4$]$^+$. Stereochemical effect due to oxidation of a completely bonding tetrahedral metal cluster system. J. Am. Chem. Soc. 94:3389–3396.

Trinh-Toan, B. K. Teo, J. A. Ferguson, T. J. Meyer, and L. F. Dahl. 1977. Electrochemical synthesis and structure of the tetrameric cyclopentadienyl-iron sulfide dication, [Fe$_4$(η^5—C$_5$H$_5$)$_4$(μ_3—S)$_4$]$^{2+}$: A metal cluster bonding description of the electrochemically reversible [Fe$_4$(η^5—C$_5$—H$_5$)$_4$(μ_3—S)$_4$]^n system (n = −1 to +3). J. Am. Chem. Soc. 99:408–416.

Yatsimirskii, K. B. 1971. Role of coordination in catalytic redox processes. Pure Appl. Chem. 27:251–264.

Nitrogen Fixation, Volume I
Edited by W. E. Newton and W. H. Orme-Johnson
Copyright 1980   University Park Press   Baltimore

Development of a Model of Nitrogenase Active Center and Mechanism of Nitrogenase Catalysis

K. R. Tsai

The importance of biological nitrogen fixation in the global nitrogen economy has spurred extensive international activities in research on nitrogenase (N_2ase) and its chemical modeling. Although important progress has been made since the 1960s, many unresolved problems still remain. Based on the known reactions of nitrogenase and the principles of coordination catalysis, we proposed, a few years ago, a twin-seated polynuclear cluster structural model of the nitrogenase active centers together with a discussion of nitrogenase catalysis (Nitrogen Fixation Research Group, 1976). Recent important breakthroughs in nitrogenase research (Gillum et al., 1977; Newton, Postgate, and Rodriguez-Barrueco, 1977; Shah and Brill, 1977; Cramer et al., 1978a, 1978b; Rawlings et al., 1978; Thorneley, Eady, and Lowe, 1978) include the successful isolation and characterization (Shah and Brill, 1977; Rawlings et al., 1978) of the iron-molybdenum cofactor (FeMoco), from the MoFe proteins of various nitrogen-fixing microorganisms and the important EXAFS work (Cramer et al., 1978a, 1978b) that gives insight into the valence state and microenvironments of the molybdenum ions, Mo(IV(III)), in the MoFe protein and FeMoco. Now, revision and development of our twin-seated polynuclear cluster structural model of the nitrogenase active centers can be undertaken, and the mechanisms of ATP-driven coupled electron and energy transports in the nitrogenase-catalyzed reactions discussed.

REVISION OF THE STRUCTURE OF THE DOUBLE-SEATED Fe₂S₂(L)(L')Mo₂ MODEL OF N₂ase ACTIVE CENTERS

Ten different types of N₂ase substrates are known: N≡N, H⁺, N≐N≐O, [N≐N≐N]⁻, [C≡N]⁻, RC≡CH, RC≡N, RN≡C, CH₂=C=CH₂, and
$\overline{CH=CH-CH_2}$ (Hardy, Burns, and Parshall, 1971; Zumft and Mortenson, 1975). From the known chemistry of these substrates and their interaction with N₂ase, a wealth of information about the structure of nitrogenase active center can be obtained. The nitrogenase-catalyzed reductive hydrogenation of an α-acetylene substrate molecule, e.g., CD≡CD, proceeds exclusively to the corresponding α-olefin with almost 100% *cis* selectivity (Hardy, Burns, and Parshall, 1971). Therefore, the RC≡CH substrate molecule is most probably coordinated double–side-on to two transition metal ions with an internuclear distance of 2.3–2.8 Å. The recent EXAFS work (Cramer et al., 1978a, 1978b) eliminates consideration of two Mo(IV(III)) atoms so close together. Also, the nitrogenase modeling systems of Schrauzer et al. (1974; also Schrauzer, 1977) make it unlikely that the α-acetylene substrate molecule can be sufficiently activated merely by double–side-on coordination to two Fe(II) atoms without the participation of Mo(IV(III)). Hence, the plausibility of the α-acetylene being coordinated double–side-on to one Mo(III(IV)) and one Fe(II) is considered first.

Like α-acetylenes, the alkyl cyanides and isocyanides also possess triple bonds. Moreover, these three types of nitrogenase substrates are all subject to the same chain length restriction for reduction, in that they must be straight chains containing no more than four carbons (Hardy, Burns, and Parshall, 1971). Thus alkyl cyanides and isocyanides are likewise coordinated double–side-on to one Mo(III(IV)) and one Fe(II) atom in approximately the same orientation as α-acetylene. However, the nitrogenase-catalyzed reduction of alkyl cyanides and isocyanides, like dinitrogen, proceeds to complete disruption of the triple bonds. Moreover, since alkyl cyanides and isocyanides strongly prefer end-on coordination to side-on coordination on transition metal ions and yet bind more strongly than CH₃C≡CH to nitrogenase (Hardy, Burns, and Parshall, 1971), alkyl cyanides and isocyanides, besides being coordinated double–side-on, are likely to coordinate end-on to another transition metal ion. This atom is most probably a Mo(IV(III)), as indicated by the peculiar reductive hydrogenation reaction of CH₃N≡C (minor amounts of C₂ and C₃ olefins and alkanes are obtained) (Hardy, Burns and Parshall, 1971), probably through molybdenum-carbene intermediates (Nitrogen Fixation Research Group, 1976). The internuclear distance between the two Mo(IV(III)) atoms is estimated to be about 3.3 Å, just about beyond the known range for the formation of intermetallic bonds in molybdenum (Cotton, 1977) and beyond the present detection limit of EXAFS (K. O. Hodgson, unpublished data). This type of end-on plus double–side-on coordination may be designated as $\omega_1/\mu_2(\eta^2)$ and is probably analogous to the bonding of the three isocyanide bridge ligands in $(Ni_4L_4)[(CH_3)_3C—N\equiv C]_3$ (Thomas et al., 1976, 1977), which span two basal Ni^0 atoms about 3.65 Å apart. Each of these isocyanide bridge ligands is also in a position to form a μ bond ($\sigma\pi$ bond) with the apical Ni^0 atom, so that each Ni^0 is just formally coordination saturated.

An α-acetylene substrate molecule coordinated double–side-on to one $Mo(III(IV))$ and one $Fe(II)$ could also form a weak, formal σ bond with the second $Mo(III(IV))$ analogous to the $\mu_3(\eta^2)$-type coordination of the acetylenic ligand in $(Ni_4L_4)(C_2\phi_2)(RNC)_2$ (Thomas et al., 1976, 1977). These $\mu_3(\eta^2)$-type coordinated acetylenes, as well as a $\mu_2(\eta^2)$-type double–side-on coordinated acetylene in a binuclear nickel complex, are hydrogenated almost exclusively to the *cis*-olefines, with a selectivity approaching that of nitrogenase, which is strong support for both the proposed model of a trinuclear active center and the proposed mode of activation.

As a nitrogenase substrate, CN^- appears to be coordinated like an isocyanide. Dinitrogen, being about the same size and bond type as CN^-, is also expected to be coordinated end-on to one Mo (III(IV)) and double–side-on to one $Mo(III(IV))$ and one $Fe(II)$. End-on–coordinated N_2 in certain mononuclear W^0 or Mo^0 complexes can be reduced either to coordinated $—N—NH_2$ or to NH_3 by internal electron transfer from W^0 or Mo^0 to ligated N_2 in conjunction with protonation (Chatt, 1975; Chatt, Pearman, and Richards, 1975, 1976). Thus, end-on coordination of N_2 to Mo^0 is very effective for back-donation of metal electrons into the antibonding $1\pi_g$ orbital of N_2, because of efficient overlap and similar energy levels. However, in nitrogenase, the presumed valence state of $Mo(IV(III))$ is considerably higher, and therefore mere end-on coordination of N_2 to one $Mo(IV(III))$ is not likely to produce sufficient back-donation of electrons, suggesting that multiple coordination of N_2 on polynuclear clusters of $Mo(IV(III))$ and $Fe(II)$ may be required for activation. A quantum-chemical theory of dinitrogen–transition metal complexes given by Tang and his students (Hsu, this volume) shows that end-on plus double–side-on coordination of N_2 on transition metal clusters is more effective in weakening the $N\equiv N$ bond and making the terminal nitrogen more negatively charged so as to serve as a point of attack in protonation. K. H. Huang (unpublished data) has recently proposed a similar mode of end-on (to one of the surface iron atoms) plus triple–side-on coordination for activation of N_2 on the (III) plane of α-Fe (ammonia synthesis iron catalysts). We have observed that the addition of a small proportion of $Fe(II)$ to the $Mo(V)$-cysteine/ BH_4^-/C_2H_2 system greatly increased the amounts of C_2H_4 and C_2H_6, while suppressing the formation of 1,3-butadiene and 1-butene (Chang and Xii, unpublished data). Thus, $Fe(II)$ probably participates with $Mo(IV(III))$ in the coordination activation of acetylene and suppresses the *cis*-insertion of acetylene into the vinyl or ethyl ligand on Mo(IV(III)). Both experimental observations and theoretical consideration indicate that activation of the N—N bond by multiple coordination seems essential.

A twin-seated, polynuclear cluster structural model ($Fe_2S_2 \cdot Mo_2L_2$) of the nitrogenase active centers similar to our previous proposal (Nitrogen Fixation Research Group, 1976) with some readjustment of the Mo—Mo and Mo—Fe internuclear distances, is shown in Figure 1. L and L' in $Fe_2S_2(L)(L')Mo_2$ denote two adjacent labile ligands, such as N_2 (or hydrogen) and H_2O. With this model, the mechanisms of reductive hydrogenation of all nitrogenase substrates can be explained (Figure 2). This figure indicates an alternative and, in fact, better explanation for the mechanism of carbene formation on Mo(III(IV)) by reductive hydrogenation of a $CH_3N{\equiv}C$ substrate molecule, the reductive hydrogenation to give CH_3NH_2, C_2H_6, C_2H_4, and the condensation of carbene with another $CH_3N{\equiv}C$ substrate molecule. Note that direct reductive hydrogenation of allene to propene

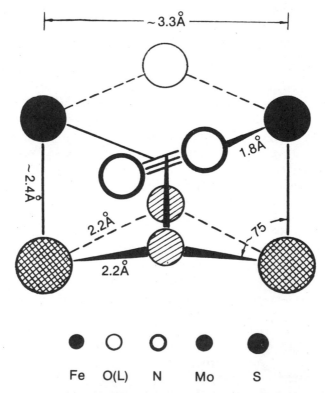

Figure 1.   A twin-seated $Fe_2S_2(L)(L')Mo_2$ cluster model with coordinated N_2 on a 2Mo—1Fe trinuclear active center.

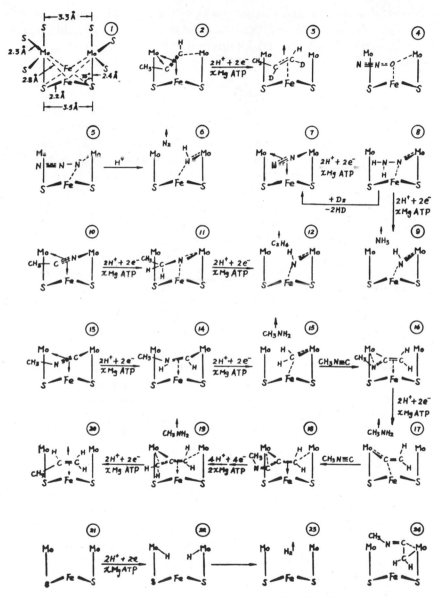

Figure 2.   Mechanisms of nitrogenase-catalyzed reactions.

without a prior isomerization to methyl acetylene has been demonstrated by Burns and Hardy (1975) and that the formation of 1-η-hydrazido ligand, $\overset{...}{N}\overset{...}{N}H_2$, rather than a bound diimide, $-NH\overset{...}{N}H-$, as a probable intermediate in reduction has recently been suggested by Thorneley, Eady, and Lowe (1978).

Since $CH\equiv CH$ is activated by double–side-on coordination, two $CH\equiv CH$ molecules, at sufficiently high pressure, might occupy both $Mo(IV(III))$—$Fe(II)$ centers in the double-seated $Fe_2S_2\cdot Mo_2L_2$ cluster, thereby inhibiting the H_2 evolution reaction. Coordination of a CO molecule to one of the twin sites on the $Mo(III(IV))$ will also inhibit the coordination of N_2, $RC\equiv CH$, $RC\equiv N$, $RN\equiv C$, CH^-, or N_3^-, or another CO molecule, to the other site, but will not inhibit the formation of a hydrido ligand there; thus, the H_2 evolution reaction is not inhibited by CO (Nitrogen Fixation Research Group, 1976). For a CO ligand to take up $2e^- + 2H^+$ at the negatively charged oxygen atom, in a mode of nitrogenase-catalyzed reductive hydrogenation like that of N_2 to $\overset{...}{N}\overset{...}{N}H_2$ and leading to complete deoxygenation of CO, would be energetically forbidden.

If reductive hydrogenation of the 1-η-hydrazido ligand successively to $=NH$ and NH_3 is thermodynamically a "downhill" process, it should require no dissipation of ATP. But both the 1-η-hydrazido and $=NH$ intermediates are stabilized by coordination, just as the 1-η-hydrazido ligand in $Mo(\overset{...}{N}\overset{...}{N}H_2)Br_2(dppe)_2$ is stabilized against further protonation by HBr (Chatt, 1975), and the reductive hydrogenation of $=NH$, produced during N_3^- reduction, is still ATP driven. These data support a stepwise reductive hydrogenation of N_2 and the other substrates by successive ATP-driven $2e^- + 2H^+$ steps.

MECHANISM OF ATP-DRIVEN COUPLED ELECTRON AND ENERGY TRANSPORT IN NITROGENASE REACTIONS

The observed EPR signal changes (Smith, Lowe, and Bray, 1973; Walker and Mortenson, 1973) in nitrogenase reactions with sufficient or deficient supply of reductant or ATP may be interpreted by a two-step ATP-driven mechanism of electron transport (Nitrogen Fixation Research Group, 1976), analogous to the two-step photon-driven reaction in photosynthesis. The essential points of this mechanism are:

1. MgATP complexes with the $Fe_4S_4^*$ cluster of the dimeric Fe protein, producing a conformational change and increasing the ligand field acting on the highly delocalized frontier orbital electron(s) of the cluster, to drive the protein-protein electron transfer with MgATP hydrolyzed immediately afterward (or simultaneously) and removed as MgADP and P_i.

2. A second ATP-driven step then lowers the redox potential of the active center on the MoFe protein to a sufficiently negative value to drive the electrons over to the coordinated substrate.
3. The complexation of MgATP with the fully reduced MoFe protein is the slow, rate-determining step common to the reduction of all substrates.
4. The huge Fe_nS_n* system of MoFe protein also serves to maintain the sufficiently negative redox potential (by a kind of condenser effect) required for substrate reduction, possibly like the $\pi^2(Pc)$ system in the ferrous phthalocyanin/K EDA catalyst, in $\pi^2(Pc)$-$d^8(Fe^0)$ state, for ammonia synthesis reaction (Sudo et al., 1969; Naito, Ichikawa, and Tamaru, 1972).
5. Reductant-independent ATP hydrolysis might be due to protein-protein electron backflow (Figure 3, via steps #10, #1, #2, and #7).

Without assuming a second step of ATP-driven electron transfer, it would be difficult to explain why, with sufficient $S_2O_4^{2-}$ but deficient ATP supply, the nitrogenase is observed (Walker and Mortenson, 1973) with the MoFe protein mostly in the semireduced, EPR-active (2s) state (Figure 3). Also, each Mo(III(IV)) serves as a one-electron valve and each dimeric Fe protein as a one-electron carrier in nitrogenase reactions. However, MgADP can strongly compete for one of the two MgATP sites on each Fe protein (Tso and Burris, 1973; Orme-Johnson et al., 1977), so that probably only one site on each Fe protein molecule is left for binding one MgATP during the ATP-driven protein-protein electron transfer. Thus, with two steps, the ATP:$2e^-$ ratio is still about 4 in the absence of electron backflow (Figure 4).

Recently, Smith et al. (1977) observed that, with the Kp1/Cp2 heterologous enzyme, the ATP-driven protein-protein electron transfer was almost as fast as with Kp1/Kp2 homologous enzyme, but that the number of ATP molecules consumed for every 2 H^+ reduced was greatly increased to about 50. Even with low ATP levels the ATP-driven protein-protein electron transfer was still much faster than the enzyme turnover, yet the enzyme activity was greatly decreased. They, too, have concluded that ATP has at least two functions in nitrogenase-catalyzed substrate reduction.

Walker and Mortenson (1973) have made the important observation that, with sufficient reductant and MgATP, a small amount of Fe protein can maintain a very large excess of MoFe protein mostly (85%) in the fully reduced, EPR-silent state, (1o); yet the enzyme activity is very much reduced. This observation indicates that, after completing the ATP-driven protein-protein electron transfer with the ensuing hydrolysis of ATP, the oxidized Fe protein can readily detach itself from the fully reduced MoFe protein and, after replenishment with one electron, may reduce another

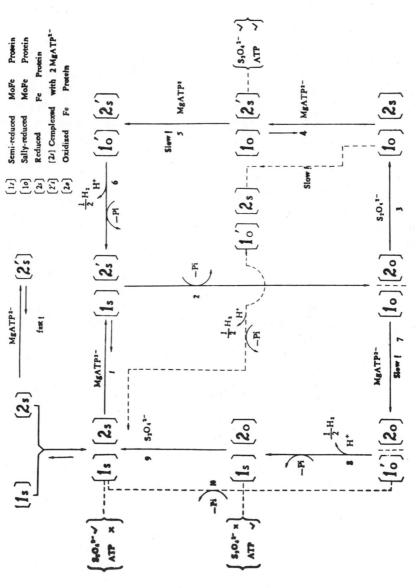

Figure 3.   Two-stepped ATP-driven electron transport.

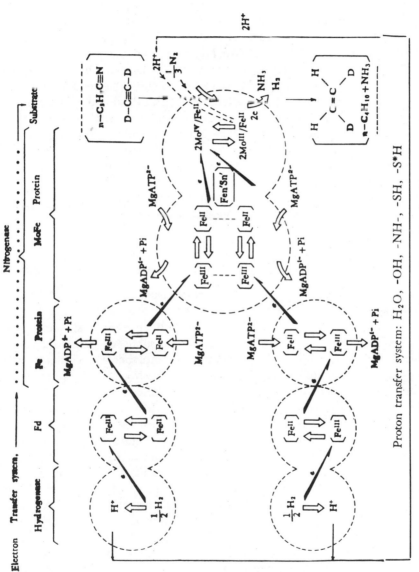

Figure 4. Electron and proton transport pathways.

Proton transfer system: H_2O, $-OH$, $-NH-$, $-SH$, $-S^*H$

semireduced MoFe protein with the help of ATP. It also indicates that the fully reduced MoFe protein in the absence of a sufficient number of bound Fe protein (in the reduced state and/or in combination with MgATP) is unable to transfer electrons to substrate molecules. As pointed out by Stiefel (1977), this may correspond to a recent observation by Ledwith and Schultz (1975). They found that, if a solution of $(Mo_2O_2(cys)_2)^{2-}$ is electrochemically prereduced to the Mo(III) stage and the potential cut off, then C_2H_2 introduced afterward is not reduced. Reapplication of the potential leads to immediate C_2H_2 reduction. Thus, a sufficiently negative redox potential, by means of Fe protein and MgATP, should be maintained at the Fe_nS_n* condenser system to make the fully reduced MoFe protein competent for substrate reduction. The critical potentials required may differ for different substrates, with the most difficult first step of N_2 reduction requiring the most stringent condition. From the results of competitive inhibition experiments with a heterologous enzyme (Kp1/Cp2) reported by Emerich and Burris (1976), at least two sites for binding Fe proteins are located on each tetrameric MoFe protein molecule.

DEVELOPMENT OF A CLUSTER STRUCTURAL
MODEL OF TWIN-SEATED Mo-2Fe TRINUCLEAR ACTIVE CENTERS

The two Fe(II) in the $Fe_2S_2 \cdot Mo_2L_2$ cluster are each singly linked through a common sulfide bridge to a Mo(III(IV)), so the Mo—Fe distance here is probably unobservable in the EXAFS, like the Fe—Fe distance in the 2Fe-2S cluster $[Fe_2S_2(S_2\text{-oxyl})_2]^{2-}$ (Hodgson, unpublished data). Based on EXAFS studies of MoFe protein and the iron-molybdenum cofactor, two cluster structural models were proposed with each Mo(III(IV)) surrounded by four sulfide ligands and two to three iron atoms (Cramer et al., 1978b). Taking one of these two models as a clue, each Mo(III(IV)) in the Fe_2S_2* (L)(L')Mo_2 cluster is linked to three Fe(II(III)) through sulfide bridges (S*) forming a $[Fe_3MoS_4*]$ cluster. This structure (Figure 5a) should allow favorable electron delocalization and electron transfer with the many other Fe_4S_4* clusters in the protein (Orme-Johnson et al., 1977). However, this model accounts for only eight iron atoms associated with two molybdenum atoms. Furthermore, the two $[Fe_3MoS_4*]$ clusters are only about 6 Å apart. These data are not in accord with the interpretations of the EPR and Mössbauer spectra (Orme-Johnson et al., 1977).

We now examine the plausibility of α-acetylene being coordinated double–side-on to 2Fe(II(III)) instead of to 1Mo(III(IV)) and 1Fe(II(III)). Following the previous line of reasoning, we can readily infer that α-acetylenes, alkyl cyanides and isocyanides, CN^-, and N_2 are all activated by double–side-on coordination to 2Fe(II(III)) plus end-on (or pseudo end-on in the case of $RC{\equiv}CH$) coordination to Mo(III(IV)), and that twin-seated

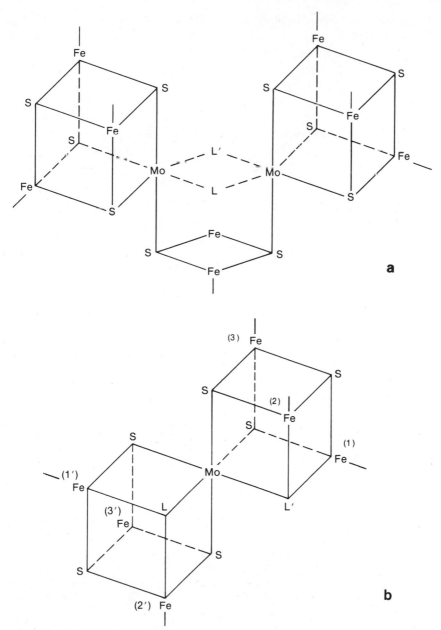

Figure 5. (a) Twin-seated 2Mo—1Fe trinuclear active centers in $Fe_2S_2(L)(L')Mo_2[S_3{}^*Fe_3S^*]_2$ cluster. (b) Twin-seated Mo—2Fe trinuclear active centers in $S^*Fe_3S_2{}^*(L)Mo(L')S_2{}^*Fe_3S^*$ cluster.

Mo—Fe-3, Mo—Fe-3' 2.7 Å

Mo—Fe-1, Mo—Fe-1'⎱ ~3.1 Å
Mo—Fe-2, Mo—Fe-2'⎰

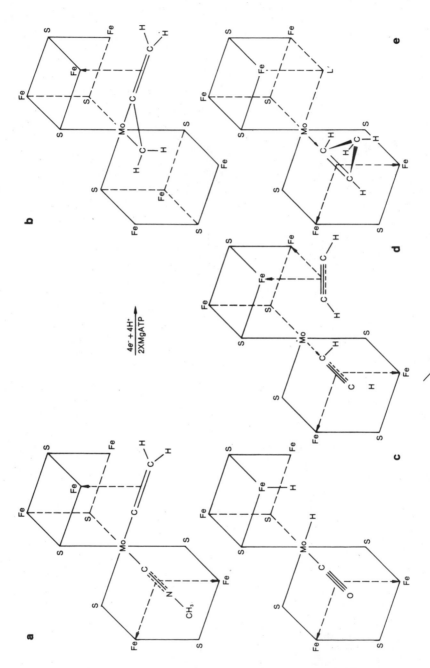

Figure 6. (a) Condensation of $CH_3N—C$ to carbene ligand, $>C=CH_2$. (b) Reductive hydrogenation of $CH_3N\equiv C$ to $CH_2=C=CH_2$ ligand and CH_3NH_2. (c) Noninhibition of H_2 evolution by CO. (d) Inhibition of H_2 evolution by $CH\equiv CH$. (e) Coordination activation of cyclopropene.

trinuclear active centers (Mo-2Fe) are required, with each Mo(III(IV)) situated at the center of an octahedral framework, $MoS_4*(L)(L')$, as in Figure 5a. This interpretation readily leads to a second model (Figure 5b): $S*Fe_3S_2(L)Mo(L')S_2*Fe_3S*$ for the twin-seated Mo—2Fe trinuclear active centers. Now associated with each Mo(III(IV)) are six Fe(II(III)) atoms, in agreement with the Mössbauer spectra (Orme-Johnson et al., 1977). A composite string bag model of nitrogenase active centers with a similar twin-seated structural feature but with Mo penta-coordinated with 4 S* and 1 —SR and a different mode of multiple–side-on plus end-on coordination of substrate has recently been proposed by C. S. Lu (this volume). However, this model of twin-seated Mo—2Fe trinuclear active centers can satisfactorily explain the mechanisms of all of the nitrogenase-catalyzed reactions (Figure 6), including the characteristic condensation reaction of RN—C and the inhibition of H_2 evolution by C_2H_2 but not by CO. Acetylene will be readily coordinated double–side-on plus pseudo end-on (a $\mu_3(\eta^2)$ type of bonding) to one of the twin-seated Mo—2Fe trinuclear active centers, and, at sufficiently high partial pressure of acetylene, a second CH—CH molecule may be coordinated double–side-on (a $\mu_2(\eta^2)$ type of bonding) to the two Fe(II(III)) of the other Mo—2Fe trinuclear active center, thus inhibiting the H_2 evolution reaction. Cyclopropene (McKenna, McKenna, and Higa, 1976) may be coordinated double–side-on to the 2Fe(II(III)) and pseudo end-on to the Mo(III(IV)) through one of the two ethylenic carbon atoms with the hydrogen of the —$\overset{\diagdown}{\underset{\diagup}{C}}$—H bond bent toward the $\overset{\diagdown}{\underset{\diagup}{C}}H_2$ group (Figure 6).

The tetrameric MoFe protein has been found to be of the $\alpha_2\beta_2$ type subunit structure with cyclic symmetry, and to have dimensions of 90 Å × 90 Å × 45 Å (Hardy, Burns, and Parshall, 1971). Computer simulations of activity titration patterns have shown (Orme-Johnson et al., 1977) that each tetrameric MoFe protein probably requires two kinetically related molecules of Fe protein to form an active enzyme. Thus, it may be inferred that the two FeMo clusters probably share the same electron transport system, consisting of about four Fe_4S_4* clusters that are interposed between the two FeMo clusters.

ACKNOWLEDGMENTS

Most members of the Nitrogen Fixation Research Group of this University have participated in the elaboration of the model, so this work is the result of group efforts. The author wishes to thank Professors Tang Au-Ching, Lu Chia-Si, W. H. Orme-Johnson, R. H. Burris, R. Holm, K. O. Hodgson, W. E. Newton, and E. I. Stiefel for preprints and helpful discussions.

REFERENCES

Burns, R. C., and R. W. F. Hardy. 1975. Nitrogen Fixation in Bacteria and Higher Plants, p. 132. Springer-Verlag, New York.

Chatt, J. 1975. Reactions of dinitrogen in its mononuclear complexes. J. Organometallic Chem. 100:17–28.

Chatt, J., A. J. Pearman, and R. L. Richards. 1975. Reduction of mono-coordinated molecular nitrogen to ammonia in a protic environment. Nature 253:39–40.

Chatt, J., A. J. Pearman, and R. L. Richards. 1976. Relevance of oxygen ligands to reduction of ligating dinitrogen. Nature 259:204.

Cotton, F. A. 1977. Molybdenum-molybdenum bonds. J. Less-Common Metals 54:3–12.

Cramer, S. P., K. O. Hodgson, W. O. Gillum, and L. E. Mortenson. 1978a. The molybdenum site in nitrogenase. Preliminary structural evidence from x-ray absorption spectroscopy. J. Am. Chem. Soc. 100:3398–3407.

Cramer, S. P., W. O. Gillum, K. O. Hodgson, L. E. Mortenson, E. I. Stiefel, J. R. Chisnell, W. J. Brill, and V. K. Shah. 1978b. The molybdenum site in nitrogenase. 2. A comparative study of Mo-Fe proteins and the iron-molybdenum cofactor by x-ray absorption spectroscopy. J. Am. Chem. Soc. 100:3804–3819.

Emerich, D. W., and R. H. Burris. 1976. Interactions of heterologous nitrogenase components that generate catalytically inactive complexes. Proc. Natl. Acad. Sci. U.S.A. 73:4369–4373.

Gillum, W. O., L. E. Mortenson, J.-S. Chen, and R. H. Holm. 1977. Quantitative extrusions of the $Fe_4S_4{}^*$ cores of the active sites of ferredoxins and the hydrogenase of *Clostridium pasteurianum*. J. Am. Chem. Soc. 99:584–595.

Hardy, R. W. F., R. C. Burns, and G. W. Parshall. 1971. Biochemistry of N_2 fixation. Adv. Chem. Series 100:219–247.

Ledwith, D. A., and F. A. Schultz. 1975. Catalytic electrochemical reduction of acetylene in the presence of a molybdenum-cysteine complex. J. Am. Chem. Soc. 97:6591–6593.

McKenna, C. E., M.-C. McKenna, and M. T. Higa. 1976. Chemical probes of nitrogenase. 1. Cyclopropene. Nitrogenase-catalyzed reduction to propene and cyclopropane. J. Am. Chem. Soc. 98:4657–4659.

Naito, S., M. Ichikawa, and K. Tamaru. 1972. Catalytic activity and electron configuration of the EDA complexes of phthalocyanins with alkali metals. J. C. S. Faraday Trans. 68:1451–1461.

Newton, W. E., J. R. Postgate, and C. Rodriguez-Barrueco. (eds.) 1977. Recent Developments in Nitrogen Fixation. Academic Press, Inc., New York.

Nitrogen Fixation Research Group, Department of Chemistry, Amoy University. 1976. A model of nitrogenase active center and the mechanism of nitrogenase catalysis. Scientia Sinica 19:460–474.

Orme-Johnson, W. H., L. C. Davis, M. T. Henzl, B. A. Averill, N. R. Orme-Johnson, E. Münck, and R. Zimmerman. 1977. Components and pathways in biological nitrogen fixation. In: W. Newton, J. R. Postgate, and C. Rodriguez-Barrueco (eds.), Recent Developments in Nitrogen Fixation, pp. 131–179. Academic Press, Inc., New York.

Rawlings, J., V. K. Shah, J. R. Chisnell, W. J. Brill, R. Zimmerman, E. Münck, and W. H. Orme-Johnson. 1978. Novel metal cluster in the iron-molybdenum cofactor of nitrogenase. J. Biol. Chem. 253:1001–1004.

Schrauzer, G. N. 1977. Nitrogenase model systems and the mechanism of biological nitrogen reduction: Advances since 1974. In: W. E. Newton, J. R. Postgate, and

C. Rodriguez-Barrueco (eds.), Recent Developments in Nitrogen Fixation, pp. 109–118. Academic Press, Inc., New York.

Schrauzer, G. N., G. W. Kiefer, K. Tano, and P. A. Doemeny. 1974. Chemical evolution of a nitrogenase model. VII. Reduction of nitrogen. J. Am. Chem. Soc. 96:641–652 and references therein.

Shah, V. K., and W. J. Brill. 1977. Isolation of an iron-molybdenum cofactor from nitrogenase. Proc. Natl. Acad. Sci. U.S.A. 74:3249–3253.

Smith, B. E., R. R. Eady, R. N. F. Thorneley, M. G. Yates, and J. R. Postgate. 1977. Some aspects of the mechanism of nitrogenase. In: W. E. Newton, J. R. Postgate, and C. Rodriguez-Barrueco (eds.), Recent Developments in Nitrogen Fixation, pp. 191–204. Academic Press, Inc., New York.

Smith, B. E., D. J. Lowe, and R. C. Bray. 1973. Studies by electron paramagnetic resonance on the catalytic mechanism of nitrogenase of Klebsiella pneumoniae. Biochem. J. 135:331–341.

Stiefel, E. I. 1977. Mechanisms of nitrogen fixation. In: W. E. Newton, J. R. Postgate, and C. Rodriguez-Barrueco (eds.), Recent Developments in Nitrogen Fixation, pp. 69–108. Academic Press, Inc., New York.

Sudo, M., M. Ichikawa, M. Soma, T. Onishi, and K. Tamaru. 1969. Catalytic synthesis of ammonia over the electron donor-acceptor complexes of alkali metals with graphite or phthalocyanins. J. Phys. Chem. 73:1174–1175.

Thomas, M. G., E. L. Muetterties, R. O. Day, and V. W. Day. 1976. Metal clusters in catalysis. 5. Four-electron η^2-ligand bonding in clusters and catalytic intermediates. J. Am. Chem. Soc. 98:4645–4646.

Thomas, M. G., W. R. Pretzer, B. F. Beier, F. J. Hirsekorn, and E. L. Muetterties. 1977. Metal clusters in catalysis. 6. Synthesis and chemistry of $Ni_4[CNC(CH_3)_3]_7$ and related clusters. J. Am. Chem. Soc. 99:743–748.

Thorneley, R. N. F., R. R. Eady, and D. J. Lowe. 1978. Biological nitrogen fixation by way of an enzyme-bound dinitrogen-hydride intermediate. Nature 272:557–558.

Tso, M.-Y. W., and R. H. Burris. 1973. Binding of ATP and ADP by nitrogenase components from Clostridium pasteurianum. Biochim. Biophys. Acta 309:263–270.

Walker, M., and L. E. Mortenson. 1973. Oxidation reduction properties of nitrogenase from Clostridium pasteurianum W5. Biochem. Biophys. Res. Commun. 54:669–676.

Zumft, W. G., and L. E. Mortenson. 1975. The nitrogen-fixing complex of bacteria. Biochim. Biophys. Acta 416:1–52.

Index

Acetylene
 active site selective probe, 234
 denitrification block by, 3, 38–40
 FeMoco reduction of, 244
 inhibition of hydrogenase by, 95
 substrate for nitrogenase, 11, 96,
  104, 142, 152, 224–227, 230,
  355–356, 377–378, 382–385
Actinomycete-alnus symbioses, 19
Activating factor, 142, 147
 see also R. rubrum nitrogenase
Activation of N_2, requirements for
 comparison with related molecules,
  344–346
 molecular orbitals and energies,
  344–346
 structural criteria, 348–350
Active site model
 cluster modification, 382–385
 model development, 374–376
 substrate interactions, 376–378
Alcaligenes spp., 29, 37
Aldehyde oxidase, 244
Ammonia-from-coal process facilities,
 47, 48
Ammonia production, commercial
 coal feedstock, 55
 economics of, 48
 electrolytic hydrogen feedstock, 56
 fuel oil feedstock, 54
 naphtha feedstock, 53
 natural gas feedstock, 52
 requirements of, 45–47

Aspergillus nidulans, cnx mutants of,
 249
ATP binding and hydrolysis
 electron microscopy studies, 201
 electron transfer relationships,
  201–204, 358, 378
 in membrane energization, 113–114
 in oxidative phosphorylation,
  120–122
 N_2 binding site relationship, 204
 NMR studies of, 200–204
 spin labels for, 200–201
Azidodithiocarbonate, inhibition of
 N_2 reduction by, 196
Azotobacter chroococcum, 20, 95
Azotobacter vinelandii, 20, 111, 145,
 153, 160, 195, 211, 227, 237,
 271, 285

Bacillus polymyxa, 145, 172, 240
Blue-green algae, 105

Carbon cycle, 12
Carbonyl cyanide *m*-chlorophenylhy-
 drazone (CCCP), 114–116
 see also Membrane energization
CCCP, *see* Carbonyl cyanide *m*-chloro-
 phenylhydrazone
Chemical probes, rationale for,
 223–227

An essential reference for botanists and plant scientists...

CO$_2$ METABOLISM AND PLANT PRODUCTIVITY

Edited by **R. H. Burris, Ph.D.**, Professor of Biochemistry, University of Wisconsin (Madison) and **C. C. Black, Ph.D.**, Professor of Botany and Biochemistry and Head, Department of Botany, University of Georgia.

For scientists studying ways to increase crop production this volume presents large amounts of new data from biochemical and physiological studies of CO$_2$ metabolism combined with evaluations of current work on plant metabolism. It is an essential reference source for botanists, biochemists, plant physiologists, plant breeders and agronomists.

The book examines complex interactions of photosynthesis and assesses, from several viewpoints, central problems concerning manipulation of CO$_2$ assimilation by environmental, chemical, and genetic means. A section is included on new methods that can supplement classical techniques in plant breeding programs to improve productivity.

This volume contains material not available elsewhere on photosynthesis, photorespiration, and dark respiration, with special consideration of the pentose phosphate and C$_4$ pathways, glycolate biosynthesis, glycine and serine metabolism, C$_3$ plants, and crassulacean acid metabolism (CAM).

Its thorough and authoritative coverage of the subject make it valuable reading for graduate students of plant physiology, plant biochemistry, crop production, and crop physiology as well as a reference for all workers and advanced students concerned with plant metabolism and crop improvement.

416 pages   *Illustrated*   *1976*

An outstanding introductory student textbook...

NITROGEN FIXATION

By **John Postgate, F. R. S.,** Professor of Microbiology and Assistant Director of the A.R.C. Unit of Nitrogen Fixation, University of Sussex

Specially designed and written for today's undergraduate biology student, this attractively priced text is a publication in the widely popular Studies in Biology Series. As with all volumes in this series, this book was developed under the prestigious auspices of the Institute of Biology, London.

CONTENTS

72 pages　　　　　*Illustrated*　　　　　*Paperback*　　　　　*1978*
Studies in Biology No. 92